Mineral Engineering: Mining and Technology

Mineral Engineering: Mining and Technology

Edited by **Beth Thorpe**

SYRAWOOD
PUBLISHING HOUSE

New York

Published by Syrawood Publishing House,
750 Third Avenue, 9th Floor,
New York, NY 10017, USA
www.syrawoodpublishinghouse.com

Mineral Engineering: Mining and Technology
Edited by Beth Thorpe

International Standard Book Number: 978-1-68286-149-3 (Hardback)

Printed in the United States of America.

Contents

Preface

Mineral Engineering is an emerging field of study that deals with the process of mining for minerals from different sources using innovative techniques and processing technologies. The topics included in this book elucidate on diverse concepts such as exploration and mining geology, advances in mineral analytical techniques, exploration of different mineral resources, industrial applications of minerals, etc. It contains contribution by eminent experts and scientists. This book focuses on the various advancements in this field and will prove to be immensely beneficial to students and researchers in this field.

After months of intensive research and writing, this book is the end result of all who devoted their time and efforts in the initiation and progress of this book. It will surely be a source of reference in enhancing the required knowledge of the new developments in the area. During the course of developing this book, certain measures such as accuracy, authenticity and research focused analytical studies were given preference in order to produce a comprehensive book in the area of study.

This book would not have been possible without the efforts of the authors and the publisher. I extend my sincere thanks to them. Secondly, I express my gratitude to my family and well-wishers. And most importantly, I thank my students for constantly expressing their willingness and curiosity in enhancing their knowledge in the field, which encourages me to take up further research projects for the advancement of the area.

Editor

Field Application of Accelerated Mineral Carbonation

Brandon Reynolds [1], K. J. Reddy [1],* and Morris D. Argyle [2]

[1] Ecosystem Science and Management, University of Wyoming, 1000 E. University Ave, Laramie, WY 82071, USA; E-Mail: brandonr@uwyo.edu

[2] Chemical Engineering, Brigham Young University, Provo, UT 84602, USA; E-Mail: mdargyle@byu.edu

* Author to whom correspondence should be addressed; E-Mail: katta@uwyo.edu or kjreddy@physics.harvard.edu

---

Abstract: Globally, coal-fired power plants are the largest industrial source of carbon dioxide (CO_2). CO_2 emissions from flue gas have potential for direct mineralization with electrostatic precipitator fly ash particles in the field. Demonstration scale accelerated mineral carbonation (AMC) studies were conducted at the Jim Bridger Power Plant, a large coal fired power plant located in Wyoming, USA. AMC produces kinetically rapid conditions for increased rates of mineralization of CO_2, sulfur dioxide (SO_2) and mercury (Hg) on fly ash particles. Control and AMC reacted fly ash particles were investigated for: change in carbon (expressed as $CaCO_3$), sulfur (expressed as SO_4^{2-}), and mercury (Hg) contents; topology and surface chemical composition by scanning electron microscope/energy dispersive X-ray spectroscopy analysis; chemical distribution of trace elements; and aqueous mineral solubility by the toxicity characteristic leaching procedure. Results of the AMC process show an increase in C, S, and Hg on AMC fly ash particles suggesting that multiple pollutants from flue gas can be removed through this direct mineral carbonation process. Results also suggest that the AMC process shifts soluble trace elements in fly ash to less leachable mineral fractions. The results of this study can provide insight into potential successful field implementation of AMC.

Keywords: accelerated mineral carbonation; fly ash particles; carbon dioxide; flue gas; coal fired power station; trace element; solubility; chemical fractionation

---

1. Introduction

In 2012, coal fired power plants were responsible for 37% of electrical generation in the United States [1]. Approximately 70% of China's [2] and 40% of India's [3] total energy were also supplied by coal in 2012. The size of these three countries highlights the importance coal plays in the domestic and global energy portfolio. Total global carbon emissions from the consumption of energy in 2011 were 35,578 Mt [4]. The products of the coal combustion process are flue gas and solid byproducts (e.g., fly ash particles, bottom ash, and boiler slag). Flue gas CO_2 emissions from coal fired power plants totaled 1718 Mt in the USA in 2011 [4]. Total annual solid byproducts in the USA are approximately 130 Mt, of which roughly 70 Mt is fly ash particles. Nationally, 35% of fly ash particles were used for beneficial purposes such as concrete additives. The remaining fly ash particles were disposed of in landfills or holding ponds. The use of fly ash particles for beneficial purposes varies worldwide; for example, it is 30% in India and China [5,6], whereas it is nearly 100% in Germany and The Netherlands [7]. Fly ash is used in construction most widely as a substitute for Portland cement in concrete as well as in wallboard, bricks, and other materials. It can be also be utilized as fill material in mining.

Both fly ash particles and flue gas contain trace elements. The levels of these trace elements depend on: the source of coal used [8], the specific conditions of the combustion process, the physical [9] and chemical properties [10,11] of the fly ash particles, the type of fly ash particle collection system [12], and the flue gas desulfurization system [13]. Based on the U.S. Resource Conservation and Recovery Act [14] of 1976, coal combustion products (CCP) are classified as non-hazardous materials and can be disposed of in lined impoundments. This classification has come under increased scrutiny following several incidents in which fly ash particles were inadvertently released into the environment, most notably the fly ash slurry holding pond spill at a Tennessee Valley Authority power plant in 2008 [15]. Numerous studies have examined the presence, the water solubility, and chemical leaching potential of major and trace elements in fly ash particles [10,12,16–30]. A recent review of both major and trace elements by Izquierdo *et al.* [31] provides a comprehensive reference for research conducted on acidic and alkaline fly ash elemental mobility across a wide range of pH values.

In the United States, proposed regulations on flue gas emissions from coal powered electrical generation, combined with increasing awareness of the potential effect these emissions have on anthropogenic climate change, have led to increased research on technologies to reduce and remove them before being emitted to the atmosphere. Early research focused on removal of flue gas SO_2, a major component of acid rain. This research led to the flue gas desulphurization techniques such as wet gas scrubbing with caustic solutions, which are widely employed today. The recent research in this field has focused on CO_2, the largest of the greenhouse gas emissions from coal-fired power plants. Technologies that have been researched are generally distinguished into separation/capture technologies and storage technologies [32,33]. Separation and capture technologies include membrane separation, sorbent capture, solvents (monoethanolamine) and other integrated processes. Storage processes include geologic sequestration and enhanced oil and gas recovery, which are arguably not a permanent storage solution. After all the efforts and research applied to this issue, few economically sound, industry-wide applications have resulted. This is due to many factors, including, but not limited to, the costs of: separation of CO_2 from the flue gas stream, capture of the CO_2 once it is separated, transportation from the source to the disposal site, and appropriate long term disposal or storage solutions for the CO_2 [32]. One method that has the potential to directly capture and store CO_2 is mineral carbonation. Mineral carbonation differs from many other methods of CO_2 capture and storage by changing flue gas CO_2 into carbonate minerals. Mineral carbonation of CO_2 (e.g., atmospheric or flue gas) with different sources of minerals (e.g., natural [34], spent oil shale [35] or fly ash [36–39]) have been studied. However, most of these studies were conducted at laboratory scale.

The focus of the research in this paper is aimed at reducing emissions from coal-fired power plants using a novel method of mineral carbonation of flue gas CO_2 with power plant fly ash particles. Fly ash from coal-fired power plants is found in many developed and developing countries. As mentioned previously, the amount of fly ash put to beneficial use varies by country drastically. For example, Germany and the Netherlands use nearly all of their fly ash for beneficial use. In the United States, China, and India it is around 30%–35%, resulting in large quantities of fly ash that is disposed in landfills or placed in holding ponds. The potential of fly ash for carbonation worldwide was reported to be ~7 Mt C/year by Renforth *et al.* [40]. The novel process is termed accelerated mineral carbonation (AMC). The AMC process relies on generating optimal conditions for nearly instantaneous mineral carbonation of flue gas CO_2 onto the surface of the fly ash particles. This process differs from other studies because it is a direct mineralization process for multiple pollutants from flue gas with short reaction times. A small pilot scale AMC process was successfully tested at a large coal power plant prior to the current demonstration scale AMC process [41]. The AMC process has shown promise not only in removing flue gas CO_2, but also SO_2 and Hg, as well as minimizing the mobility and leaching of trace elements in the fly ash [42–44]. However, for this current study, we made several modifications to the demonstration apparatus (Figure 1). These modifications included addition of a screw conveying system to accurately deliver the required amount of fly ash from the electrostatic precipitator hopper to the fluidized bed reactor and installation of better controls for reactor temperature and pressure, and moisture of flue gas leaving the humidifier into the fluidized bed reactor (FBR). We conducted three experiments with 0–120 min reaction time, at 60 °C heater/humidifier temperature, ~21 kPa gauge pressure, and 16% of flue gas moisture. Control and flue gas CO_2 reacted fly ash samples were analyzed for carbon, sulfur, and mercury and subjected to studies of scanning electron microscopy/energy dispersive X-ray spectroscopy (SEM/EDS), toxicity characteristic leaching, and mineral fractionation. Data from these studies were used to discuss the direct mineralization of flue gas CO_2, SO_2, and Hg by fly ash particles using the AMC process.

2. Materials and Methods

2.1. Demonstration Scale Accelerated Mineral Carbonation Field Testing

The AMC study was conducted at the Jim Bridger Power Plant (JBPP), a 2120 MW coal fired electrical generation plant, owned by PacifiCorp, located near Point of Rocks, Wyoming, USA. JBPP consumes approximately 19,900 t of low sulfur sub-bituminous coal per day to feed its four independent power generating units, each unit capable of producing 530 MW. Post-combustion flue gas sulfur removal at JBPP is accomplished by a wet Na_2CO_3 scrubber.

The AMC demonstration scale process (Figure 1) is located at Unit 2 of JPBB. In the AMC process, a continuous slipstream of flue gas is piped from the outlet of the flue gas stack (after all scrubbing/treatment processes) through a moisture reducing (or knockout) drum (MRD) to a blower (model 1602 Atlantic Blower (Carrollton, TX, USA), producing approximately 0.094 m^3/s, STP), which slightly pressurizes the flue gas. From the blower, the flue gas is routed through a water reservoir in the heater/humidifier (H/H) drum, before finally fluidizing and reacting with a batch of ash in the fluidized bed reactor (FBR). The fly ash particles for each batch experiment conveyed by a screw conveyor (UCC Corp., Waukegan, IL, USA) directly from the electrostatic precipitator hopper above the reactor. The temperature of the fly ash particles entering the FBR is approximately 70 °C. The fly ash bed depth was approximately 45 cm (~300–350 kg) in each experiment.

Figure 1. Schematic of accelerated mineral carbonation (AMC) process design (not to scale).

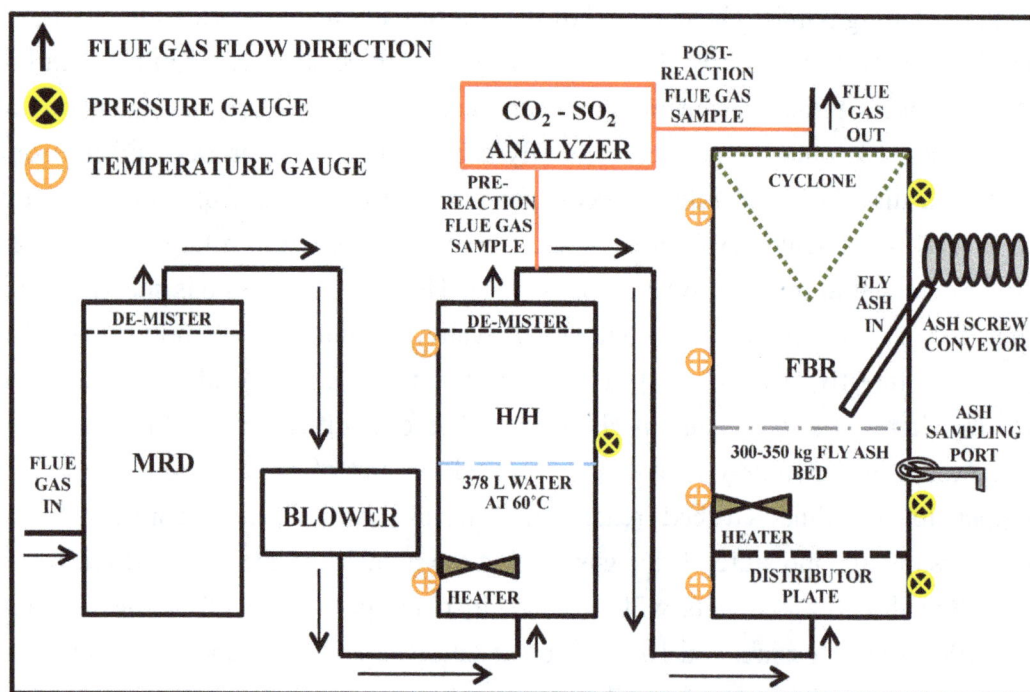

The FBR has an operating section of 0.9 m diameter and 3.0 m length with a perforated distributor plate containing 9 mm diameter holes to distribute the flue gas to provide proper fluidization of the fly ash particles, a heater (ASB Heating Elements Ltd., Toronto, ON, Canada), a cyclone to separate the fly ash particles from the flue gas as it leaves the FBR, and a pressure control valve with controller (Leslie Controls, Inc, Tampa, FL, USA). The flue gas temperature entering the MRD is approximately 50 °C. Both the MRD and H/H drums contain demister screens to remove entrained water droplets from the flue gas to protect the blower and the FBR, respectively. The H/H drum is filled with 378 L of plant process water, which is heated (ASB Heating Elements Ltd.) to condition the flue gas to the desired humidity. The temperature of the H/H in this study was 60 °C, which is approximately 16 mol % moisture. Pressure (Ashcroft gauges), temperature (Omega Engineering thermocouples), and CO$_2$ and SO$_2$ concentrations (Horiba continuous emissions flue gas analyzer) are continuously monitored at strategic locations noted in Figure 1. However, we did not report flue gas measurement of CO$_2$ and SO$_2$ for two reasons: (1) the Horiba instrument presented problems with calibration and measurement due to condensation of water in the sample lines during the testing; and (2) the flue gas supplied was constantly replenished with CO_2 and SO_2 during the reaction, which rendered the small differentials in concentration not meaningful to analyze in the gas samples. For these reasons, the removal of CO_2 and SO_2 from the flue gas was calculated based on C and S contents of control and treated fly ash samples. A detailed description of the demonstration scale AMC process is available in Reddy *et al.* [42,43]. Fly ash solubility and fractionation data from a pilot scale AMC process at JBPP, smaller than the current demonstration scale conducted at JBPP, is detailed in Bhattacharya *et al.* [44].

Experimental conditions in the fluidizing bed were as follows: fly ash temperature of 70 °C, flue gas temperature of 60 °C corresponding to ~16 mol % moisture, and ~21 kPa gauge pressure. Three separate experiments were conducted on different days, each with the same experimental conditions. Fly ash samples were collected directly out of the FBR before the reaction (control), during the reaction at 10 min and 30 min through a sampling valve, and directly out of the FBR at 120 min or the end of the experiment (final). The samples were sealed immediately after sampling and transported to the lab for analysis. The chemical composition of control fly ash samples from Unit 2 electrostatic precipitator hopper was gathered from JBPP.

2.2. Total Carbon, Sulfur, and Mercury

Samples from three separate tests at 60 °C humidifier water temperature were analyzed for total carbon, sulfur, and mercury by the Wyoming Department of Agriculture Analytical Services Laboratory, Laramie, WY, USA. Analysis was accomplished using a CE Elantech EA 1112 elemental combustion analyzer (Thermo Scientific, Waltham, MA, USA) for carbon and sulfur. Mercury analysis was performed by an Agilent 7500ce inductively coupled plasma mass spectrometer (ICP-MS). All other analyses described next were performed on ash samples from one of the 60 °C humidifier water temperature tests.

2.3. Scanning Electron Microscopy-Energy Dispersive X-Ray Spectroscopy

Scanning electron microscopy (SEM) and energy dispersive X-ray spectroscopy (EDS) were used to determine the texture and composition of the surface of the control and AMC fly ash particles. Analysis was conducted at Colorado State University, Ft. Collins, CO, USA on a JSM-6500F (JEOL, Tokyo, Japan) SEM coupled with a NORAN System SIX (Thermo Fisher Scientific, Waltham, MA, USA) EDS. Samples were prepared by coating, mounting, and inserting into the JSM-6500F. Analysis was performed by examining SEM images and performing EDS measurements on areas of interest.

2.4. Sequential Chemical Fractionation of Trace Elements

Sequential chemical fraction of the trace elements (Al, Cr, Mn, Cu, Zn, As, Se, Pb, Cd, and Hg) in the fly ash was conducted following a slightly modified method of Tessier *et al.* [45] at the Water Quality Laboratory, College of Agriculture and Natural Resources, University of Wyoming, Laramie, WY, USA. The sequences of extraction are water soluble, exchangeable, carbonate bound, oxide bound, organic matter bound, and the residual fraction, as detailed in Table 1. The modified methodology uses 1 g of fly ash with the following extraction fluids in 50 mL centrifuge tubes:

Table 1. Sequential chemical fraction procedure, modified from Tessier *et al.* [45].

Fraction	Extractant Fluid	Procedure
Exchangeable	0.5 M Mg(NO$_3$)$_2$	30 min shaking at 120 rpm
Carbonate	1 M NaOAc	5 h shaking at 120 rpm
Oxide	0.08 M NH$_2$OH·HCl	6 h shaking at 120 rpm
Organic Matter	0.02 M HNO$_3$ and H$_2$O$_2$	2 h shaking at 120 rpm at 85 °C
Residual	Concentrated HNO$_3$	2 h shaking at 120 rpm

Recovery of the extractant fluid was performed through centrifugation at 10,000 rpm. Supernatant was removed and analyzed. To wash the remaining residue of the previous extractant fluid, samples were rinsed with 8 mL deionized water between every step, centrifuged again for 30 min at 10,000 rpm, and supernatant disposed. Analysis of the leachate fluid was performed using ICP-MS for cations and ion chromatography (IC) for anions at Wyoming Department of Agriculture Analytical Services Laboratory, Laramie, WY, USA.

2.5. Toxicity Characteristic Leaching Procedure (TCLP)

Toxicity characteristic leaching procedure (TCLP) was conducted at the ALS Environmental laboratory in Ft. Collins, CO, USA. TCLP is one of the measurements used to determine if a disposal product is classified as a hazardous or non-hazardous waste based on the leaching of certain elements of concern. The TCLP is often used when wastes are disposed of in a landfill. Slightly more than half of JBPP fly ash is disposed of in a local landfill, with the remaining portion being used in cement production. The current study examined leachability of trace elements including As, Ba, Cd, Cr, Pb, Se, Ag and Hg. The purpose of the TCLP test is to help predict the mobility and leaching potential of these trace elements from fly ash before and after the AMC process.

3. Results and Discussion

3.1. Elemental Composition and Distribution on Fly Ash Particles

The composition of fly ash particles and the distribution of elements on their surfaces and interiors are important when discussing elemental solubility and fractionation, with an emphasis on control fly ash samples and of changes on AMC fly ash samples. A model for fly ash particles has been proposed [46], with the particles composed of a thin exterior surface of differing composition from the interior of the fly ash particle. Subsequent studies featuring SEM-EDS and XRD analysis [25,47,48] indicate general agreement with their findings. The exterior is proposed to be made up of inorganic oxides and hydroxides and reactive aluminosilicate glass, while the interior is primarily a less reactive aluminosilicate matrix [24]. The exterior shell has been found to contain elements, both trace and major, that undergo phase changes during the combustion process, form small particles upon cooling, and can attach to the surface of the aluminosilicate matrices that form the larger fly ash particles [49]. The interior of the particle is less reactive and composed primarily of aluminosilicates with a distribution of various cations and anions. Major cations, anions, and trace elements are not all evenly distributed between the silicate and non-silicate phases of the fly ash particles [24]. The fly ash particles in this study were from low sulfur sub-bituminous coal with an initial carbon content of

35–45 wt %. The combustion of the coal results in fly ash particles with inorganic oxides and hydroxides and amorphous aluminosilicates [26,50]. The approximate distribution of chemical species in JBPP fly ash particles is shown in Table 2.

Table 2. Distribution of chemical species on Jim Bridger Power Plant (JBPP) fly ash particles (wt %).

Component	wt %
SiO_2	60.04
Al_2O_3	19.67
CaO	5.86
Fe_2O_3	4.66
MgO	3.85
K_2O	2.00
N_2O	1.00
Loss on Ignition	0.60

The amorphous silicates are primarily Si, Al, and Ca minerals [25,47]. The pH of the fly ash is 11.5–12.5. The physical size of the fly ash is approximately 40 μm, spherical in shape, with very low moisture content. The flue gas is relatively low in SO_2 (20–25 ppm) and NO_x (100–120 ppm) concentrations, contains 11–13.5 mol % CO_2, 66–70 mol % N_2, 10–20 mol % moisture and trace amounts of Hg, As, and various other trace elements. The pH of the flue gas is 3.3 [41,42].

3.2. Demonstration Scale Studies

During demonstration scale AMC testing, fly ash samples collected were: control, 10 min, 30 min, and 120 min (or final). The controlling factors in the AMC process are the chemical and physical properties of the fly ash and flue gas and the conditions in which they are mixed. Three factors influencing the chemical composition of the fly ash are the combustion process, the mineralogy and chemical composition of parent coal, and the ash collection method.

The fly ash particles and flue gas exiting the combustion chamber are reactive due in part to their respective intrinsic properties; fly ash particles with alkaline pH, high metal oxide content, and warm temperature and flue gas with high CO_2 content, presence of SO_2 and NO_x, low pH, optimized moisture content from the H/H, and warm temperature. In addition to these properties, the conditions created within the process to facilitate AMC include increased in pressure to 21 kPa guage in the FBR, mixing via the fluidizing bed in the FBR, and conditioning of the flue gas via the H/H. The combination of these factors results in an increased rate of mineral carbonation of fly ash inorganic oxides and hydroxides and flue gas CO_2 and SO_2. As an example, Summers *et al.* [51] investigated the combined effect of CO_2 and SO_2 in mineral carbonation. Results from these studies show that small fraction of SO_2 in the flue gas in fact enhances the mineral carbonation of CO_2. The formation of carbonates and sulfates in fly ash through the AMC process can be explained by the following example reactions involving CaO and CO_2 and SO_2 [43]:

Carbonates:

$$CO_2 \text{ (g) (flue gas)} + H_2O \text{ (moisture in flue gas)} \leftrightarrow H_2CO_3 \text{ (carbonic acid)} \tag{1}$$

$$CaO \text{ (fly ash particles)} + H_2CO_3 \text{ (carbonic acid)} \leftrightarrow CaCO_3 \text{ (calcite)} + H_2O \qquad (2)$$

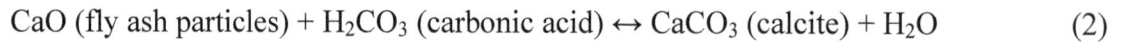

The overall Reaction (3) results from combining Reactions (1) and (2):

$$CO_2 \text{ (g) (flue gas)} + CaO \text{ (fly ash particles)} \leftrightarrow CaCO_3 \text{ (calcite)} \qquad (3)$$

Sulfates:

$$2SO_2 \text{ (flue gas)} + O_2 + 2CaO \text{ (fly ash)} \leftrightarrow 2CaSO_4 \text{ (anhydrite)} \qquad (4)$$

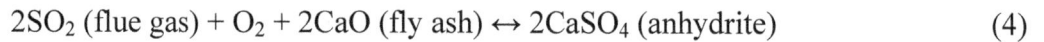

The total carbon, sulfur and mercury content of the fly ash particles after the AMC process (Table 3) increased from the control sample to the final sample. As noted in Section 2.2, C, S, and Hg were calculated by the reported methods. To obtain the values for $CaCO_3$ and SO_4^{2-}, conversion factors were multiplied by the elemental value. The conversion factor is the molecular weight of the compound divided by the molecular weight of the individual element. This equals to 8.33 for $CaCO_3$, and 3.0 for SO_4^{2-}. These data are averaged from three experiments. The control value is based on two samples collected prior to the experiments. These data are averaged from three separate experiments at ~16% moisture in the flue gas. The control value is based on two of the experiments.

Table 3. $CaCO_3$ (wt %), SO_4 (wt %), and Hg (mg/kg) content of JBPP control and AMC fly ash samples.

Sample	$CaCO_3$ (wt %)	SO_4^{2-} (wt %)	Hg (mg/kg)
Control	2.88 ±0.96	0.48 ±0.03	0.12 ±0.09
10 min	3.11 ±1.57	1.21 ±0.19	0.27 ±0.12
30 min	3.67 ±1.53	1.35 ±0.13	0.25 ±0.08
Final	3.86 ±1.28	1.42 ±0.14	0.30 ±0.11

These data indicate formation of carbonates, as the only potential source for an increase in carbon content of the fly ash in the AMC process was due to flue gas CO_2. Carbonates have been shown to form during hydrolysis of calcium oxides and hydroxides in alkaline fly ash as well as in the presence of CO_2 [46,52]. This is possibly occurring in the AMC samples due to the high relative moisture and 11.0%–13.5% CO_2 contained in the flue gas. Examination of the interaction of atmospheric and flue gas SO_2 with fly ash particles indicates that it is possible for anhydrite ($CaSO_4$), gypsum ($CaSO_4 \cdot 2H_2O$), amorphous calcium sulfate, or various other sulfate minerals to be formed on the surface of the particles of alkaline fly ashes [25,31,50]. The observed increase in total sulfur on the AMC samples is likely due to the interaction with the 20–25 ppm SO_2 in the flue gas. Tsuchiai *et al.* [53] found that SO_2 oxidized to SO_3 and then was converted to $CaSO_4$ by the presence of NO_x, which the flue gas also contains. The observed neutralization of the high pH in the control sample to the AMC sample suggests that more calcium will be available as Ca^{2+} to react with either CO_2 or SO_2. Extensive studies on mercury [13,54–60] have suggested that the majority of Hg in post desulfurization flue gas is typically in elemental form, which can be oxidized by soluble inorganic oxides (*i.e.*, CaO) or chlorine [60], and potentially adsorb to unburned carbon in the fly ash. However, since fly ash from JBPP has low or insignificant unburned carbon content, this mechanism might not explain the increase in mercury content of the fly ash. Studies have shown that mercury can also coprecipitate as $HgCO_3$ with calcite in fly ash or adsorb to the calcite surface [61], which might be happening during the AMC process.

3.3. SEM/EDS

SEM/EDS images and data (Figure 2) suggest an increase in the carbon content from the control to the AMC samples. In the control SEM image (Figure 2a), spherical fly ash particles are visible with little crystal structure formation or variation in their general shapes. The corresponding EDS data suggest that aluminum, oxygen, calcium, magnesium, and silicon are the primary elements at the point of analysis. In the 10 min (Figure 2b) and 30 min (Figure 2c) SEM images, formation of crystals and a change from the uniform shape observed in the control occurs. Correspondingly, an increase in carbon is noted in both EDS data. This increase in carbon content is attributable to the formation of amorphous or crystalline carbonate minerals. The SEM/EDS data (Figure 2c) showed significant amount of carbon and sulfur in 30 min flue gas reacted sample. From this investigation we conclude that carbonation of flue gas CO_2 takes place in the presence of SO_2 gas.

Figure 2. Scanning electron microscopy (SEM) images with corresponding X-ray spectroscopy (EDS) spectra: (**a**) control; (**b**) 10 min; (**c**) 30 min. Point in the yellow circle corresponds to the respective EDS spectra.

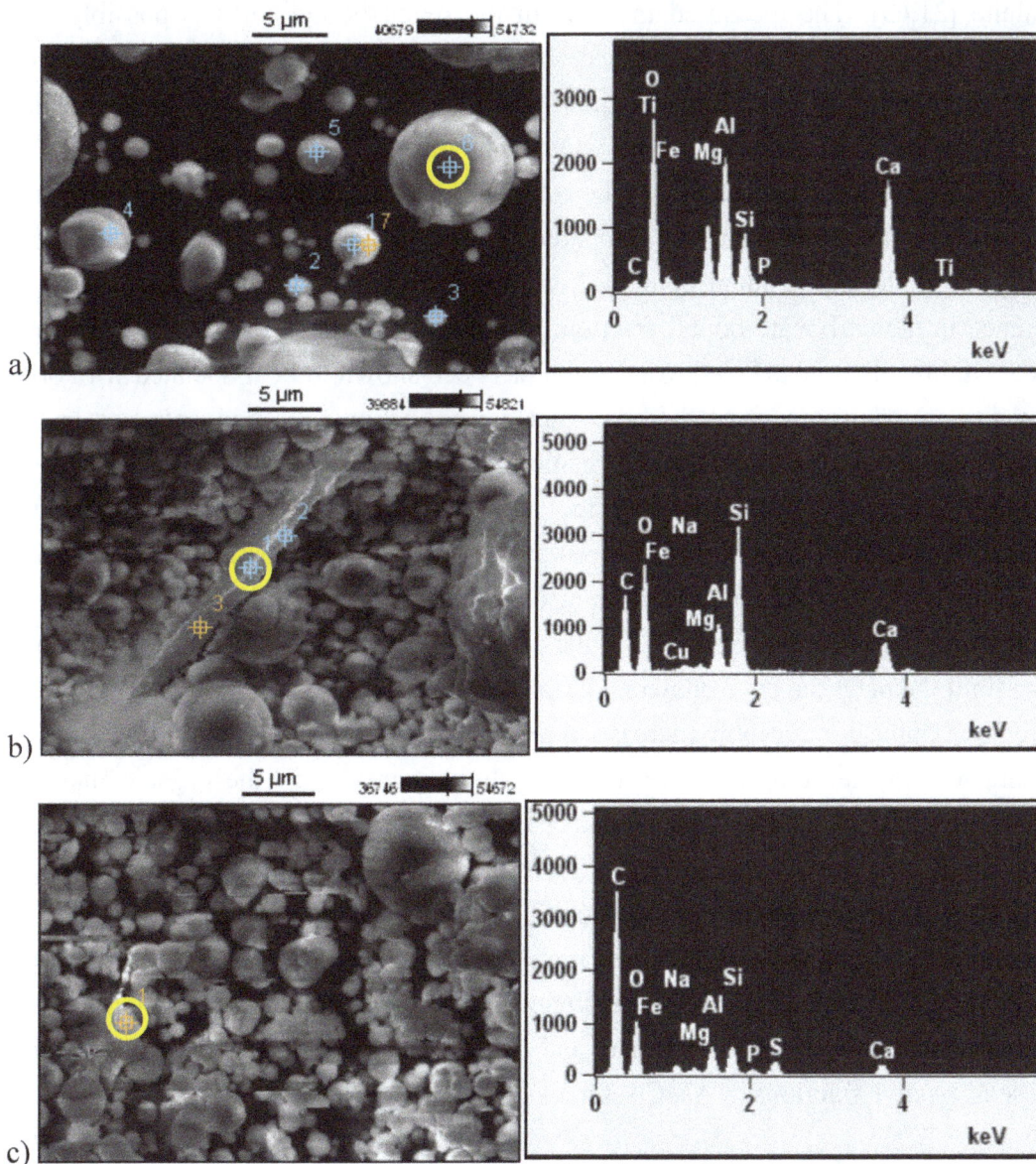

3.4. Sequential Chemical Fractionation of Trace Elements

Trace element (As, Cd, Cr, Cu, Hg, Mn, Pb, Se, Ti, Zn) fractionation data (Figure 3) suggest a large increase in the carbonate fraction and, to a lesser extent, oxide bound fraction for arsenic, chromium, copper, and zinc in the AMC process samples. An increase in the oxide fraction for manganese and aluminum was also observed. Lead, selenium, cadmium, mercury and titanium were either below instrumental detection limits or if present, were found at very low concentrations in all fractions. This suggests that their mobility would not pose a problem in either the control or AMC samples. To attempt to interpret possible effects on the distribution of lead, selenium, cadmium, and mercury as a result of the AMC process would be difficult due to the lack of definitive data.

Arsenic (Figure 3a) chemical fractions changed in distribution from the control to the AMC samples. This change results in a decrease in the exchangeable fraction and an increase in the carbonate, oxide, and residual bound arsenic, with minimal change observed in the organic matter bound from the control to the AMC samples. Narukawa *et al.* [62] found similar results for the majority of arsenic bound to carbonates and oxides in a sequential leaching procedure similar to the current study. The increase in the carbonate fraction of arsenic is possibly attributed to adsorption onto the surface of calcite [31,63]. The measured increase in the oxide bound form is possibly due to absorption with iron oxide [27] or to a lesser extent aluminum oxide [63]. The AMC samples potentially reduced the mobility of arsenic in the environment by increasing the concentrations in the less soluble carbonate, oxide, and residual fractions when compared to the control sample.

Chromium (Figure 3b) fractionation data from the AMC process show the greatest increases in the oxide and organic matter bound fraction when compared to the control samples. Lesser increases are observed in the exchangeable and residual bound AMC samples compared to the control, with little discernable difference in the carbonate bound fraction. Chromium is known to adsorb with iron and aluminum oxides at slightly alkaline pH [64]. Chromium has been shown to be associated with organic matter in the depositional environment of the coal, and depending on total concentration and coal combustion methods, can be associated with the organic matter content in the fly ash particles [31]. Chromium in the AMC samples increased in all the fractions except carbonate bound, where no difference is detectable. Although chromium has been suggested to be an element of concern in the environment, the more toxic hexavalent species in fly ash particles is readily reduced to trivalent chromium by interaction with flue gas SO_2 [31], which the AMC process promotes. This change in oxidation state can limit the potential environmental hazards of chromium.

Copper (Figure 3c) exhibits a noticeable increase in the carbonate and oxide bound fractions of the AMC samples compared to the control. There are very slight increases in the organic matter and residual fractions from AMC to control and no detectable difference in the exchangeable fraction. Copper can be adsorbed onto the aluminum oxides in fly ash particles [31,65], and this association might help explain the exhibited increase in oxide bound fractions of copper in the AMC samples. In addition to adsorption on aluminum oxides, formation of copper carbonate, oxide, or hydroxide minerals is possible [65,66]. Formation of these may increase the carbonate bound fraction of copper. The observed increases in copper in both carbonate and oxide fractions indicates that more copper is being fixed in these respective fractions in AMC samples compared to the control.

Figure 3. Sequential chemical fractionation concentrations (mg/L) for the selected trace elements: **(a)** arsenic; **(b)** chromium; **(c)** copper; **(d)** zinc; **(e)** manganese; **(f)** aluminum Fractions: Ex = exchangeable; Carb = carbonate; Ox = oxide; OM = organic matter; Res = residual.

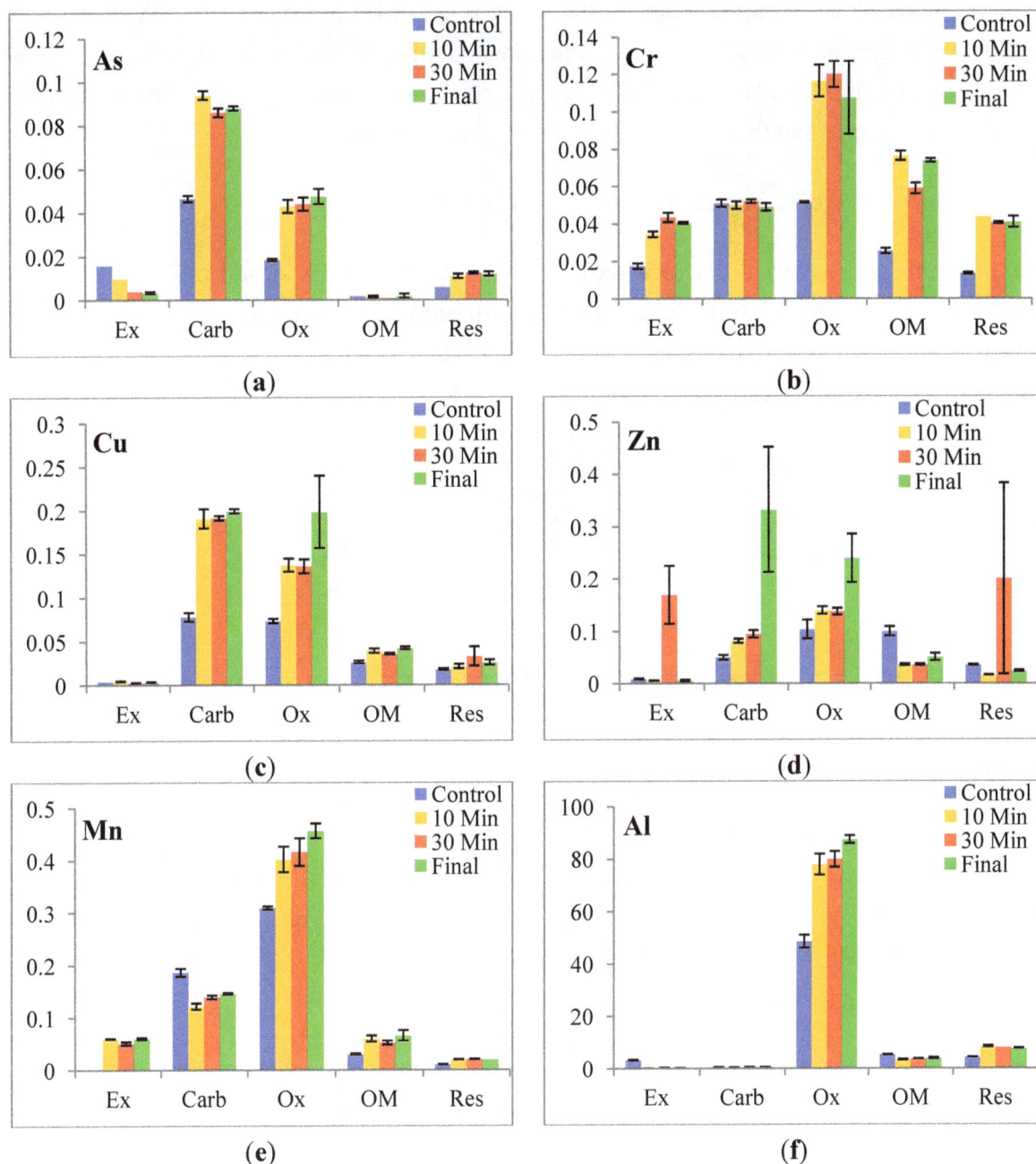

Zinc (Figure 3d) displays an increase in the carbonate and oxide bound fractions and a decrease in the organic matter fraction from the control to the AMC samples. Exchangeable and residual fractions do not exhibit an appreciable difference. Zinc has been suggested to associate with aluminum oxides [67] or be present as zinc oxide or hydroxide in fly ash [31]. The observed increase in carbonate bound zinc can possibly be explained by adsorption of zinc onto calcite [68] or some other carbonate species or perhaps the formation of a zinc carbonate species.

Manganese (Figure 3e) shows slight increases in exchangeable and oxide bound fractions from the control to AMC samples. Carbonate, organic matter, and residual fractions show slight changes, but the magnitude of the difference is very minimal. Kim and Kazonich [24] found manganese in higher quantities in the silicate bound forms. This could mean a possible association with aluminum oxides that leach from the fly ash inner matrix and are available to scavenge manganese. The manganese increase in the exchangeable fraction from control to AMC samples is difficult to explain.

Aluminum (Figure 3f) shows an increase in the oxide fraction from the control to AMC samples, with little observable change in any of the other fractions. Aluminum is a major constituent of fly ash particles, and by decreasing the pH in the AMC samples, increased amounts can possibly be dissolved from the glassy inner matrix and then form hydroxide or oxide species.

3.5. Toxicity Characteristic Leaching Procedure

TCLP experiments were carried out to determine if the AMC process would possibly altered the classification of fly ash particles from its currently non-hazardous status. Results of the TCLP are shown in Table 4. All of the constituents sampled for in the TCLP analysis are well below the limits in both the control and AMC samples.

Table 4. Toxicity characteristic leaching procedure (TCLP) concentrations (mg/L).

TCLP (mg/L)	As	Ba	Cd	Cr	Hg	Pb	Se	Ag
Reporting Limit	5.0	100	1.0	5.0	0.2	5.0	1.0	5.0
Control	*	*	*	*	*	*	0.21	*
30 min	*	*	*	*	*	*	0.18	*
Final	*	*	*	*	*	*	0.18	*

Note: * = below the lower limit of quantitation.

4. Conclusions

Results of this study suggest changes in the fly ash geochemistry as a result of the demonstration scale AMC process. Increases in C, S, and Hg were observed, suggesting a mechanism of capture from flue gas onto the surface of the fly ash particles during the AMC process. Analysis of SEM/EDS, sequential chemical fractionation, and TCLP further provided evidence for the direct mineralization of flue gas CO_2, SO_2, and Hg by fly ash particles.

SEM/EDS analysis show changes in the morphology from uniform, spherical particles observed in the control to crystal structures and less uniformity in the shape in the AMC samples, and chemical composition shown by an increase in carbon content of the AMC samples, indicative of carbonation of the fly ash particles. Sequential chemical fractionation data show an increase in the carbonate and oxide bound forms of arsenic, copper, and possibly zinc. Increases in the oxide bound forms of chromium, manganese, and aluminum were observed. Chromium and manganese were the only two trace elements with an apparent increase in the exchangeable forms. TCLP data did not indicate any difference due to AMC process.

The results of the AMC process on fly ash major and trace element mobility are important to determine if this process can potentially be implemented at an industrial scale. The conclusion of the experiments and analysis suggest direct mineralization of flue gas CO_2, SO_2, and Hg is possible in the field at a large scale through the AMC process. As a complement to the mineralization of multiple flue pollutants, this study also suggests that the AMC process possibly alters some elements of concern in the fly ash to less mobile fractions.

Future study on the AMC process on other coal-fired power plants would help to determine if the AMC process results are applicable to a wide range of conditions. Additionally, increasing the AMC reaction to industrial scale would be needed to determine if it is cost effective and efficient [42]. Fly ash and flue gas conditions vary between different power plants and studies with differing conditions would elucidate if the AMC process works elsewhere. Potential beneficial end uses of the carbonated fly ash include use as an amendment for naturally or anthropogenically created sodic soils. The effect of the AMC process on the pozzolanic properties of fly ash needs to be studied if the beneficial use as construction material, and especially as a replacement for Portland cement, is to be retained. Additional studies could examine application in any industrial setting where an alkaline solid waste and acidic flue gas are produced, including but not limited to: steel industry, cement industry, and solid waste incinerators.

Acknowledgments

The authors would like to thank Dr. Ajay Kumar Sankara Warrier for his contributions in sampling and analysis for the project. We would also like to thank Ryan Taucher, Roger Warner, Dan Miller, Jim Sedey, and Jerry Dickson at Jim Bridger Power Plant, PacifiCorp, and the University of Wyoming Office of Research and Economic Development. Funds for this project were contributed by PacifiCorp, University of Wyoming Ecosystem Science and Management, and the Department and Office of Research and Economic Development.

Author Contributions

K. J. Reddy conceived the original idea. K. J. Reddy and Morris D. Argyle designed and developed AMC rector skid. Brandon Reynolds, K. J. Reddy, and Morris D. Argyle all participated in conducting experiments. Brandon Reynolds collected the data. All authors participated in writing of the manuscript.

Conflicts of Interest

The authors declare no conflict of interest.

References

1. U.S. Energy Information Administration. Energy in Brief: What is the Role of Coal in the United States? Available online: http://www.eia.gov/energy_in_brief/article/role_coal_us.cfm (accessed on 21 September 2013).
2. U.S. Energy Information Administration. China. Available online: http://www.eia.gov/countries/ cab.cfm?fips=CH (accessed on 21 September 2013).
3. U.S. Energy Information Adminiatration. India. Available online: http://www.eia.gov/countries/ cab.cfm?fips=IN (accessed on 21 September 2013).

4. U.S. Energy Information Administration. Monthly Energy Review October 2013. Available online: http://www.eia.gov/totalenergy/data/monthly/pdf/sec12_9.pdf (accessed on 5 October 2013).

5. Asokan, P.; Saxena, M.; Asolekar, S. Coal combustion residues-environmental implications and recycling potentials. *Resour. Conserv. Recycl.* **2005**, *43*, 239–262.

6. Pei-Wei, G.; Xiao-Lin, L.; Hui, L.; Xiaoyan, L.; Jie, H. Effects of fly ash on the properties of environmentally friendly dam concrete. *Fuel* **2007**, *86*, 1208–1211.

7. Barnes, D.; Sear, L. *Ash Utilisation from Coal-Based Power Plants*; Report COAL-R-274 for UK Quality Ash Association; Hatterrall Associates: Cheltenham, UK, 2004.

8. Mardon, S.; Hower, J. Impact of coal properties on coal combustion by-product quality: Examples from a Kentucky power plant. *Int. J. Coal Geol.* **2004**, *59*, 153–169.

9. Seames, W. An initial study of the fine fragmentation fly ash particle mode generated during pulverized coal combustion. *Fuel Process. Technol.* **2003**, *81*, 109–125.

10. Coles, D.; Ragaini, R.; Ondov, J.; Fisher, G.; Silberman, D.; Prentice, B. Chemical studies of stack fly ash from a coal-fired power plant. *Environ. Sci. Technol.* **1979**, *13*, 455–459.

11. Hower, J. Petrographic examination of coal-combustion fly ash. *Int. J. Coal Geol.* **2013**, *92*, 90–97.

12. Davison, R.; Natusch, D.; Wallace, J.; Evans, C. Trace elements in fly ash. Dependence on concentration of particle size. *Environ. Sci. Technol.* **1974**, *8*, 1107–1113.

13. Galbreath, K.; Zygarlicke, C. Mercury speciation in coal combustion and gasification flue gases. *Environ. Sci. Technol.* **1996**, *30*, 2421–2426.

14. Resource Conservation and Recovery Act (Public Law 94-580). *Code of Federal Regulations*, Chapter 139, Title 42, 1980.

15. Ruhl, L.; Vengosh, A.; Dwyer, G.; Hsu-Kim, H.; Deonarine, A. Environmental impacts of the coal ash spill in Kingston, Tennessee: An 18-month survey. *Environ. Sci. Technol.* **2010**, *44*, 9272–9278.

16. Talbot, R.; Anderson, M.; Andren, A. Qualitative model of heterogeneous equilibria in a fly ash pond. *Environ. Sci. Technol.* **1978**, *12*, 1057–1062.

17. Dudas, M. Long-term leachability of selected elements from fly ash. *Environ. Sci. Technol.* **1981**, *15*, 840–843.

18. Wadge, A.; Hutton, M. The leachability and chemical speciations of selected trace elements in fly ash and coal combustion refuse incineration. *Environ. Pollut.* **1987**, *48*, 85–99.

19. Fernandez-Turiel, J.; de Carvalho, W.; Cabanas, M.; Querol, X.; Lopez-Soler, A. Mobility of heavy metals from coal fly ash. *Environ. Geol.* **1994**, *23*, 264–270.

20. Querol, X.; Umana, J.; Alastuey, A.; Ayora, C.; Lopez-Soler, A.; Plana, F. Extraction of soluble major and trace elements from fly ash in open and closed leaching systems. *Fuel* **2001**, *80*, 801–813.

21. Nugteren, H.; Janssen-Jurkovicova, M.; Scarlett, B. Removal of heavy metals from fly ash and the impact on its quality. *J. Chem. Technol. Biotechnol.* **2002**, *77*, 389–395.

22. Smeda, A.; Zyrnicki, W. Application of sequential extraction and the ICP-AES method for study of the portioning of metals in fly ashes. *Microchem. J.* **2002**, *72*, 9–16.

23. Kim, A.; Kazonich, G. Relative solubility of cations in Class F fly ash. *Environ. Sci. Technol.* **2003**, *37*, 4507–4511.

24. Kim, A.; Kazonich, G. The silicate/non-silicate distribution of metals in fly ash and its effect on solubility. *Fuel* **2004**, *83*, 2285–2292.

25. Kim, A. The effect of alkalinity of Class F PC fly ash on metal release. *Fuel* **2006**, *85*, 1403–1410.

26. Kutchko, B.; Kim, A. Fly ash characterization by SEM-EDS. *Fuel* **2006**, *85*, 2537–2544.

27. Jegadeesan, G.; Al-Abed, S.; Pinto, P. Influence of trace metal distribution on its leachability from coal fly ash. *Fuel* **2008**, *87*, 1887–1893.

28. Yang, C. Leaching characteristics of metals in fly ash from coal-fired power plant by sequential extraction procedure. *Michrochim. Acta* **2009**, *165*, 91–96.

29. Ahmaruzzaman, M. A review of the utilization of fly ash. *Prog. Energy Combust. Sci.* **2010**, *36*, 327–363.

30. Bhattacharyya, P.; Reddy, K.; Attili, V. Solubility and fractionation of different metals in fly ash of the Powder River Basin coal. *Water Air Soil Pollut.* **2011**, *220*, 327–337.

31. Izquierdo, M.; Querol, X. Leaching behavior of elements from coal combustion fly ash: An overview. *Int. J. Coal Geol.* **2012**, *94*, 54–66.

32. Pacala, S.; Socolow, R. Stabilization wedges: Solving the climate problem for the next 50 years with current technologies. *Science* **2004**, *305*, 968–972.

33. Huijgen, W.; Comans, R. *Carbon Dioxide Sequestration by Mineral Carbonation: Literature Review*; Report ECN-C-03-016 for Energy Research Centre of The Netherlands: Petten, The Netherlands, 2003.

34. Oelkers, E.; Gislason, S.; Juerg, M. Mineral carbonation of CO_2. *Elements* **2008**, *4*, 333–337.

35. Reddy, K.J.; Lindsay, W.L.; Boyle, F.W.; Redente, E.F. Solubility relationships and mineral transformations associated with recarbonation of retorted oil shales. *J. Environ. Qual.* **1986**, *15*, 129–133.

36. Reddy, K.; Drever, J.; Hasfurther, V. Effects of a CO_2 pressure process on the solubilities of major and trace elements in oil shale solid wastes. *Environ. Sci. Technol.* **1991**, *25*, 1466–1469.

37. Tawfic, T.; Reddy, K.; Gloss, S. Reaction of CO_2 with clean coal technology ash to reduce trace element mobility. *Wate Air Soil Pollut.* **1994**, *84*, 385–398.

38. Wee, J. A review on carbon dioxide capture and storage technology using coal fly ash. *Appl. Energy* **2013**, *106*, 143–151.

39. Reddy, K. Coal Fly Ash Chemistry and Carbon Dioxide Infusion Process to Enhance Its Utilization. In *Biogeochemistry of Trace Elements in Coal and Coal Combustion Byproducts*; Sajwan, K., Alva, A., Keefer, R., Eds.; Springer: New York, NY, USA, 1999; pp. 133–143.

40. Renforth, P.; Washbourne, C.; Taylder, J.; Manning, D. Silicate production and availability for mineral carbonation. *Environ. Sci. Technol.* **2011**, *45*, 2035–2041.

41. Attili, V. Capture and Mineralization of Carbon Dioxide from Coal Combustion Flue Gas Emissions. Ph.D Thesis, University of Wyoming, Laramie, WY, USA, 2009.

42. Reddy, K.; Weber, H.; Bhattacharyya, P.; Argyle, M.; Taylor, D.; Christensen, M.; Foulke, T.; Fahlsing, P. Instantaneous Capture and Mineralization of Flue Gas Carbon Dioxide: Pilot Scale Study. Available online: http://precedings.nature.com/documents/5404/version/1 (accessed on 17 March 2014).

43. Reddy, K.; John, S.; Weber, H.; Argyle, M.; Bhattacharya, P.; Taylor, D.; Christensen, M.; Foulke, T.; Fahlsing, P. Simultaneous capture and mineralization of anthropogenic carbon dioxide (CO_2). *Energy Procedia* **2010**, *4*, 1574–1583.

44. Bhattacharyya, P.; Reddy, K. Effect of flue gas treatment on the solubility and fractionation of different metals in fly ash of Powder River Basin coal. *Water Air Soil Pollut.* **2012**, *223*, 4169–4181.

45. Tessier, A. Sequential extraction procedure for the speciation of particulate trace metals. *Anal. Chem.* **1979**, *51*, 844–851.

46. Warren, C.; Dudas, M. Formation of secondary minerals in artificially weather fly ash. *J. Environ. Qual.* **1985**, *14*, 405–410.

47. Medina, A.; Gamero, P.; Querol, X.; Moreno, N.; DeLeon, B.; Almanza, M.; Vargas, G.; Izquierdo, M.; Font, O. Fly ash from a Mexican mineral coal I: Mineralogical and chemical characterization. *J. Hazard. Mater.* **2010**, *181*, 82–90.

48. Reardon, E.; Czank, C.; Warren, C.; Dayal, R.; Johnston, H. Determining controls on elemental concentrations in fly ash leachate. *Waste Manag. Res.* **1995**, *13*, 435–450.

49. Jones, D. The Leaching of Major and Trace Elements from Coal Fly Ash. In *Environmental Aspects of Trace Elements in Coal*; Swaine, D., Goodzari, F., Eds.; Springer: Berlin, Germany, 1995; Volume 2, pp. 221–262.

50. Dudas, M.; Warren, C. Submicroscopic model of fly ash particles. *Geoderma* **1987**, *40*, 101–114.

51. Summers, C.; Dahlin, D.; Ochs, T. *The Effect of SO_2 on Mineral Carbonation in Batch Tests*; DOE/ARC-2004-022; Office of Fossil Energy (FE), U.S. Department of Energy: Washington, DC, USA, 2004.

52. Schramke, J. Neutralization of alkaline coal fly ash leachates by $CO_2(g)$. *Appl. Geochem.* **1992**, *7*, 481–492.

53. Tsuchiai, H.; Ishizuka, T.; Ueno, T.; Hattori, H.; Kita, H. Highly active absorbent for SO_2 removal prepared from coal fly ash. *Ind. Eng. Chem. Res.* **1995**, *34*, 1404–1411.

54. Serre, S.; Silcox, G. Adsorption of elemental mercury on the residual carbon in coal fly ash. *Ind. Eng. Chem. Res.* **2000**, *39*, 1723–1730.

55. Dunham, G.; DeWall, R.; Senior, C. Fixed-bed studies of the interactions between mercury and coal combustion fly ash. *Fuel Process. Technol.* **2003**, *82*, 197–213.

56. Niksa, S.; Naoki, F. Predicting extents of mercury oxidation in coal-derived flue gases. *J. Air Waste Manag. Assoc.* **2005**, *55*, 930–939.

57. Gale, T.; Lani, B.; Offen, G. Mechanisms governing the fate of mercury in coal-fired power systems. *Fuel Process. Technol.* **2008**, *89*, 139–151.

58. Goodzari, F.; Hower, J. Classification of carbon in Canadian fly ashes and their implications in the capture of mercury. *Fuel* **2008**, *81*, 1949–1957.

59. Shah, P.; Strezov, V.; Prince, K.; Nelson, P. Speciation of As, Cr, Se, and Hg under coal fired power station conditions. *Fuel* **2008**, *87*, 1859–1869.

60. Hower, J.; Senior, C.; Suuberg, E.; Hurt, R.; Wilcox, J.; Olson, E. Mercury capture by native fly ash carbons in coal-fired power plants. *Prog. Energy Combust. Sci.* **2010**, *36*, 510–529.

61. Noel, J.; Biswas, P.; Giammar, D. Evaluation of a sequential extraction process used for determining mercury binding mechanisms to coal combustion byproducts. *J. Air Waste Manag. Assoc.* **2007**, *57*, 856–867.

62. Narukawa, T.; Takatsu, C.; Riley, K.; French, D. Investigation on chemical species of arsenic, selenium and antimony in fly ash from coal fuel thermal power stations. *J. Environ. Monit.* **2005**, *7*, 1342–1348.

63. Cornelis, G.; Johnson, A.; VanGerven, T.; Vandecasteele, C. Leaching mechanism of oxyanionic metalloid and metal species in alkaline solid wastes: A review. *Appl. Geochem.* **2008**, *23*, 955–976.

64. Rai, D.; Eary, L.; Zachara, J. Environmental chemistry of chromium. *Sci. Total Environ.* **1989**, *86*, 15–23.

65. Meima, J.; Comans, N. The leaching of trace elements from municipal solid waste incinerator bottom ash at different stages of weathering. *Appl. Geochem.* **1999**, *14*, 159–171.

66. Meima, J.; van der Weijden, R.; Eighmy, T.; Comans, R. Carbonation processes in municipal solid waste incinerator bottom ash and their effect on the leaching of copper and molybdenum. *Appl. Geochem.* **2002**, *17*, 1503–1513.

67. Meima, J.; Comans, N. Application of surface complexation/precipitation modeling to contaminant leaching from weathered solid waste incinerator bottom ash. *Environ. Sci. Technol.* **1998**, *32*, 688–693.

68. Zachara, J.; Kittrick, J.; Harsh, J. The mechanism of Zn^{2+} adsorption on calcite. *Geochem. Cosmochim. Acta* **1988**, *52*, 2281–2291.

Computational Redox Potential Predictions: Applications to Inorganic and Organic Aqueous Complexes, and Complexes Adsorbed to Mineral Surfaces

Krishnamoorthy Arumugam and Udo Becker *

Department of Earth and Environmental Sciences, University of Michigan, 1100 North University Avenue, 2534 C.C. Little, Ann Arbor, MI 48109-1005, USA; E-Mail: arumugam@umich.edu

* Author to whom correspondence should be addressed; E-Mail: ubecker@umich.edu

Abstract: Applications of redox processes range over a number of scientific fields. This review article summarizes the theory behind the calculation of redox potentials in solution for species such as organic compounds, inorganic complexes, actinides, battery materials, and mineral surface-bound-species. Different computational approaches to predict and determine redox potentials of electron transitions are discussed along with their respective pros and cons for the prediction of redox potentials. Subsequently, recommendations are made for certain necessary computational settings required for accurate calculation of redox potentials. This article reviews the importance of computational parameters, such as basis sets, density functional theory (DFT) functionals, and relativistic approaches and the role that physicochemical processes play on the shift of redox potentials, such as hydration or spin orbit coupling, and will aid in finding suitable combinations of approaches for different chemical and geochemical applications. Identifying cost-effective and credible computational approaches is essential to benchmark redox potential calculations against experiments. Once a good theoretical approach is found to model the chemistry and thermodynamics of the redox and electron transfer process, this knowledge can be incorporated into models of more complex reaction mechanisms that include diffusion in the solute, surface diffusion, and dehydration, to name a few. This knowledge is important to fully understand the nature of redox processes be it a geochemical process that dictates natural redox reactions or one that is being used for the optimization of a chemical process in industry. In addition, it will help identify materials that will be useful to design catalytic

redox agents, to come up with materials to be used for batteries and photovoltaic processes, and to identify new and improved remediation strategies in environmental engineering, for example the reduction of actinides and their subsequent immobilization. Highly under-investigated is the role of redox-active semiconducting mineral surfaces as catalysts for promoting natural redox processes. Such knowledge is crucial to derive process-oriented mechanisms, kinetics, and rate laws for inorganic and organic redox processes in nature. In addition, molecular-level details still need to be explored and understood to plan for safer disposal of hazardous materials. In light of this, we include new research on the effect of iron-sulfide mineral surfaces, such as pyrite and mackinawite, on the redox chemistry of actinyl aqua complexes in aqueous solution.

Keywords: redox potential calculations; density functional theory (DFT) methods; semiconducting minerals; actinides; continuum solvation; mineralogy

---

1. Introduction

One fundamental type of process that control energy fluxes in nature is redox processes, which involves electron transfer reactions that relate to a number of scientific fields, such as chemistry, biology, geochemistry, and mineralogy.

Reduction of hazardous toxic elements such as Cr(VI) and As(V) by redox active minerals where the role of redox chemistry is not well understood. The toxic Cr(VI) exhibits as CrO_4^{2-} ion and this is relatively more toxic to the environment and regarded as one of the dangerous carcinogenic agent [1], however the reduction of Cr(VI) into Cr(III) process makes it as less toxic than the former one. Molecular level details for redox mechanisms for the Cr(VI) into Cr(III) process are still unexplored and not properly studied [1], except a recent computational study on the cytochrome mediated enzymatic reduction of chromate [2]. In addition, the semiconducting redox-active surface mediated redox processes of these toxic elements are under-studied and molecular level details are lacking [3–5]. As in the environment exists as arsenate As(V) and arsenite As(III). This is also toxic to human environment and lead to several health problems in human. The penta-valent As is more toxic than the tri-valent As [6]. Surface mediated redox process of As(V) and As(III) on Fe-sulfide mineral surfaces, such as pyrite [7] and mackinawite [8], are unexplored and not well understood.

An(VI) (An = U, Np, and Pu) is one of the most stable oxidation state for the actinides and exists as a linear oxo cation, in particular for the U and Pu. In contrast, for the Np the stable oxidation state is V for the actinyl species. These actinyl ions are soluble in aqueous environment and highly mobile; this makes these ions for their active transport process into the geosphere. These are radioactive materials, highly hazardous and lead to long term severe contaminations. Reduction process of these highly mobile and soluble actinyl ions are often catalyzed by the redox-active semiconducting minerals, for example iron sulfides (pyrite and mackinawite) [9–13] or oxides (hematite and magnetite) [9,12,14] present in the environment, in the presence of reductants, for instance, quinones, and Fe(II) species in solution, *etc.* Although the reduction process is a more complicated process and it was found that there were different mechanistic pathways were proposed depending on the reaction conditions and the presence of chelating ligands, for instance, hydroxyl, carbonate, chloride, and bicarbonate are available in the environmental conditions. These ligands would more likely to form chelation and coordination around the equatorial plane of these actinyl ions, which in turn alters the redox behavior of these actinyl ions in solution [15,16].

After the reduction process, either by the redox-active mineral surface, or by any organic reductants, for example quinones, or radicals, for example OH radicals, the actinyl(V) reduced species is formed, which is unstable with respect to disproportionation in aqueous environment. This disproportionation process takes place via the formation of a cation-cation intermediate and followed by protonation steps as computationally proposed by Steele et al. [17]. Eventually this lead to the formation of actinyl(VI) and An(IV) species. The formed An(IV) species are stable and insoluble in aqueous solution and precipitate as uraninite [UO_2] or colloidal precipitated species. If there is a chelating ligand in solution, for instance carbonate ligands lead to form U(IV) carbonate complex rather than uraninite [14,18].

In addition, the redox-active mineral surface plays an important role in the reduction process and acts as a template and adsorbs the formed product on its surface and retains more often. However, there are different possibilities for the reduction to take place; either it could take place in solution, or on the surface adsorbed complexes. If the reduction takes place in solution, then there are two possibilities here, it could be adsorbed onto mineral surface and then either go for disproportionation or second reduction. Even disproportionation reaction for the actinyl(V) may possible in solution itself even before adsorption. Another interesting possibility is that the An(VI) species might be first adsorbed on the redox active mineral surface and then the reduction process, followed by either surface mediated disproportionation or the proton coupled second direct reduction. These various possibilities and complicated redox mechanistic pathways are challenging to density functional theory (DFT) methods, since treating a surface process using small cluster models has its limitations. However, this could give a deeper understanding to these fundamental redox processes that have been taking place for hundreds of years perhaps millions of years. Understanding these semiconductor redox active minerals surface mediated redox process of the environmentally toxic and hazardous actinide materials will help us to design effective remediation processes for immobilization and eventually even reused and recycled. This prevents long-standing and long term serious contamination to the environment.

Aqueous sediments contain a variety of dissimilatory metal-reducing bacteria (DMRB), which are often involved in reduction processes of heavy metals [19]. The cell walls of DMRBs carry redox-active proteins, for instance cytochromes. Numerous reports are available for the reduction of actinides by bacterial strains and this is bioremediation technique [19–22]. Despite the large number of experimental studies, the actual mechanism for the reduction process is still unknown or controversial. These studies agree that the redox-active proteins present in the bacterial strains are responsible for the electron transfer and reduction. Recently, reduction of actinyl(VI) by cytochrome enzymes, followed by the disproportionation of actinyl(V) through the formation of a cation-cation intermediate were computationally examined [23]. In addition, electron transfer and reduction of actinyl(VI) by protonated mackinawite surfaces and the subsequent disproportionation were modeled using a DFT computational approach [24]. The results of these studies confirm that the enzymes and semiconducting redox-active minerals are playing a vital role in the redox and immobilization process, in general. As a consequence, the optimization of effective bio-remediation methods may profit from an atomic-scale knowledge gained from the computational modeling of enzymes and semiconducting redox active minerals mediated redox process of actinides [23,24].

Another interesting aspect in understanding redox processes are E_H-pH or Pourbaix diagrams in which the electrode potentials of relevant species are plotted against the pH of the solution. Depending on the pH of the solution, different species often exists in equilibrium with each other. By combining, for example Fe with As or Fe with organic compounds will help us to understand the redox active behavior of hazardous pollutants in the environmental conditions.

One-electron redox process of a redox reaction can be simply defined in terms of a half-cell reaction as follows:

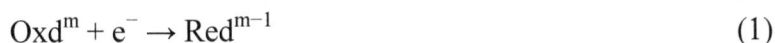

$$Oxd^m + e^- \rightarrow Red^{m-1} \tag{1}$$

where Oxd is an oxidized species and Red is a reduced species. During a redox reaction, the Oxd species gains an electron and forms the reduced species, Red. Overall, this reaction is called as redox half-cell reaction. Then, two half-cell reactions can be combined to obtain a complete redox reaction.

The accurate prediction of redox potentials using appropriate computational approaches can help us understand redox mechanisms of geochemical reactions and aid us in designing and optimizing redox-sensitive remediation techniques. In addition, the controlled modification of geochemical or industrial redox processes can optimize the design of redox-active catalytic agents, which can be utilized in large-scale industrial applications.

In this paper, we are going to give an overview and summarize the available computational methods for predictions of redox potentials in solution. Advantages and disadvantages of different methods reported in the literature are being examined and a suitable approach for the prediction of redox potentials of semiconducting mineral surfaces will be recommended. In addition, we will describe the prediction of reduction potentials for redox reactions involving organic materials, transition metal complexes, and actinides in solution. However, our main intention is, after having evaluated computational approaches for the prediction of reduction potentials from the literature, to develop and apply a reliable computational approach and use it to investigate the redox chemistry of semi-conducting minerals and redox-active surface-mediated chemical transformations.

Several computational settings have to be taken into account when tackling the computation of reduction potentials, which include the basis sets used, solvation models, free energy corrections, zero point energy (ZPE) corrections, standard state corrections, spin-orbit coupling interactions, and relativistic effects. Although, to a certain degree, relativistic effects can be included into pseudopotentials (PP) of actinide elements, the number core electrons replaced by the pseudopotential, which is 78 electrons for large-core PPs and 60 electrons for small-core ones. Using small-core PPs in combination with DFT methods was found to reproduce experimental reduction potentials in aqueous solution more closely, which will be described in more detail later.

In the introduction section, we have demonstrated the motivation for this article based on the wider applications of redox processes in various fields such as chemistry, biology, and mineralogy. The remaining sections of this article are organized as follows. We will introduce the essential background for calculating redox potentials followed by describing the importance of the reference electrode potential and the thermodynamic cycle in various redox potential prediction methods. This is followed by how these computational tools are applied to redox potentials predictions of organic compounds, transition metal complexes, actinides, and semiconducting materials in solution.

2. Theory of Redox Potential Predictions

After describing the computational treatment of reference electrodes for aqueous and non-aqueous solutions, we will briefly explain the relation between the thermodynamic cycle and the calculation of redox potentials. Finally, we will illustrate various methods to predict the Gibbs free energy of redox processes in solution, which is the essential part of calculating redox potentials. In addition, we will outline the absolute potentials of reference electrodes and their values in aqueous and non-aqueous solutions.

2.1. Aqueous Solutions

Standard reduction potentials are typically referenced to the standard hydrogen electrode (SHE), whose reaction is given as:

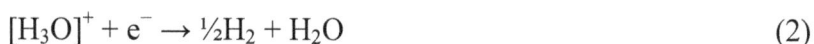

$$[H_3O]^+ + e^- \rightarrow \tfrac{1}{2}H_2 + H_2O \tag{2}$$

When the SHE is used as a reference electrode in cyclic voltammetry experiment, its value is defined to be zero at pH = 0 and atmospheric pressure of H_2 = 1 bar. Similar to the SHE, there are other reference electrodes such as the saturated calomel electrode (SCE) and $Ag^+/AgCl$ electrode that are in practice easier to handle in aqueous solution than the SHE. Although the experimentalist would determine the redox potentials of complexes with respect to any suitable and practical reference electrode, the determined redox potentials can then be converted with respect to the SHE or any other reference electrode of interest. However, care should be taken when converting experimental redox values determined with respect to one reference electrode *vs.* another, the liquid junction potential is often problematic. For instance, to convert redox potentials determined against SCE to SHE, 0.24 V have to be subtracted from the determined values.

Typically, experimental redox potentials are referenced or reported with respect to reference electrodes, for instance the SHE, in literature. Although during the experiment the value of the SHE is considered zero, its absolute value is not zero; a range of absolute reference values for the SHE have been reported from 4.24 to 4.73 eV [25–29]. However, the IUPAC had recommended a value of 4.44 eV as the absolute value of the SHE in 1986 [26], a value that has been recently confirmed by an experimental study [27]. It should be noted that the absolute value for the SHE has been in debate for years. Later in this article, we will provide the various absolute values determined for the SHE in aqueous and non-aqueous solutions.

2.2. Non-Aqueous Solutions

In some cases, it may be necessary to use non-aqueous solutions, e.g., if the reactants are not stable or soluble in water. The ferrocene redox couple has been widely used as a reference redox couple for non-aqueous solutions. The ferrocene molecule is a metallocene type transition metal organo-metallic complex, in which the Fe(II) ion is sandwiched between two cyclopentadienyl (Cp) anionic ligands. Depending on the Cp ring conformations, two structural isomers with different symmetries are possible, eclipsed (d5h, Figure 1) and staggered (d5d). However, the barrier is small between these two conformers [30].

Figure 1. Ferrocene (eclipsed) structure in vertical (**a**) and horizontal (**b**) views.

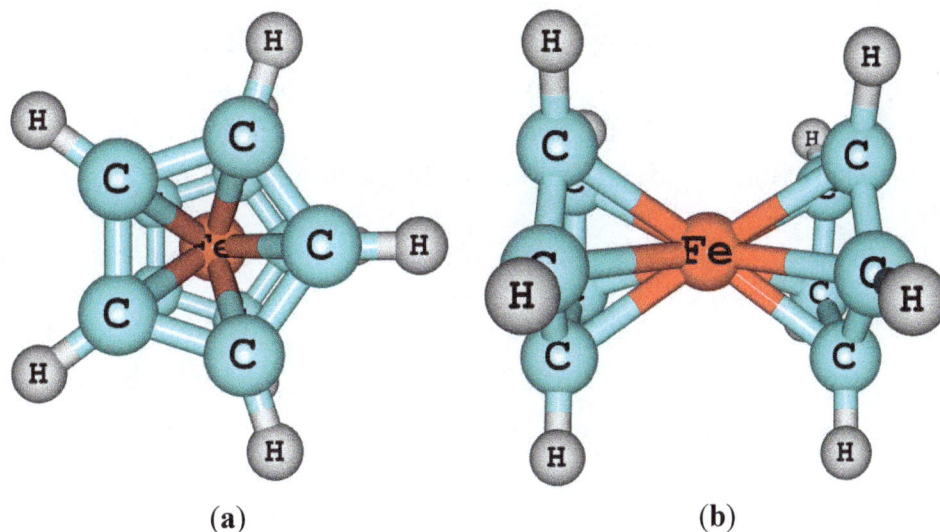

(**a**)   (**b**)

The ferrocene/ferrocenium ion (Fc/Fc$^+$) redox system was found to show a solvent-independent redox behavior in a range of 22 non-aqueous solvents and was thus recommended as a reference electrode system by the IUPAC for electrochemical studies in non-aqueous solutions [31]. Since then, the Fc/Fc$^+$ reference electrode has been widely used and accepted as a reference electrode system for non-aqueous solutions. The half-cell reaction for Fc/Fc$^+$ reference electrode system is given as follows:

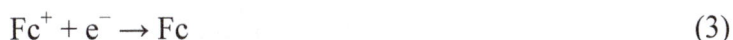

$$Fc^+ + e^- \rightarrow Fc \qquad (3)$$

Interconversion of electrode potentials between different reference electrodes is a problematic issue unless similar experimental conditions, for example electrolyte concentration, ionic strength, and solvents, are used during the redox potential determinations [32,33]. To illustrate the effect of ionic-strength on the Fc/Fc$^+$ potentials, different electrolyte concentrations were investigated and reported. Depending on the electrolyte concentration, the Fc/Fc$^+$ redox potential varies within a range of ~0.1–0.15 eV, and while this is significant as far as computational predictions are concerned, these uncertainties in experimental redox potentials could also introduce systematic errors in computational predictions [34].

Absolute potentials for the Fc/Fc$^+$ reference electrode system in various solvents have recently been determined computationally using a high-level G3 method (G3 method involves several post self-consistent field (SCF) calculations where the energy expression is given as, $E_0[G3(MP2)] = QCISD(T)/6\text{-}31G(d) + \Delta E(MP2) + \Delta E(SO) + E(HLC) + E(ZPE)$, where SO is the spin-orbit interaction, HLC is the "high level correction", and ZPE is the zero point energy correction). Although the computational study had used the G3 method, the gas-phase ionization potential for Fc was predicted to be 0.20 eV smaller than the experimental ionization potential. Despite this underestimation, the absolute potential for the Fc/Fc$^+$ reference electrode in dimethylsulfoxide (DMSO), acetonitrile (ACN) and dichloroethane (DCE) solvents were accurately predicted by encompassing the conductor-like polarizable-continuum (CPCM) and the charge-density based solvation model (SMD).

2.3. Thermodynamic Cycle

To get the reduction potentials from computations, the thermodynamic cycle (see Schematic 1) is used. This is generally defined as a schematic representation of gas-phase and solution phase reactions and the relation between the phases. If the reaction is a redox reaction or any other chemical transformation, the thermodynamic cycle can be used to evaluate the reaction free energy.

Schematic 1. Thermodynamic cycle for the calculation of Gibbs free energies of a one-electron reduction process.

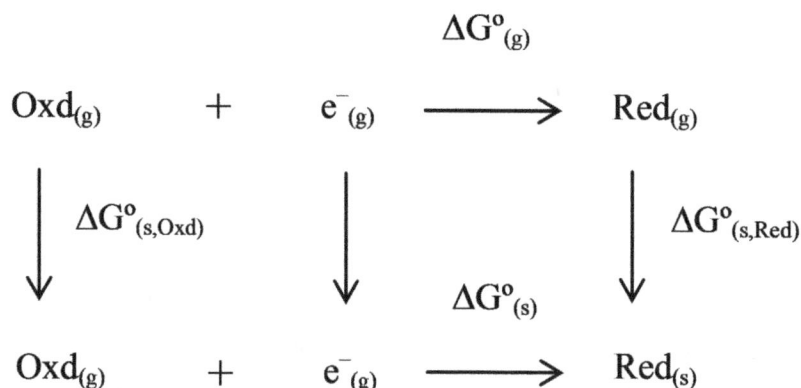

$$\Delta G^{o}_{(g)}$$

$$Oxd_{(g)} \quad + \quad e^{-}_{(g)} \quad \longrightarrow \quad Red_{(g)}$$

$$\Delta G^{o}_{(s,Oxd)} \qquad\qquad \Delta G^{o}_{(s)} \qquad\qquad \Delta G^{o}_{(s,Red)}$$

$$Oxd_{(g)} \quad + \quad e^{-}_{(g)} \quad \longrightarrow \quad Red_{(s)}$$

The thermodynamic cycle involves several terms that correspond to the gas or solution phase. The free energy of the electron is not taken into account since the addition of another half-cell reaction for the reference electrode automatically cancels the energy of the electron out. It could also be argued that the ionic convention of an electron leads to a zero value of its free energy in the gas and solution phase.

For the free electron, there are three different conventions, namely ionic, Fermi-Dirac, and Boltzmann statistics conventions (the ionic convention is based on the "gaseous ion energetics", in which the heat capacity of a free electron is assigned a value of zero. The other two conventions are based on different statistical treatments, though the resulting differences in the energy are small (0.04 eV) between these two conventions). Based on the ionic convention, the free energy of a free electron is considered zero in the gas as well as in the solution phase. In contrast, the other two conventions give slightly different heat capacity values for the free electron. However, the resulting overall free energy value for the redox reaction is not significantly different between the ionic and Fermi-Dirac conventions of the electron, hence this is typically ignored [35].

$$\Delta G^{o}(s) = \Delta G^{o}_{(gas)} + \Delta\Delta G^{o}(s) + \Delta ZPE + \Delta FEc \tag{4}$$

$$\Delta\Delta G^{o}_{(s)} = \Delta G^{o}_{(s,Red)} - \Delta G^{o}_{(s,Oxd)} \tag{5}$$

$$\Delta G^{*}_{(s)} = \Delta G^{o}_{(gas)} + [(\Delta G^{o}_{(s,Red)} + \Delta G^{1atm \rightarrow 1M})] - [(\Delta G^{o}_{(s,Oxd)} + \Delta G^{1atm \rightarrow 1M})] + \Delta ZPE + \Delta FEc \tag{6}$$

$$\Delta G^{*}_{red} = \Delta G^{*}(s) - \Delta G_{(ref.elec)} \tag{7}$$

where $\Delta G^{o}_{(g)}$ is the reduction free energy in the gas-phase, $\Delta G^{o}_{(s)}$ is the standard state reduction free energy of the redox reaction in the solution-phase, $\Delta G^{o}_{(s,Oxd)}$ is the standard state solvation free energy of the oxidized species, $\Delta G^{o}_{(s,Red)}$ is the standard state solvation free energy of the reduced species, ZPE is the zero-point energy correction obtained from frequency calculations at stationary equilibrium geometry, FEc is the free-energy correction from thermal contributions ΔG^{*}_{red} is the standard reduction free energy of the redox reaction referenced with respect to reference electrode, and $\Delta G_{(ref.elec)}$ is the free energy of the reference electrode. Here, the ($^\circ$) notation corresponds to the standard state at 1 atm and 298.15 K, whereas the (*) notation corresponds to 1 mol/L. The term $\Delta G^{1atm \rightarrow 1M}$ equals to transfer a reagent from its gas phase at 1 atm to its dissolved state at 1 mol/L, which numerically equals to 7.93 kJ/mol. In the above redox equation (see Schematic 1), the number of species on the left and right hand sides are the same such that there is no need to include the standard free energy correction term, $\Delta G^{1atm \rightarrow 1M}$, in this case. Another way to look at the Gibbs free energy of mixing is using $G = E_{elec} + ZPE + E_{trans} + E_{rot} + E_{vib} + RT - TS$, where E_{elec} is the SCF energy, ZPE the zero point energy, *i.e.*, vibrations that still exist at 0 K, the next three terms refer to translational, rotational, and vibrational energies, respectively. R is the ideal gas-constant, T temperature, and S entropy. The last five terms totally referred as free-energy correction. It should be noted that if there is a change in number of moles in the redox reaction of interest, a standard free energy correction of 7.93 kJ/mol has to be included [36] (this standard free energy correction term can be understood, e.g., for a chemical reaction, $A + B \rightarrow C$, where the free energy for this reaction in the standard state can be expressed as $\Delta G^* = \Delta G^\circ + RT \ln([C]/[A][B])$, in which * refers to 1 mol/L whereas the $^\circ$ refers to 1 atm standard states for all species. Based on the ideal-gas assumption, the concentrations of [A], [B], and [C] are defined as 1/25.4 mol/L (at 298.15 K). Inserting these values into the above equation lead to $\Delta G^* = \Delta G^\circ + RT \ln(25.4)$ or $\Delta G^* = \Delta G^\circ + 7.93$ kJ/mol.

According to the Nernst equation, the free energy of a reduction reaction is related to an experimentally determined reduction potential.

$$E_0 = -\Delta G^*_{red}/nF \tag{8}$$

where n is the number of electrons transferred in the redox reaction, F is the Faraday constant (96.48 kJ mol^{-1} V^{-1} or 96485 C mol^{-1}) and E_0 is the experimentally determined redox potential (in V).

2.4. Methods Used to Calculate the Gibbs Free Energy and Redox Potential of a Redox Reaction

In this section, different computational methods that were employed to predict the reduction potentials are briefly described. Two frequently used methods for reduction potential predictions are the direct and isodesmic reaction method.

2.4.1. Direct Method

The reference-electrode half-cell reaction, for instance the half-cell reaction of the SHE or Fc/Fc$^+$ reference electrode, can be included into the redox reaction of interest and the reference electrode value would be calculated in a way similar to the half-cell reaction of the reactant of interest. Combining the redox half-cell reaction Equations (1) and (2) will, e.g., produce the following complete redox reaction.

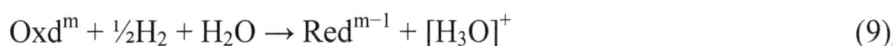

$$Oxd^m + \tfrac{1}{2}H_2 + H_2O \rightarrow Red^{m-1} + [H_3O]^+ \tag{9}$$

The reduction free energy for the overall redox reaction (Equation (9)) can then be calculated according to the expression in Equation (6). The overall redox equation (Equation (9)) already contains the half-cell reaction of the reference electrode (SHE) in the overall redox reaction, hence the

$\Delta G_{(ref\text{-}elec)}$ term vanishes. The free energy value of the reference electrode half-cell reaction is calculated using the relevant species involved in the redox reaction of the reference electrode system. There are experimentally determined absolute values available for these reference electrodes in different solvents; for example, the absolute values of the SHE in H_2O, methanol, ethanol, acetonitrile, and dimethylsulfoxide (DMSO) are different, these absolute values were derived from thermodynamic parameters [6,7,26,27,35,37] (see Table 1). In addition, the absolute value of the Fc/Fc^+ reference electrode with respect to the SHE is also known. By knowing the correct interconversion factors, the determined redox potential can be converted into the reference electrode of interest. This issue of interconversion has been recently addressed [33]. Typically, the absolute value of the reference electrode potential can be taken for redox potential predictions, which minimizes systematic errors compared to calculating the absolute value of the reference electrode, for example the SHE, at the same level of theory employed for the redox half-cell reaction of interest. If the reference electrode potential is not calculated accurately, the redox potential estimated with the erroneous reference potential will introduce errors that lead the predicted redox potentials to deviate from experimentally-determined redox potential values.

Table 1. Absolute potentials of the standard hydrogen electrode (SHE) (V) in different solvents and from different sources.

Solvents	Trassatti	Fawcett	Kelly [a]	Kelly [b]
Water	4.44	4.42	4.24	4.28
Methanol	4.19	4.17	4.34	4.38
Ethanol	4.21	4.24	-	-
Acetonitrile	4.60	4.59	4.48	4.52
Dimethylsulfoxide	-	3.83	3.92	3.96

Notes: [a] the integrated heat capacity and entropy values for the free electron were based on the Fermi-Dirac statistics; [b] the integrated heat capacity and entropy values for the free electron were based on the Boltzmann statistics.

2.4.2. Isodesmic Method

In the isodesmic model, rather than using a reference electrode reaction in the overall redox reaction, calculated redox potentials are referenced with respect to the redox half-cell reaction of a reference complex (Equation (10)). The fact that the inclusion of the redox potential of the reference complex automatically determines the reference potential with respect to the reference electrode as to the determined experimental redox potential. Error cancellations lead to minimize systematic errors in the redox potential prediction. Moreover calculating reference electrode potentials accurately is difficult, which in turn introduces systematic errors to redox potential predictions.

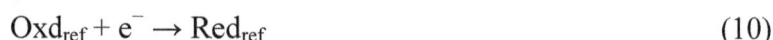

$$Oxd_{ref} + e^- \rightarrow Red_{ref} \tag{10}$$

Combining Equations (1) and (10) gives an overall redox reaction,

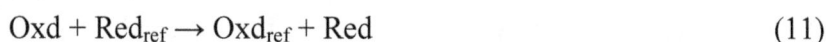

$$Oxd + Red_{ref} \rightarrow Oxd_{ref} + Red \tag{11}$$

Free energies of the species involved in the isodesmic redox reaction (Equation (3)) are calculated using the chosen DFT method. The reduction free energy ($\Delta G^*_{(s)}$) is calculated for the isodesmic model redox reaction according to Equation (11) using the thermodynamic cycle scheme as explained earlier (see Schematic 1).

$$\Delta G^*{}_{red} = \Delta G^*{}_{(s)} - \Delta G_{(ref)} \tag{12}$$

where the $\Delta G_{(ref)}$ is the experimental redox free energy of the reference complex used as isodesmic model. This method had been successfully applied not only to organic compounds [38–41] but also for transition metal complexes and reproduced experimentally determined redox potentials accurately within ~0.1 eV [42–44]. Moreover, this isodesmic method was employed to predict the redox potentials of actinyl (VI/V) in solution and calculated redox potentials were in good agreement with the experimentally determined redox potentials [45].

3. Computational Methods

This section summarizes computational methods used to calculate redox potentials, with an overview on the suitability of different density functional theory (DFT) methods, basis sets, and solvation methods in the first part and an introduction to thermodynamic integration at the end.

3.1. DFT or Wave Function Based Methods

DFT is based on the two theorems proposed by Kohn and Hohenberg [46,47]. We are not going to provide any rigorous treatment and derive DFT equations here, which is beyond the scope of this paper, instead, we urge the readers to refer the literature on DFT methods [46–50]. The general form of energy as a function of density within the DFT formalism is given below,

$$E[\rho(r)] = T_e[\rho(r)] + V_{ne}[\rho(r)] + V_{ee}[\rho(r)] + E_{XC}[\rho(r)] \tag{13}$$

In Equation (13), $T_e[\rho(r)]$ is the kinetic energy of electrons, $V_{ne}[\rho(r)]$ is the nuclear-electron interaction, $V_{ee}[\rho(r)]$ is the Coulomb repulsion, and $E_{XC}[\rho(r)]$ is the exchange-correlation functional. The Kohn-Sham one electron orbital equations as a function of density are iteratively solved using a self-consistent field algorithm and the overall energy is minimized according to the variational principle.

There are several ways to define the exchange-correlation functional term ($E_{XC}[\rho(r)]$): the first approximation introduced was the local density approximation (LDA, simplifies electron-electron interaction by the interaction between an electron and the charge density of the other electrons) followed the local spin density approximation (LSDA), which allowed for the calculation of systems with unpaired spins. In addition to this, the generalized gradient approximation (GGA, simplifies electron-electron interaction by the interaction between an electron and the charge density and its gradient of the other electrons) and hybrid forms between Hartree-Fock approaches and DFT functionals (e.g., B3LYP [51–53], and HSE [54]) were introduced later. The following expression gives the functional form for the most widely used hybrid DFT B3LYP in computational studies,

$$E_{XC}^{B3LYP} = aE_X^{HF} + (1-a)E_X^{LSDA} + b\Delta E_X^{B88} + (1+c)E_C^{LSDA} + c\Delta E_C^{LYP} \tag{14}$$

E_X^{HF} is the Hartree-Fock exchange energy, E_X^{LSDA} the local spin density approximation, B88 the Becke's exchange functional, a, b, and c are semi-empirical parameters and the corresponding values are 0.20, 0.72, and 0.81, respectively. The LYP term corresponds to the Lee Yang Parr correlational functional [53]. The subscript C refers to correlation whereas the X refers to exchange.

3.1.1. Choice of DFT Functional

Unless the choice of DFT functional produces accurate geometries as compared to high-quality experiments, other predicted properties are likely to be inaccurate, for instance the redox potential. It is worth to be cautious in choosing appropriate DFT functionals for redox potential predictions. Although the B3LYP hybrid (Hartree-Fock-DFT) functional has been found to accurately predict redox potentials for organic compounds, for transition metal complexes, the GGA DFT functional BP86 performed better than the B3LYP one [55–57]. Despite these results for transition metals [34], the B3LYP functional has reproduced redox potentials for actinyl complexes in agreement with experimental values in solution. In addition, the M06 and PBE0 functionals were also found to give good agreement with experimental redox potentials [45,58] and recently, the M06L functional was found to give very small mean unsigned error (MUE) error values, e.g., 0.04 eV for actinyl(VI/V) redox potentials predictions in aqueous solution with respect to experimental redox potentials [59].

3.1.2. Basis Set Choice

In general, basis sets such as the double or triple zeta basis with additional diffuse and polarization functions are sufficient to capture the energetics of redox processes [60,61]. Since the redox process may lead to anionic species, diffuse functions are crucial to smear the electron density over the space. As noted by Baik *et al.*, larger basis sets with additional diffuse functions improve the predicted redox potentials for organics and transition metal complexes in solution [61]. This study used a 6-31G** basis set for geometry optimizations and single-point energy evaluations were done with the cc-pVTZ(-f) and cc-pVTZ(-f)++ basis sets. The average deviations for the calculated reduction potentials with respect to experiments for these three types of basis sets were 0.425, 0.171, and 0.120 eV, respectively. In addition, for actinyl(VI/V) reduction potential calculations in solution, the B3LYP/SC-SDD/6-31G* (SC-SDD for An-atoms with An = U, Np, and Pu, and 6-31G* basis set for non-metal atoms) level of theory was used for geometry optimizations, later energy evaluations were done in gas and aqueous-phase with the inclusion of additional polarization and diffusion functions for non-metal atoms (6-311+G** basis set) [16,45,58]. Although a medium-size double zeta basis set, for example, 6-31G* (for non-metals), is adequate for geometry optimizations, energy evaluations with additional diffuse and polarization functions result in more accurate thermodynamic properties. Often, reduction of a neutral molecule results in an anionic species formation, hence, it is suggested to use an appropriate basis set containing diffuse and polarization functions to define atoms present in the molecule for the effective modeling of redox process in solution.

3.2. Solvation

Solvation is a crucial factor in redox potential determination. Solvation can be modeled by static or dynamic simulations of explicit water molecules around the ion or complex, by replacing the water by a homogenous dielectric fluid, or by a combination of the above, e.g., by calculating the first or first two hydration spheres with explicit water and the surroundings by a dielectric fluid. Several solvation models for the dielectric fluid approach are available in the literature and incorporated in some quantum-mechanical programs, such as PCM (polarizable continuum model) [62], CPCM (conductor-like polarizable continuum model) [63–65], IEF-PCM (integral equation formalism-polarizable continuum model) [66,67], SMD (solvation model density) [68], COSMO (conductor like screening model) [63], COSMO-RS (conductor like screening model for real solvents) [69,70], and PB (Poisson-Boltzmann) finite element model [71–73]. Among these solvation methods, CPCM solvation has been one of the most widely used solvation method to study solvation effects. A more rigorous treatment of solvation models can be found in a review by Tomasi *et al.* [74].

In CPCM solvation calculations, first, a cavity, mimicking the water-free region around the aqueous complex, is built by placing interlocking spheres around each atom or group of atoms of solute. Then, the surface around the cavity is mapped by small regions, called tesserae. Inside the cavity the solute is placed in a vacuum whereas outside the cavity the value of the dielectric constant is equal to the solvent of interest, for instance, $\varepsilon = 78.39$ for water. However, for the solute-cavity description, different radii are available in some programs) (e.g., in *Gaussian09* [75]). These can be based on atomic or ionic radii, or on iso-charge-density surfaces and as a result, different solute cavities can be described. Recently, Takano *et al.* have performed benchmarking studies for the solute-cavity descriptions within the CPCM solvation model for the prediction of acidity constants. They found that the UAKS cavity definition, which uses united atom radii optimized at the PBE1PBE/6-31G(d) level of theory, estimates the solvation free energies within ~10.5 kJ/mol of experimental solvation free energies [76].

Implicit solvation models such as PCM, CPCM, and the Poisson-Boltzmann finite (PBF) element method were benchmarked for the predicted standard redox potentials of eighty Ru-based complexes in solution using the DFT-HF B3LYP hybrid method [77]. Within the PCM and CPCM solvation models, different cavity definitions were examined. In contrast to Takano *et al.*'s [76] results, this study reported that the Bondi radii (van der Waals radii published by Bondi [78]) for the solute cavity description were found to give better hydration energies than the UAKS cavity radii. However, this study did not distinguish the PCM and CPCM solvation models since both produced the same MUE values using the Bondi solute cavity definition with respect to experimental standard redox potentials. Moreover, the PBF solvation method was found to give slightly better results compared to the PCM and CPCM solvation models [77].

Explicit solvation by adding explicit water molecules around the solute can describe the hydrogen bonding environment more accurately [60,79] than a pure dielectric fluid approach. However, this method is computationally expensive, and therefore less commonly used because the CPCM solvation captures the solvation process comparatively well. The standard reduction potential for Ru(3+/2+) aqueous redox couple was examined using 37 DFT functionals with five different basis sets and the solvation effects were modeled using the SM6 solvation model [60]. In addition, the effect of the second hydration shell on the predicted standard reduction potential was investigated. The total number of water molecules modeled in the first and second hydration sphere was 18. It was found that including the second hydration shell was important for the accuracy of predicted redox potentials. However, interestingly, including the second hydration shell underestimated the redox potential for the Ru(2+/3+) redox couple. In contrast, just including the first hydration shell around the metal center tended to overestimate the redox potential. Even though this study does not recommend any specific DFT functional for the redox potential prediction in solution (though it emphasizes not to rely on a single DFT functional [60]), in general, care should be taken to choose a suitable DFT functional for the redox potential prediction. Short-range effects, such as local-hydrogen bonding effects and solute-solvent interaction can be effectively described with the addition of explicit solvation shells. However, to account for the long-range effects, a dielectric continuum model solvation approach can be used. Combination of the explicit and implicit solvation models is currently the most promising approach [80]; however, one should be cautious since additional hydration spheres can result in accurate reduction potential, but this approach of explicit hydration is computationally expensive [60].

In transition metal coordination chemistry, the field strength of the coordinating ligand on the metal center and the coordination number are the determining factors for the amount of crystal field splitting. The field exerted by the ligands splits up the degeneracy of the metal d-orbitals for TM coordination complexes. The higher this energy difference between the energy levels of orbital is (the former d orbitals now have so-called t_{2g} or e_g symmetry), the most likely is a low-spin (LS) configuration of the transition metal. In contrast, for degenerate orbitals or small amounts of crystal-field splitting, Hund's rule calls for a high-spin (HS) configuration [81].

Conjugated double bonds between organic multi-dentate ligands that complex metal centers are often involved in redox reactions or change the reduction potential of the metal by modifying their immediate local electronic environment. This phenomenon is termed non-innocence. The fact that the incoming electrons tends to occupy low-lying delocalized ligand-based orbitals leads to either under- or over-estimation of redox potentials. While for experimentalists, the nature of the redox process of metal-bound ligands is often difficult to define, be it either metal-based or ligand-based, the source and sink of the electron density can be more easily and quantifiably tracked using calculations. Spin cross over and ligand non-innocence in redox potential predictions are discussed by Hughes et al. [81].

3.3. Spin-Orbit Coupling

Spin-orbit coupling (SO) effect is usually not taken into account in most transition metal redox chemistry calculations, as this effect is generally insignificant (and only the spin-orbit coupling difference between the oxidized and reduced complex plays a role). Redox potentials of M(2+/3+) (M = Ru and Os) complexes were investigated using the CASSCF approach. The calculated SO coupling on the predicted redox potentials were found to be −0.07 and −0.30 eV for the Ru and Os complexes, respectively. This shows that even for the second row transition metal redox chemistry, the SO effect is not very significant since SO effect value found for the Ru(2+/3+) redox is −0.07 eV, which is negligible. However, for the fourth row transition metal redox chemistry, SO effects have to be taken into account since the SO value found for the Os(2+/3+) redox is −0.30 eV, which is significant [82]. For elements with unpaired f-electrons, spin orbit coupling can even be more significant, for instance, its value for f^1 electron of the uranyl(V) ion is about −0.31 eV [83].

3.4. Molecular Dynamics Simulations

Using quantum-mechanical molecular-dynamics simulations, the free energy of a redox process can be calculated. In order to calculate the free energy of a redox process, the thermodynamic integration method [84–95] is employed. The relation between the reduced and oxidized species of the redox reaction can be expressed using a coupling parameter (η) as shown below.

$$E(\eta) = E_{Oxd}(1 - \eta) + E_{Red}(\eta) \tag{15}$$

$$E(\eta = 0) = E_{Oxd} \tag{16}$$

$$E(\eta = 1) = E_{Red} \tag{17}$$

where η is the coupling or integration parameter, E_{Oxd} is the energy of the oxidized species (where $\eta = 0$) and E_{Red} is the energy of the reduced species (where $\eta = 1$). The derivative of the energy of the redox reaction with respect to the integration parameter, η can be written in terms of the vertical energy gap, where the vertical energy gap is defined as the difference between the E_{Oxd} and E_{Red} terms.

$$\partial E(\eta)/\partial \eta = E_{Red} - E_{Oxd} = \Delta E \tag{18}$$

$$\langle \frac{\partial E(\lambda)}{\partial \lambda} \rangle = \langle \Delta E \rangle \tag{19}$$

Using the canonical ensemble formalism, integration of the expectation value $\langle \frac{\partial E(\lambda)}{\partial \lambda} \rangle$ with respect to the integration parameter, η taking values from 0 to 1 produces the free energy difference between the reactant and the product, here this free energy difference corresponds to the reduction free energy of the redox reaction.

$$\Delta G = E(\eta = 1) - E(\eta = 0) = \int_0^1 \partial \lambda \langle \frac{\partial E(\lambda)}{\partial \lambda} \rangle_\lambda = \int_0^1 \partial \lambda \langle \Delta E \rangle_\lambda \tag{20}$$

More often, only the initial state ($\eta = 0$) corresponds to the oxidized state and the final state ($\eta = 1$) corresponds to the reduced state are studied according to linear response approximation. However, this has to be validated; on the other hand one could argue that the intermediate states are chemically meaningless entities. Thus, the free energy difference for the redox process can be thermally averaged over the energies of the oxidized and reduced species corresponding to the η values, 0 and 1, respectively.

$$\Delta G = 1/2(\langle \Delta E \rangle_0 + \langle \Delta E \rangle_1) \tag{21}$$

In addition, there are studies that investigated the redox transformations in which the redox process involves protonation or deprotonation, where three states for the integration parameter (η) are investigated ($\eta = 0, 0.5$ and 1) [89,96,97]. Using the Simpson rule based on the quadratic interpolation, the free energy difference can be determined when the intermediate state considered for the η in addition to the 0 and 1 states.

$$\Delta G = 1/6(\langle \Delta E \rangle_0 + \langle \Delta E \rangle_1) + 2/3(\langle \Delta E \rangle_{0.5}) \tag{22}$$

According to Marcus theory of electron transfer, the oxidized and reduced species attain a certain configuration favorable for the electron transfer, after the electron transfer these species tend to relax themselves to their equilibrium state. Corresponding free energies for these relaxation processes are reorganization free energies (λ). These reorganization free energies (λ) for the oxidized and reduced species can also be deduced from the calculated reduction free energy of the redox process. The relevant expression to compute the reorganization free energies for the oxidized and reduced species are shown below (Equations (23) and (24)).

$$\lambda_0 = \Delta G - \langle \Delta E \rangle_0 \tag{23}$$

$$\lambda_1 = \Delta G - \langle \Delta E \rangle_1 \tag{24}$$

After getting the reduction free energy from Equation (22), this can be further modified into reduction potential according to the Nernst equation (see Equation (8)) as explained earlier. Then, the obtained reduction potential has to be referenced with respect to the reference electrode potential, for instance the SHE and this makes the calculated reduction potential to be directly comparable with the experimental reduction potential.

4. Applications

4.1. Organics

Neutral organic molecules occur in a wide variety of chemistries, such as aliphatics, aromatics, phenols, quinones, amines, and nitro compounds. The redox chemistry of these compounds is interesting in terms of the described approaches, since most organic transformations take place by electron transfer in solution. Electrochemical organic transformations are often used as an efficient way to perform complicated organic syntheses. Various aspects involved in the electrochemical synthesis of organic compounds, for example, mechanism of redox processes, kinetics of electrode reactions, homogeneous or heterogeneous electron transfers, coupled electron transfer processes, have been reviewed [98–101]. Redox processes also help design organic photovoltaic materials, which are cost effective compared to metal-based photovoltaic or fuel cell materials. Organic solar cells are promising since carbon is one of the most abundant elements on Earth. In addition, certain organic photosensitive materials can be environmentally friendly, such that finding and optimizing such materials for solar-cell applications would be very helpful for the advancement of clean energy.

Molecular and mechanistic-level details of reduction of carbon dioxide (CO_2) by organic compounds, for instance by pyridine that reduces CO_2, is of fundamental interest. Once these redox processes are properly understood using computational approaches, this knowledge can be applied to reduce CO_2 emission [102].

One approach to grasp the theory of one-electron transfer processes is Marcus theory. Based on this theory, the one-electron redox process happens adiabatically and as soon as the electron is transferred from the oxidized species to the reduced species, relaxation of the complex and surrounding solvent molecules are expected to happen. By applying this, several authors have predicted the adiabatic electron affinity of different species, which is equivalent to calculating the energy of oxidized species with an additional electron. However, subsequent structural relaxation was not taken into account. The calculated adiabatic ionization potentials (IP) were also correlated with respect to the available experimental redox potentials.

Quinones are often involved in biological electron transfer and redox reactions. These organic molecules undergo one and two-electron transformations. Redox potentials of quinones were calculated using both the direct and isodesmic method of redox potential predictions. One-electron redox potentials of eight quinones with different substituents were determined using the DFT/B3LYP/PCM approach. A correlation ($E_{red} = -2.115 - 12.845E_{HOMO}$) between the calculated HOMO orbital energy to the experimental redox values was obtained. Redox potentials were predicted to be within a MUE of 0.03 eV of the experimental values [38]. Furthermore, a high level G3 method combined with the CPCM solvation model were used to calculate one-electron reduction potential of thirteen quinones with respect to the SCE potential (4.67 eV) in acetonitrile solution, and the average error was 0.07 eV [103]. Similarly, there have been few other studies, in which redox potentials for a number of quinone derivatives (isoindole-4,7-diones-(IIDs)) in ACN solvent were calculated with respect to the SHE reference potential (4.44 eV) and the MUE was determined to be 0.03 eV. The calculated reduction potentials of the IIDs compounds showed a linear correlation to the Hammett constant values of the ring substituents [104,105]. In addition, two-electron redox potentials for eight substituted quinones in acetonitrile solvent were also accurately calculated by using the B3LYP method including the PCM solvation model for solvation effects [40]. Recently, IPs and HOMO orbital energies were calculated for a number of substituted aryl imidazoles using the B3LYP method and plotted against the experimental oxidation potentials. In addition, this study found a linear correlation between the Hammett constant values and experimental oxidation potentials (for triarylimidazoles, E_{ox} (in eV) = 0.949 + 0.134$\Sigma\sigma$, R^2 = 0.973, where σ is the Hammett constant of the substituent). This study claims that using these empirical correlations, the redox potentials of unknown arylimidazole derivatives and the effect of various substitutions on the redox behavior can be obtained with reasonable accuracy [106].

One-electron reduction potentials of 116 (*para*- and *ortho*)-quinones in DMSO and ACN solvents [41] and oxidation potentials in DMSO solvent [39] were calculated. These studies used the B3LYP/PCM method for the redox potential calculations in combination with the isodesmic model [41]. The calculated one-electron redox potentials of the o- and p-quinones correlated with respect to the corresponding Hammett constant values of the substituent groups. The obtained empirical relation is E_{ox} (in eV) =1.66$\Sigma\sigma$ + 0.54. Then using this empirical relation, a large number of substitution effects on the reduction and oxidation potentials of quinones can be obtained without much further computational effort. In addition, the calculated electron affinities were plotted against the Hammett constants of the substituents and excellent correlations were obtained [39,41].

By employing the B3LYP/PCM method, redox potentials of 270 different organic compounds in acetonitrile solvent were calculated [107]. Of these 270 compounds, this study calculated adiabatic ionization potentials of 160 organic compounds, for which the experimental ionization potentials were accurately known. The calculated IPs were plotted with respect to experimental IPs, and the plot produced an intercept value of 0.28 eV, which corresponds to underestimation of IPs [107]. This was added as a correction to predict the redox potentials of the complete investigated set with respect to the SHE reference potential. Surprisingly, this approach predicted the redox potentials with a MUE of 0.17 eV to experimental redox potentials. Using the direct method for the redox potentials prediction, another computational study calculated the redox potentials of 250 distinct organic compounds in DMSO solution [108] by employing the B3LYP DFT functional and including solvation effects using the integral equation formalism polarizable continuum model (IEFPCM) solvation. Interestingly, in this study, the calculated ionization potential at the gas-phase by the B3LYP method were corrected by 0.28 eV, as obtained from an earlier study [107]. Using this correction, the direct method applied for predicting redox potentials of different organic compounds in DMSO solution were within a MUE of 0.11 eV with respect to the experimental redox potentials [108].

Reduction potentials for 74 different organic (cyano aromatics, quinones, flexible pi molecules, polyaromatic hydrocarbons (PAH), heterocyclic amines, and *N*-methyl heterocyclic aromatic cations) compounds in acetonitrile solution utilizing the B3LYP method with the CPCM/UAKS solvation model were modeled [109]. Calculated solution phase energy differences between the neutral and reduced species in acetonitrile solution were plotted against the experimental reduction potentials and an empirical correlation was obtained. Using this empirical correlation equation, redox potentials were determined within a MUE of 0.07 eV of experimental reduction potentials [109]. Similar types of correlation were obtained for the oxidation potentials of a number of pyridylhydroxyl amines calculated in ACN solution by applying the B3LYP/CPCM method. This study revealed an excellent direct correlation between the Hammett constant values of the pyridyl ring substituents and the experimental redox potentials [110]. Using the B3LYP/SMD method, redox potentials for 51 PAHs were calculated with respect to the Fc/Fc$^+$ reference potential (5.22 eV) in acetonitrile solution. The estimated MUE for the calculated redox potentials was 0.03 eV. The calculated absolute redox potentials plotted against the experimental redox potentials produced an excellent straight line correlation fit and an accurate Fc/Fc$^+$ reference potential (5.17 eV) was obtained [111].

Reduction potentials of 25 cyclic nitroxide (pyrrolidine, piperidine, isoindoline, and azaphenalene) organic compounds in acetonitrile and water solution were calculated using ab-initio methods (G3 and B3LYP) incorporating solvation effects using the PCM solvation model [112]. The calculated reduction potentials were referenced with respect to the SHE reference potential. Surprisingly, calculated redox potentials for nitroxide compounds were in excellent agreement (MUE 0.05 eV) with the experimental redox potentials, except for the azaphenalene nitroxide systems (MUE 0.60 eV). This discrepancy was attributed to chemical reactions followed by electron transfer that were not taken into account and studied [112]. For example, redox potentials of similar nitroxide compounds calculated in water solvent exhibited protonation states, which explains the complication of nitroxide redox reactions. Although for acetonitrile solvent, protonation is not possible, such side reactions may be the reason for the observed discrepancy. However, the calculated overall MUE for the redox potentials predicted in water was 0.04 eV for the examined cyclic nitroxides [113]. Moreover, a linear correlation between the Hammett constant values of the substituents and the calculated redox potentials was obtained.

Implicit solvation models, such as SM8, SMD, CPCM, IEFPCM, and COSMO-RS combined with the CBS-QB3 method were examined for the redox potential predictions to obtain a suitable solvation model [114]. This study calculated the redox potentials of 27 different neutral organic compounds in ACN and water solvents with respect to the SHE reference potential. The estimated MUEs for the calculated redox potentials were within ~0.2–0.50 eV for the examined implicit solvation models. Main sources for the solvation free energy error may have been due to the insufficient description of radical-cation solvation. This study recommended the SMD solvation model over the other investigated implicit solvation models for the redox potential calculations of neutral organic compounds in solution [114].

Another important field for the application of redox chemistry is DNA bases because the nucleotide bases are the fundamental constituents of DNA. These bases are of two categories, namely purine- and pyrimidine-type bases, in which adenine and guanine are purine bases, whereas cytosine, thymine, and uracil are pyrimidine bases. One-electron oxidation potentials of these DNA bases were estimated computationally using DFT B3LYP and complete basis set (CBS-QB3) methods incorporating solvation effects using the solvation model density (SMD). However, both of these DFT and CBS-QB3 methods were found to underestimate the one electron oxidation potentials of nucleotide bases in acetonitrile solvent by MUEs of 0.33 and 0.21 eV, respectively [115]. Although the predicted oxidation potentials for these bases with respect to the SHE reference electrode, it should be noted that the absolute potential value used for the SHE was 4.28 eV as reported by Kelly *et al.*, underestimated the experimental values by ~0.2–0.3 eV. On the other hand, the oxidation potentials predicted relative to adenine were found to be in excellent agreement with the experimental redox potential values within the MUE of 0.07–0.10 eV. Interesting thing to note here is that the redox reaction of nucleic bases in aqueous solution would more likely to exhibit different tautomeric forms; this further complicates the redox potential prediction in aqueous solution. This problem was overcome by introducing ensemble redox potential, which is actually estimated based on the relative energies of the tautomers in the oxidized and reduced states [115].

Recently, the oxidation mechanism of guanine and the redox potentials of intermediates along the proposed mechanistic pathways were determined using the DFT B3LYP and CBS-QB3 methods [116]. Moreover, redox potentials for these DNA constituents were investigated at various DFT levels [117–120]. An atomic-level understanding of the redox behavior of these fundamental building blocks of DNA is important because it will help to detect DNA mutation problems which may be the reason for a lot of genetic disorders in human. One-electron redox processes of these DNA base molecular units lead to form radicals, which are dangerous and eventually forming mutations and disorders in DNA.

Flavonoids are another important class of organic compounds who play the role of antioxidants and biological one- and two-electron catalysts. Reduction potentials of 28 flavins in water solvent with respect to the SHE reference potential (4.44 eV) were calculated using the B3LYP method including solvation effects by employing the CPCM solvation model. The estimated MUE for the calculated redox potentials was 0.06 eV. In addition, substitution effects were systematically evaluated for the redox behavior of flavins. Linear correlations were obtained for the Hammett substitution constants when plotted against the computed and experimental redox potentials. In addition, the calculated HOMO orbital energies with respect the calculated redox potentials revealed an excellent linear correlation [121]. In another recent study, for a number of flavonoids, one- and two-electron redox potentials were calculated using the M06-L DFT method combined with the SM6 solvation model. The SHE (4.28 eV) reference electrode potential was used as the reference. The redox potentials were determined within a MUE of 0.042 eV of experimental redox potentials. Empirical linear correlations were obtained by plotting the calculated one electron-reduction potentials against the Hammett constants of the substituents groups [122]. Using these empirical linear relations and without applying computer-intensive DFT computations for all investigated flavonoids, redox potentials of unknown flavonoids can be predicted accurately. This implies that the electronic effects of the substituents have a strong influence on the redox behavior of the flavonoids. By introducing relevant substituents with either electron-donating or drawing nature, flavonoids with appropriate redox behavior can be fine-tuned [123].

4.2. Inorganic Compounds

In this section, we summarize the previous computational investigations on redox potential calculations for inorganic compounds, for example, carboranes and oxo acids, such as chloro-, bromo-, and nitro-oxo acids in solution. Compared to redox calculations on organic compounds, published reports on computational redox calculations of inorganic compounds are few in number.

Carboranes are a type of inorganic cluster compounds which contain carbon, boron, and hydrogen atoms, and often H-atoms are substituted by different groups, for instance, chloride, methyl, *etc*. The carboranes are classified into different types depending on the skeletal structural variation, for example, closo, nido, and arachno [124,125]. Oxidation potentials for 31 icosahedral carborane anions [$1\text{-}X\text{-}12\text{-}Y\text{-}CB_{11}Me_{10}^-$] were calculated at three different levels of theoretical methods, such as the RI-B3LYP, RI-BP86, and RI-HF (RI is the resolution of identity, an algorithm to speed-up HF and DFT calculations). Solvation effects were treated using the COSMO solvation model. The isodesmic model was used for the calculation of oxidation potentials and the standard deviation for the calculated oxidation potentials was 0.02 eV. The calculated EAs in solvent were found to produce a linear correlation with respect to the experimental oxidation potentials. Furthermore, substitution effects were examined and a linear relationship revealed between the Hammett constants and the calculated oxidation potentials [126]. Likewise, using the PBE0 method, gas-phase ionization potentials and electron affinities for a series of 1-carba-closo-dodecaborate anions were calculated, and these IPs and EAs were plotted against the experimental oxidation and reduction potentials, respectively and excellent linear correlations were obtained [127]. Similarly, another study predicted oxidation potentials with respect to the Fc/Fc^+ reference potential in acetonitrile solvent for few carboranes using the B3LYP/PCM method, for which the predicted oxidation potentials were in agreement with the experimental oxidation potentials.

Boron hydrides are structurally similar to carboranes. Recently, reduction potentials for a series of hypercloso boron hydrides, B_nH_n ($n = 6\text{--}13$) and $B_{12}X_{12}$ ($X = F, Cl, OH,$ and CH_3) [128] in aqueous solution were calculated using the G4, B3LYP, and M06-2X methods incorporating solvation effects using the CPCM and SMD solvation models. The G4 method was used to calculate the gas-phase free energies. The reference electrode potential used was the SHE (4.28 eV). The calculated reduction potentials at the G4/M06-2X method combined with the Pauling energy/solvation-cavity method were in agreement with the experiments within 0.20 eV. For one-electron reductions, a linear correlation was obtained between the electron affinities and the experimental reduction potentials; however, deviations were observed for two-electron reduction [128].

Employing the hybrid DFT B3LYP method in combination with the PCM method for solvation effects, reduction potentials for chloro-, bromo-, and nitro- oxo acids in acidic and basic environments were calculated with respect to the SHE reference potential [129]. The calculated reduction potentials were in excellent agreement with the experimental values and the MUE was 0.10 eV. This study proposed a decomposition scheme to interpret the chemistry behind the redox reaction of the above mentioned oxo acids in aqueous solution. The decomposition scheme consists of different terms in the overall reduction potential, for example, the electrophilicity, protonation, and formation/dissociation of water. This scheme explained the quantitative contribution of different terms into the chemistry of the redox reactions of the above mentioned oxo acids in aqueous solution [129].

4.3. Metal Complexes

In this section, we will discuss computational reduction potential predictions for transition metal complexes, from the 3d to 5d series, and actinyl complexes (5f series) in aqueous and non-aqueous solution.

4.3.1. Transition Metal Complexes

First redox potential calculations of the 3d transition metal aqua complexes are discussed. Uudsemaa *et al.* [79] computationally determined the reduction potentials of transition metal (3+/2+) aqua complexes using the DFT BP86 method, water solvation effects modeled by the implicit COSMO solvation model, and the SHE as a reference potential. The predicted reduction potentials for these TM aqua complexes including the water molecules of the first solvation sphere (six first hydration sphere water molecules) were overestimated by 1.3 eV compared to the corresponding experimental reduction potentials. Results were improved by adding additional explicit water molecules (twelve second hydration sphere water molecules in total) around the metal-bound water molecules. This explicit solvation (hydration) approach (six (first hydration) + twelve (second hydration) = eighteen water molecules) predicted the reduction potentials within a MUE of 0.29 eV of experimental redox potentials, a significant improvement by 1 eV over the previous approach. The local hydrogen bonding effects were important in order to obtain more accurate reduction potentials [79,130]. This confirms that the solvation process is usually an important aspect of the reduction potentials of TM complexes. Similar to this, redox potentials of the Ru(2+/3+) couple were calculated using different DFT functionals, the local solvation effects were modeled using explicit water molecules [60], as explained earlier under the solvation section.

Reduction potentials for organic compounds and TM complexes (including metallocenes and coordination complexes) were calculated with respect the SCE reference potential (4.188 eV) using the PCM solvation model for solvation effects combined with the hybrid DFT B3LYP method [61]. The obtained results were in excellent agreement with experimental redox potentials, the MUE was 0.15 eV. These authors pointed out that the diffuse basis sets were important for accurate redox calculations, while the effects of zero point energy (ZPE), and free-energy (FE) corrections were negligible, though improved the calculated redox potential. However, this study used an erroneous absolute value for the SCE as the reference potential. The actual absolute value for the SCE is 4.60 eV, not the 4.188 eV, as noted recently by Castro *et al.* [55].

The hybrid DFT B3LYP method with the solvation effects included using the PB continuum solvation model predicted the reduction potentials of 95 octahedral 3d TM complexes coordinating with different ligands [81]. Systematic improvement of the calculated reduction potentials were obtained by introducing a correction term, in which the B3LYP-predicted redox potentials were corrected by applying an empirical correction, the so-called D-block Localized Orbital Correction (DBLOC). The addition of this correction improved the predicted redox potentials. The MUEs calculated for the predicted redox transitions with the addition of DBLOC terms is 0.12 eV and 0.40 eV without this correction.

The hybrid DFT B3LYP functional combined with the IEF-PCM (integral equation formalism-polarizable continuum model) solvation model resulted in a good prediction of reduction potentials for a series of TM complexes (including metallozenes, metallozenedichlorides, bipyridyine complexes, carbonyl complexes, and maleonitriledithiolate complexes) in non-aqueous solution such as dichloromethane (DCM), acetonitrile, and dimethylformamide (DMF) only if corrected for an inaccurate reference potential [56]. The redox potentials were predicted with a relatively high MUE of 0.54 eV compared to the experimental reduction potentials [56]. The reference used was the calculated Fc/Fc^{+} reference potential. In contrast, the GGA BP86 and the PBE functionals performed better than the hybrid DFT functional and the MUE of the predicted reduction potentials was ~0.30 eV smaller than the MUE calculated for the B3LYP predicted reduction potentials. All the predicted B3LYP reduction potentials in this study had to be shifted by a constant value of 0.43 eV to improve the linear correlation between the calculated and the experimental redox potentials [56]. In addition, for a number of iron model complexes of the hydrogenase enzyme, the B3LYP method combined with the PCM solvation model calculated oxidation and reduction potential in CH_3CN solvent were underestimated and systematic shifts of -0.82 and -0.53 eV had to be added to reproduce the experimental oxidation and reduction potentials, respectively. In contrast, the GGA BP86 method predicted the redox potentials with a MUE of 0.12 eV [57]. Another recent study found a superior performance of the BP86 method over B3LYP, in which the one-electron reduction potentials of oxo iron-porphyrin complexes are calculated using the DFT B3LYP, BP86 and M06-L methods and the solvation effects are comprised with the CPCM solvation model [55]. However, another current study improved the DFT B3LYP method coupled with the CPCM solvation model, and predicted reduction potentials by employing the isodesmic model. In addition, this study suggests that the reference complex, as usually employed instead the reference electrode potential in this method, should belong to the same period of the periodic table. This often minimizes systematic errors (to a certain degree by error cancellations using species from the same row) and the redox potentials in solution can be predicted within a MUE of 0.06 eV of experimental redox potentials [34].

Recently, the reduction potentials of group eight (Fe, Ru, and Os) metal octahedral complexes were calculated using two different DFT functionals, PBE and M06L combined with the COSMO-RS and SMD solvation models, respectively [131]. The reference electrode used was the SHE (4.28 eV). Interestingly, the predicted redox potentials for the negatively charged complexes $(3-/4-)$ were significantly underestimated, which could be due to large solvation errors. Even the effect of explicit water molecules did not show any significant improvement for the computed redox potentials of the cationic complexes whereas significant improvement was observed for the anionic complexes, the latter one in agreement with an earlier study [79]. In addition, the QM/MM model approach was employed in which the thermodynamic integration method was used to predict the free energy of redox processes, and then the reduction potentials were predicted with respect to the SHE reference potential for TM complexes in aqueous solution. It is worth noting that the calculated root mean square deviation (RMSD) for the QM/MM method predicted redox potential is 0.36 eV, which is close to the experimental redox potentials. In contrast, the B3LYP DFT method predicted redox potentials using the PCM solvation model, in which the TM aqua complexes had only the first solvation sphere and the RMSD is 1.5 eV. In an implicit solvation model, there is no way to include local hydrogen bonding effects, for instance the solute-solvent hydrogen bonding interactions, and these issues were addressed [84].

A new approach for the redox potential calculation of TM complexes has been recently proposed. This approach is called as the pseudo counter-ion solvation, which is basically described by adding an oppositely charged sphere $(-q)$ around the actual charge (q) of the solute cavity sphere and a correction to the solvation energy based on the generalized Born model is added. This has been found to improve the predicted redox potentials from 0.5 to 0.17 eV for the B3LYP functional. However, this method has only been tested for a set of 39 transition metal complexes, applying this approach to predict redox potentials for a larger data set of transition metal complexes in solution would reveal the reliability of this approach [132].

To elucidate the ligand additive effects on the redox properties using the B3LYP method for computations [133], the oxidation (2+/3+) process of transition metal (Ru, Os, and Tc) carbonyl (CO) complexes were investigated by substituting the metal-bound CO ligands with the CN^-, Cl^-, H_2O, CH_3CN, and N_2 ligands. The linear regression fitting method is then applied to correlate the adiabatic oxidation energy, calculated for the oxidation reaction (Equation (25)), with the number of CO ligands and the straight line expression is shown below (Equation (26)).

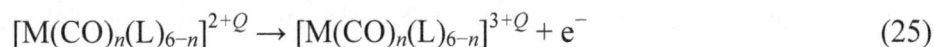

$$[M(CO)_n(L)_{6-n}]^{2+Q} \rightarrow [M(CO)_n(L)_{6-n}]^{3+Q} + e^- \qquad (25)$$

where, M (Ru, Os, and Tc) is the transition metal cation, n equals to 6 and the term Q is defined as, $Q = (6 - n)q$, in which the q is the charge of the ligand, L (CN^-, Cl^-, H_2O, CH_3CN and N_2).

$$\Delta E_{adiabatic} = I + S \cdot [nCO] \qquad (26)$$

where, the $\Delta E_{adiabatic}$ term refers to the adiabatic energy difference for the oxidation process, the term "I" refers to the intercept, the term "S" refers to the slope of the straight line and the term n refers to the number of CO ligands present in the complex. The obtained slope values from these fitted lines showed a non-dependent behavior with respect to the metals, this confirms that the determined parameters are ligand specific [133].

Redox potentials for 30 octahedral tungsten-alkylidyne complexes with a variety of different coordinating ligands were calculated by employing the hybrid B3LYP DFT method [134]. Interestingly, the orbital participates in the one electron oxidation process is the d_{xy} (HOMO) orbital of W atom and the calculated HOMO orbital energies of these various complexes are linearly correlated with the experimental redox potentials. By altering the coordinating ligands, the electronic properties of the metal center can be fine-tuned and the expected redox properties can be obtained [134]. Similar to this study, the calculated redox potentials using the B3LYP method in water and DMSO solvents, produced excellent correlations with respect the HOMO and LUMO orbital energies for a number of Cu complexes [135]. DFT methods are powerful tools to design redox active complexes and catalysts.

4.3.2. Actinides

Actinides are 5f elements. Except the protactinium (Pa), uranium (U) and thorium (Th) elements, the remaining elements are manmade elements. The interests to study the redox chemistry of actinides have been growing over the years [16,45,58,59,83,136–140]. The reason for this high emphasis is that these elements, in particular U, Np, and Pu, are used in nuclear reactors for power generation. However, the highly radioactive wastes produced during the nuclear reactions are dangerous to dispose. Proper and more precise understanding of the chemistry of these elements and their redox behavior in solution is necessary and inevitable to store in geological repositories.

More importantly, interactions of these elements with the geologically most abundant minerals, transport, speciation, precipitation, and migration behavior have to be understood and indeed these processes are complicated processes. Because minerals present in geo-sphere can promote different reactions such as precipitation, adsorption, reduction, and surface mediated chemical reactions. In general, mineral surfaces are more chaotic and actively involved in bio mineralization processes and surface mediated reactions. Redox active minerals will help us to design more effective remediation strategies for these actinide elements, recycling process and reuse in future.

Reduction potentials of actinyl (U, Np, and Pu) aqua complexes were calculated using the DFT B3LYP method with respect to the calculated SHE reference. The An (An = U, Np, and Pu) atoms were described by large core (LC) PPs and basis sets, this lead to an inadequate description of valence states of these elements and resulted in an overestimation reduction potentials by ~2.5 eV relative to experimental values. Even though the calculated reduction potentials are overestimated, spin-orbit coupling interaction and multiplet effects were found to be significant to get the experimental trend [83]. Later with the use of small core (SC) PPs and basis sets for An atom of these actinyl aqua complexes in the DFT B3LYP redox potential calculations were found to improve the reduction potentials and the experimental redox potentials were reproduced within the MUE of 0.40 eV, a significant change. Moreover, the use of SC PPs for An (U, Np, and Pu) atoms were found to predict better properties such as vibrational frequencies and thermochemistry. All electron (AE) and zeroth order regular approximation (ZORA) methods have also predicted similar reduction potentials like the SC PPs relativistic methods for these actinyl aqua complexes [137,139,141].

High level *ab initio* studies for the actinyl redox potential prediction were also reported. The CASPT2 method reproduced the experimental redox potentials within the MUE of 0.20 eV. In addition, the reduction of uranyl(VI) by Fe(II) was also studied in which QM/MM approach employed. The redox reaction is: $[UO_2]^{2+} + Fe(II) \rightarrow [UO_2]^+ + Fe(III)$ [140]. Recently, using DFT B3LYP and M06L functionals the actinyl redox potentials were calculated with the inclusion of PCM solvation effect. This study used the experimental SHE potential, 4.44 eV as the reference potential. In this study, the calculated potentials were adiabatic electron affinities, even the structural optimization of reduced actinyl species was not taken into account, and the reason was stated that according to the Marcus theory of electron transfer the relaxation is believed to take place after the electron transfer. ZPE and FE correction were not included since the authors argued that these effects were minimal and hence negligible. The calculated MUEs for the predicted reduction potentials were 0.13 and 0.04 eV, respectively for the B3LYP and M06 functionals [59].

The actinyl(VI, V) aqua complexes are linear oxo-cations with water molecules coordinating to the metal center in the equatorial plane whereas the An(IV, III) aqua complexes do not have axial oxygen atoms and the coordination number can go up from 8 to 10. The following overall redox reactions including the SHE reference were employed to calculate the actinyl(V) to An(IV) and An(IV) to An(III) redox potentials [139].

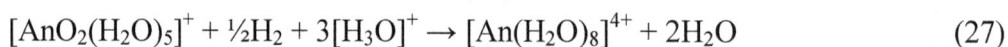

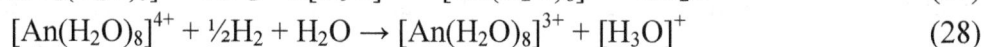

$$[AnO_2(H_2O)_5]^+ + \tfrac{1}{2}H_2 + 3[H_3O]^+ \rightarrow [An(H_2O)_8]^{4+} + 2H_2O \qquad (27)$$
$$[An(H_2O)_8]^{4+} + \tfrac{1}{2}H_2 + H_2O \rightarrow [An(H_2O)_8]^{3+} + [H_3O]^+ \qquad (28)$$

Austin *et al.* [16,58] have predicted $[AnO_2]^{2+/+}$ (An = U, Np, and Pu) and $[AnO_2(L)_n]^m$ (where L = H_2O, OH^-, Cl^-, AcO^-, and CO_3^{2-})) redox potentials in aqueous solution with respect to the calculated SHE reference electrode potential using DFT methods. Solvation effects were incorporated with the CPCM solvation model and the solute cavities were described using the universal force field (UFF) radii [142,143]. Despite the presence of an additional explicit water molecules (hydration shell) around the actinyl-bound neutral and anionic ligands, aqueous actinyl(VI/V) redox potentials were predicted within ~1.3 eV (MUE) of experimental redox potential values [16]. Reasons for these larger deviations were attributed to solvation effects. However, an important point to note is that the calculated absolute SHE reference potential values are 5.56 and 5.61 eV for the B3LYP and BP86 methods, respectively, and these values are ~1.1 eV higher than the IUPAC SHE value (4.44 eV). This explains the relatively high MUE (~1.3 eV) of calculated reduction potentials.

Moreover, a series of DFT functionals ranging from GGA to hybrid DFT functionals and recently developed M0x functionals from Truhlar *et al.* [49] were examined for the redox potentials prediction, in particular for actinyl(VI/V) aqua complexes. The calculated mean unsigned error (MUE) with respect to the experimental reduction potentials from these authors reveal that the hybrid B3LYP functional predicts the reduction potentials of actinyl aqua complexes within the MUE of ~0.40 eV of experimental redox potential values whereas the M06 functional was found to be accurate than the B3LYP and the MUE for the M06 functional was reported to be 0.20 eV less than the B3LYP functional predicted reduction potentials [58].

Reduction potentials for actinyl(VI/V) in aqueous solution and uranyl(VI/V) complexed with organic multi-dentate ligands in a range of non-aqueous solutions, such as DMSO, dimethylformamide (DMF), dichloromethane (DCM), acetonitrile (ACN), and pyridine, using the DFTB3LYP and M06 functionals combined with the CPCM/UAKS solvation model were investigated. The importance of reference electrode potential, solute cavity description and the effect of explicit solvation were elucidated. In addition, the effect ionic strength on the redox potentials was studied, however this effect was found to be negligible. The effect of counter ions on the calculated redox potentials were also studied, because of strong binding of counter ions with the metal-bound anionic ligands did not produce any noticeable trend. The reason for this could be attributed to the tight binding and the employed static model, whereas in solution the counter ions are dynamic and solvated all the time. Both the direct and isodesmic methods applied for the redox potentials predictions in solution. The results suggests that the reduction potentials of actinyl complexes in solution can be predictable with in ~0.1–0.2 eV of experimental values using the hybrid DFT B3LYP/CPCM/UAKS method. The isodesmic model of redox potential prediction gets upper hand over the direct method; this is mainly due to cancellation of errors. It should be noted that the reference electrodes considered were the SHE and Fc/Fc^+ for aqueous and non-aqueous solutions, respectively. Moreover, the uranyl(VI/V) redox was found to be altered by introducing relevant substitutions into the periphery of the uranyl bound multi-dentate ligand. Electron-releasing substituents increase the electron density around the metal center which in turn increases the uranyl(VI/V) redox potentials, while electron-drawing substituents decrease the reduction potential. A direct correlation between the calculated redox potentials and the Hammett parameters of the peripheral ligand substituent groups were established, which could be used for ligand design, e.g., to control separation chemistry of actinides [45]. Other studies include the calculation of redox potentials of actinyls complexed with multi-dentate ligands in non-aqueous solutions such as DCM [138] and tetrahydrofuran (THF) [136].

Electron affinities of a series of organometallic complexes, $[Cp^*_2UX_2$ (X = BH_4, Me, and NEt_2Cl), $Cp3UX$ (X = Cl, BH_4, SPh, S^iPr, and O^iPr), $L_2U(BH_4)$ (L = Cp_2, tmp_2, $tBuCp_2$, Cp^*tmp, and Cp^*_2), and L_3UCl (L = Cp, MeCp, TMSCp, tBuCp, and Cp^*)] were calculated using the DFT BP86/ZORA method and the solvation effects were included with the COSMO solvation model. The calculated electron affinities in THF solvent were highly correlated with experimental U(IV/III) reduction potentials. Similarly, the calculated LUMO and HOMO orbital energies showed linear correlation with experimental reduction potentials. This implies that the electronic nature of the ligand plays a significant role in altering the redox properties of U(IV) organometallic complexes in THF solution [144–147].

4.4. Actinyl(VI/V), Pyrite-Actinyl(VI/V), and Mackinawite-Actinyl(VI/V) Redox Potential Calculations

In Section 4.3.2, we have discussed the previous computational studies reported in the literature on redox potential calculations of various actinides, for example, actinyl aqua complexes, uranyl complexes of various organic ligands, and U(IV) organometallic complexes in aqueous and non-aqueous solutions. Herein, we describe our current investigation of reduction potential calculations of actinyl aqua complexes, and semiconducting mineral surfaces such as pyrite and mackinawite surface-adsorbed actinyl aqua complexes in aqueous solution by employing cluster models within the DFT/CPCM computational approach.

4.4.1. Computational Details

Previous calculations have used the hybrid DFT B3LYP [51–53] method in combination with the continuum solvation model to calculate reduction potentials of actinyl aqua complexes in aqueous solution. We employ the B3LYP functional combined with the Conductor-like Polarizable Continuum Model (CPCM) solvation [63–65], in which the solute cavities are described using the UAKS cavity definition. Small-core (60 core electrons) pseudopotentials and respective double-zeta basis sets for An atoms produced accurate redox properties in earlier studies on redox potentials [16,58,59,137]. Thus, in this study we use the small-core SDD pseudopotential and basis set [148] for U, Np, and Pu-atoms. For Fe, O, and H-atoms LANL2DZ basis sets, and for S-atom LANL2DZdp are employed.

4.4.2. Hydration of $[AnO_2(H_2O)_5]^{2+/+}$ Complexes (An = U, Np, and Pu)

Actinyl ion aqua model complexes contain five water molecules in their equatorial plane. Although, four and six water molecules in the equatorial plane of actinyl ions were computationally proposed [149], the predominant species is the penta-aqua complex. Here, we have considered only the first solvation water molecules, adding explicit water molecules is computationally time consuming and expensive. Recent calculations have showed that the redox potentials for actinyl(VI/V) aqua complexes can be determined accurately with just the first solvation water molecules using computed adiabatic electron affinity at the DFT M06-L method [59] in combination with the PCM solvation model. In addition to the usual redox potential calculations, we will apply this adiabatic ionization calculation approach to determine redox potentials, and discuss the consequences. One-electron reduction half-cell reaction for actinyl aqua complexes is shown below in Equation (29).

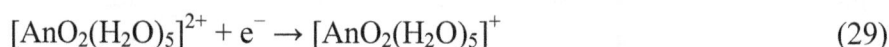

$$[AnO_2(H_2O)_5]^{2+} + e^- \rightarrow [AnO_2(H_2O)_5]^+ \tag{29}$$

4.4.3. Actinyl (An = U, Np, and Pu) Adsorption/Reduction to Small Pyrite Clusters $[Fe_4S_8-AnO_2(H_2O)_5]^{2+/+}$

Pyrite is an iron-disulfide mineral with a bulk formula of FeS_2. A small stoichiometric cubic pyrite cluster of molecular formula Fe_4S_8 is used to examine pyrite-actinyl interactions and effects of pyrite on the reduction potentials of actinyl aqua complexes are investigated. The Fe_4S_8 pyrite cluster model has four Fe(II) ions, and one of the Fe(II) is attached to a $[AnO_2(H_2O)_5]^{2+}$ aqua complex through one of the actinyl oxygen atom, a so-called cation-cation type interaction. Using these cluster models, reduction potentials for the pyrite attached actinyl aqua complexes are calculated. In our calculations, the spins of these cluster models are treated as high-spin (HS) configurations, because the Fe(II) ions present in the pyrite fragment are coordinatively unsaturated. The Fe_4S_8 coordinates of the super-cluster (pyrite-actinyl) were kept frozen in all optimizations in order to minimize the computational cost, and the structural relaxation of the pyrite moiety is expected not to have significant impact on the calculated redox potentials of pyrite-actinyl complex cluster models. A half-cell reaction used to calculate the reduction potentials of pyrite-actinyl cluster is given in Equation (30) and the optimized structure of $[Fe_4S_8\text{-}UO_2(H_2O)_5]^{2+}$ cluster model is shown in Figure 2.

$$[Fe_4S_8\text{-}AnO_2(H_2O)_5]^{2+} + e^- \rightarrow [Fe_4S_8\text{-}AnO_2(H_2O)_5]^+ \tag{30}$$

Figure 2. DFT-optimized geometry of pyrite-uranyl cluster ($[Fe_4S_8\text{-}UO_2(H_2O)_5]^{2+}$) model.

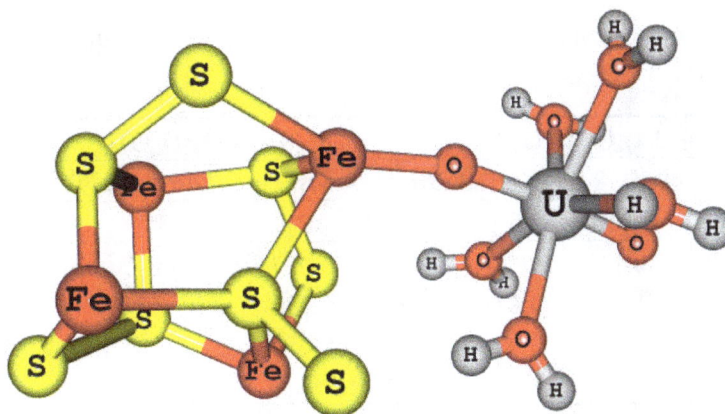

4.4.4. Actinyl (An = U, Np, and Pu) Adsorption/Reduction on Small Mackinawite Clusters $[Fe_8S_8\text{-}AnO_2(H_2O)_5]^{2+/+}$

Mackinawite is a layered iron sulfide, its general formula is FeS. In this study, a stoichiometric sheet-like cluster of molecular formula Fe_8S_8 is used as a model for mackinawite semiconducting mineral. The Fe(II) atoms in the cluster are treated as low spins as in bulk mackinawite. Although this cluster model contains coordinatively undersaturated Fe(II) atoms at corners, we have not explored yet the relative thermodynamics of high-*versus*-low spin setups for each site. A half-cell reaction for one-electron reduction is shown in Equation (31) and the optimized structure of $[Fe_4S_8\text{-}UO_2(H_2O)_5]^{2+}$ cluster model is shown in Figure 3.

$$[Fe_8S_8\text{-}AnO_2(H_2O)_5]^{2+} + e^- \rightarrow [Fe_8S_8\text{-}AnO_2(H_2O)_5]^+ \tag{31}$$

By applying the thermodynamic cycle scheme, free energies for the above one-electron reduction half-cell reactions (Equations (29)–(31)) can be evaluated in gas and aqueous solution phases. From the reduction free energies, utilizing the Nernst relationship between the free energy and electrode potential, absolute reduction potentials were obtained. The calculated absolute reduction potentials are then referenced with respect to the standard hydrogen electrode (SHE) reference potential (4.44 eV). The calculated reduction free energies are not adjusted by zero-point energy (ZPE) and free-energy (FE) corrections, since these additions have only a minor influence on the potential. Computationally obtained one-electron reduction potentials in aqueous solution for different model complexes are collated in Table 2.

Figure 3. DFT-optimized geometry of mackinawite-uranyl cluster ($[Fe_8S_8-UO_2(H_2O)_5]^{2+}$) model.

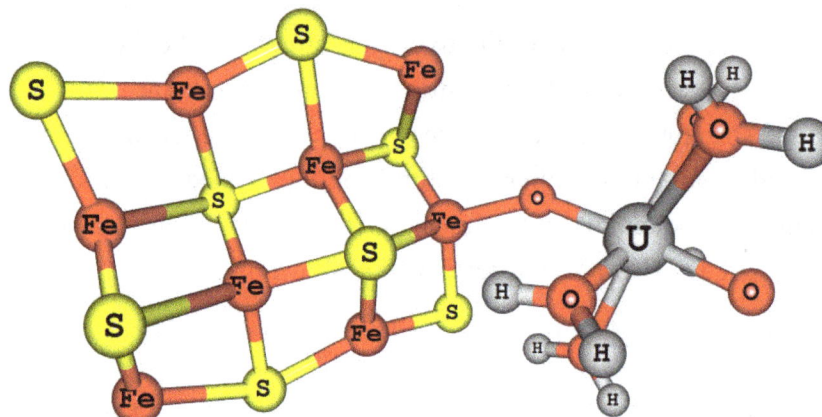

Table 2. Experimental and calculated reduction potentials of actinyl aqua model complexes in aqueous solution (eV).

Models	Experimental	$[AnO_2(H_2O)_5]^{2+/+}$				$[Fe_4S_8-AnO_2(H_2O)_5]^{2+/+}$		$[Fe_8S_8-AnO_2(H_2O)_5]^{2+/+}$	
		Opt. [a]		Adiabatic [b]		Opt.	Adiabatic [b]	Opt.	Adiabatic [b]
		E_0	E_0+SO^c	E_0	E_0+SO^c	E_0	E_0	E_0	E_0
U	0.088	−0.173	0.137	−0.504	−0.194	0.017	0.145	−0.256	0.234
Np	1.159	0.820	1.210	0.471	0.861	0.036	0.154	−0.267	−0.699
Pu	0.936	1.332	1.422	0.975	1.065	0.036	0.163	3.174	−0.658
MUE	-	0.33	0.20	0.44	0.24				

Notes: [a] reduction potentials calculated based on optimized geometries; [b] reduction potentials calculated based on adiabatic approach; [c] spin-orbit(SO) interaction corrections are −0.31, −0.34, and −0.09 eV for uranyl, neptunyl, and plutonyl redox reactions, respectively taken from Hay *et al.* [83].

By applying a similar approach to Steele *et al.* [59], we do not take multiplet effects into account; however, we do correct our calculated reduction potentials by spin-orbit interaction correction values reported by Hay *et al.* [83] for actinyl f electrons. Despite not including the multiplet effects, the actinyl(VI/V) reduction potentials are reproduced within ~0.2–0.24 eV of experimental values (see Table 2). According to Marcus theory of electron transfer, it is an adiabatic process. The calculated reduction potentials, based on the adiabatic process, are also in good agreement with the experiments. However, this approach underestimated the reduction potentials, because structural relaxations of the reduced species are not included.

Calculated redox potentials for pyrite-actinyl cluster models revealed a change in redox potentials for all the actinyl aqua complexes. Interestingly, the difference between the adiabatic and the full optimization redox calculation do not show significant variation, the values only differ by ~0.1 eV. Our results confirm that the redox-active semiconducting pyrite mineral surface plays a critical role in altering the redox potential of actinyl aqua complexes. Although there is no experimental proof for the redox potentials change in the presence of minerals for Np and Pu complexes, recent powder-micro electrode study by Renock *et al.* [9] shows that the VI/V redox potential of uranyl aqua complex is altered by the pyrite mineral. This investigation also proposed various surface-mediated processes, for example disproportionation of the reduced uranyl(V) on the surface. The experimentally determined uranyl(VI/V) reduction potential on pyrite mineral surface is 0.003 eV (*vs.* standard hydrogen electrode, SHE) which is in excellent agreement with our calculated reduction potential of uranyl-pyrite cluster, 0.017 eV, though this value does not include the approximate spin-orbit interaction correction from Hay *et al.* [83]. Even though the experiment has showed a small change (0.088 to 0.003 eV (85 mV)) for the uranyl(VI/V) redox on the pyrite mineral surface, this is a result of the alteration of the electronic environment of uranyl ion near the mineral surface. In contrast, for the neptunyl and plutonyl aqua complexes, the calculated change in redox for pyrite-actinyl adduct is ~1 eV. This huge difference of change in reduction potential will play a key role in altering the redox behavior of these complexes when they are in contact with redox-active mineral surfaces. Although, these three actinyl aqua complexes exhibit different VI/V reduction potentials in aqueous solution, it should be noted that the reduced penta-valent uranyl and plutonyl aqua complexes would more likely undergo disproportionation. In contrast, the neptunyl(V) is relatively stable against disproportionation. Despite their different redox behavior, the pyrite mineral surface brings the reduction potential of these aqua complexes down to near zero. This confirms that altering the redox environment of actinyl aqua complexes will result in different redox behavior and this can have profound implications on immobilizations and remediation techniques and safer disposal.

The semiconducting mackinawite surface has shown a significant influence on the redox behavior of actinyl aqua complexes. The adiabatic method results in a reduction potential of 0.234 eV for the mackinawite-uranyl cluster. In contrast, the full optimization method produced −0.256 eV, which is ~0.5 eV lower than the former one, which is due to the relaxation of the reduced species. For the mackinawite-neptunyl cluster, the scenario is different. The adiabatic method produces a ~0.5 eV smaller reduction potential compared to the full optimization method. However, the full optimization method results in similar redox potentials (−0.256 and −0.267 eV) for both the mackinawite adsorbed uranyl and neptunyl cluster models. The aqueous plutonyl complex on mackinawite behaves differently than the U and Np equivalents. The full optimization method produces a reduction potential of 3.17 eV for the mackinawite-plutonyl cluster model which is ~2 eV higher than the plutonyl(VI/V) reduction potential. However, the adiabatic method produced a reduction potential value of −0.658 eV which is close to the value obtained for the mackinawite-neptunyl reduction potential, −0.699 eV. The reason for these diverging results of aqueous actinyl complexes on mackinawite is unclear. When the role of pyrite and mackinawite in altering the redox behavior of actinyl aqua complexes are compared, mackinawite has significantly more impact (~0.2–0.3 eV) than the pyrite and this inference is based on the negative reduction potentials obtained for the mackinawite adsorbed actinyl aqua complexes.

We showed that the stoichiometric cluster model approach can be applied to study the interaction semiconducting mineral surfaces on the redox properties of surface-adsorbed actinyl aqua complexes. Our computational approach precisely reproduced the experimental uranyl(VI/V) reduction potential on pyrite mineral surface (experimental value is 0.003 eV and our calculated value is 0.017 eV). Computational investigations are ongoing to underpin the surface-mediated redox process of adsorbed actinyls on semiconducting mineral surfaces such as iron-sulfides and iron-oxides.

4.5. Redox Potentials of Li Ion Battery Materials, Semiconductors, and Surfaces

Li-based battery materials have been widely used in electronics applications. Myriad of Li-based materials have been synthesized and the important properties of these materials such as the electric conductivity, chemical composition, thermal stability, diffusion, efficiency, and service life have been characterized over the years [150,151]. However, in order to improve the efficiency, performance, and cost effectiveness of these battery materials, numerous studies are ongoing, not just the experimental synthesis and characterization, but also computational studies, for instance to predict the redox properties, especially the Li intercalation potentials [151]. Li^+ ion intercalation potentials for Li ion battery materials, such as layered Li_xMO_2 (M = Co and Ni), Li_xTiS_2, olivine-structured Li_xMPO_4 (M = Mn, Fe, Co, and Ni), and spinel-like $Li_xMn_2O_4$, $Li_xTi_2O_4$ were determined using *ab initio* DFT-based periodic boundary calculations [152–154].

Insertion of Li into a transition metal oxide can be expressed as:

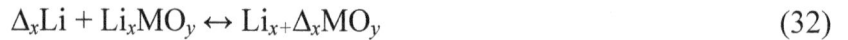

$$\Delta_x Li + Li_x MO_y \leftrightarrow Li_{x+\Delta_x} MO_y \tag{32}$$

MO is the transition-metal oxide material. The relation between the voltage of the cell and the Li chemical potential is given by the following equation.

$$V(x) = -(\mu_{Li(x)}^{cathode} - \mu_{Li}^{anode})/F \tag{33}$$

where μ_{Li}^{anode} is the chemical potential of Li. Integrating the above equation ($x1 = 0$ and $x2 = 1$) gives the average voltage $\langle V \rangle$ for the Li insertion between $Li_{x1}MO_y$ and $Li_{x2}MO_y$ compositions. From the total energies of the $Li_{x1}MO_y$ and $Li_{x2}MO_y$ and Li species, the average voltage $\langle V \rangle$ can be determined using the expression shown below:

$$\langle V \rangle = -\left[E\left(Li_{x2}MO_y\right) - E\left(Li_{x1}MO_y\right) - (x2 - x1)E\left(Li\ metal\right)\right]/(x2 - x1)F \tag{34}$$

Conventional GGA DFT studies were found to underestimate the experimental Li intercalation potentials for lithium oxide materials, such as Li_xMO_2 (M = Co and Ni), Li_xTS_2, and Li_xMPO_4 (M = Mn, Fe, Co, and Ni), and the calculated MUEs for the predicted Li^+ ion intercalation potentials were 0.75 eV less than the experimental Li^+ ion intercalation potentials. This underestimation was attributed to the lack of cancellation of self-interaction. In contrast, the use of the DFT+U method was found to predict the Li^+ ion intercalation potentials for these materials to be within 0.15 eV of the experimental intercalation potential. Moreover, a recent study using the newly-developed hybrid DFT functional HSE06 has predicted the Li^+ ion intercalation potentials for the above mentioned materials as accurate as the DFT+U methods and reproduced the experimental intercalation potentials within a MUE of 0.19 eV [153,154]. Using the reliable *ab initio* DFT methods, the battery materials can be designed and properties can be predicted before the synthesis, for instance the Li ion intercalation potentials for the $LiNiPO_4$ and Li_2CoSiO_4 materials were first calculated using the DFT+U approach and later synthesized and characterized by experimental studies. The theoretically predicted Li ion intercalation potentials have been found to be in excellent agreement with the experiments, which shows that the material design using prior *ab initio* methods is a powerful tool [151]. In recent years, high-throughput computing approaches for materials design are gaining rapid attention, in particular the Li ion battery materials, and this method has been applied to screen a large number of battery materials by applying DFT+U *ab initio* methods [155,156].

Similar to the Li based materials, Ce(4+/3+) reduction free energies of ceria were calculated using classical molecular-dynamics simulations [157–160]. Using empirical potentials, Ce (4+/3+) reduction free energies were predicted. This method involves creating defects and eventually the reduced species in terms of excess negative charge, for instance the reduction of an M^{n+} species forms $M^{(n-1)+}$, and the overall free energy difference for the formation of reduced species can be calculated. The authors have used a Buckingham potential and shell-models to describe the interactions in CeO_2. Using the following reaction scheme, the free energy for the Ce (4+/3+) reduction process was estimated.

$$2Ce_{Ce}^{\times} + O_O^{\times} \leftrightarrow 2Ce_{Ce}^{'} + V_O^{\cdot\cdot} 1/2 O_{2(g)} \tag{35}$$

According to the Kröger–Vink notation, $Ce_{Ce}^{\times}$ is the Ce^{4+} ion in its lattice position and $O_O^{\times}$ is the oxygen atom in its lattice position. On the right hand side of the equation, the first term designates Ce^{3+} and the next term designates an oxygen vacancy. Moreover, the Ce(4+/3+) reduction free energies in mixed oxides such as CeO_2–MO (M = Zr, Ca, Mn, Ni, and Zn) and CeO_2–M_2O_3 (M = Sc, Mn, Y, Gd, and La) [157,160] were modeled using this approach. The Ce (4+/3+) reduction free energies were found to decrease when the content of divalent metal increased in CeO_2–MO mixed oxides. Similarly, in CeO_2–M_2O_3 mixed oxides, an increase in trivalent metal was found to predict a decrease in the Ce(4+/3+) reduction free energies. Although reduction free energies for the Ce (4+/3+) reduction process in cerium oxide materials were modeled using classical molecular dynamics approaches, these studies did not include reference electrode systems.

Oxidation and reduction potentials of photo catalytic semiconductors were recently determined using DFT methods [131]. The reduction and oxidation potentials were calculated with respect to the experimentally well-known reduction and oxidation half-cell reactions, for instance the redox half-cell reactions of water (Equations (36) and (37)):

$$2H^+ + 2e^- \rightarrow H_2 \qquad (Reduction) \tag{36}$$

$$O_2 + 4H^+ + 4e^- \rightarrow 2H_2O \quad (Oxidation) \tag{37}$$

The thermodynamic parameters, for example the Gibbs free energies of the water redox reaction species such as H_2, H^+, O_2, and H_2O are available in literature [161,162]. Using the reported Gibbs free energies, the redox potential for overall water redox reaction can be calculated. The reduction reaction corresponds (Equation (36)) to the SHE and its value is assigned to zero (pH = 0). Oxidation potential of water with respect to the SHE is well-known and it is reported to be 1.23 V with respect to the SHE.

The next task is to calculate the reduction and oxidation potentials for a semiconductor with respect to these well-known water redox reactions. As an example, naturally-occurring and semiconducting zinc sulfide (ZnS) [131] is used and the respective reduction and oxidation reaction are given below (Equations (38) and (39)):

Reduction:

$$ZnS + H_2 \rightarrow Zn + H_2S \tag{38}$$

Oxidation:

$$ZnS + H_2O \rightarrow ZnO + S + H_2 \tag{39}$$

The conceptual basis used in this study is relatively simple, "obtain the Gibbs free energies of formation from already-available databases and calculate the free-energy (ΔG) for oxidation and reduction reactions with respect to the above water redox reactions (Equations (36) and (37))". It should be noted that the oxidation and reduction reaction proposed with these well-known oxidation and reduction reaction of water should produce thermodynamically stable compounds. Otherwise obtaining thermodynamic parameters from experimental data bases is not possible, and the relevant energies have to be calculated from *ab initio* methods. The reduction and oxidation potentials of a range of semiconductors that are relevant to photo-voltaic applications have been determined using the first principle DFT methods and experimental free energies of formation. Pourbaix diagrams (pH dependent redox potentials) were obtained, and thermodynamic stabilities of photocatalytic semiconductors, for example oxidative and reductive photocorrosion, were explored for thirty photocatalytic semiconductors [131].

Now, we discuss computational studies on the calculation of redox potentials of surface-grafted substituted ferrocenes (SFCs). In order to make the SFCs, ferrocene was functionalized with ethyl, vinyl or ethynyl groups and then attached to Si(100) surfaces. Using electrochemical methods, redox potentials of these adsorbed SFCs were determined. The DFT B3LYP method was used to calculate redox potentials of the surface-functionalized SFCs in CH_3CN solution, and the solvation effects were included using the PCM solvation method. The Si(100) surface-functionalized SFCs were modeled using mono and di hydrogenated Si surface cluster models. The calculated redox potentials are in good agreement with the experiments [163,164]. In a similar way, redox potentials were successfully calculated for the SFCs functionalized with different organic linkers on Si(100) surfaces [165,166].

5. Conclusions

In this review article, we have tried to provide a perspective view on various computational redox potentials predictions methods and outlined the frequently used methods available in the literature and briefly discussed the methods and the applicability and transferability of the approaches to other complexes, for instances, to transition-metal complexes, actinides, semiconductors, mineral surfaces, and surface-bound species. In order to give some guidance on the suitability of computational settings for the calculation of reduction potentials, we summarize some of these computational parameters in the following.

Basis Sets: Diffuse and polarization function containing basis sets recommended since the redox process involves anionic or radical species which requires to be stabilized.

DFT Functional: Despite the discrepancies reported for the B3LYP hybrid DFT functional, in particular for the redox potential predictions of Fe oxo-porpyrine systems and for few other open-shell systems, this functional has been reported to reproduce experimental redox potentials accurately. However, relying on a single DFT method is often problematic and it is emphasized that so far, there is not one DFT functional that can be universally applied to all kinds of computational redox-chemistry problems. More often, a DFT functional that might work for certain redox processes may fail for others.

Solvation Model: Although, there are several solvation models widely used to study the solvation effects in computational methods, based on the previous bench mark studies as discussed earlier, either the CPCM or SMD solvation methods are recommended for redox potential calculations. However, we prefer the CPCM method over the SMD since the CPCM method has been widely tested for actinides and transition metal redox potentials predictions. Using explicit solvation is promising to produce accurate redox potentials and mimics the local hydrogen effects effectively. In contrast, it is computationally expensive and obtaining the explicit solvation configuration may require pre-computations from low-level molecular mechanics methods. Unless very accurate redox potentials are expected from computations, for example benchmarking purpose, it is preferable to choose one of the implicit continuum solvation models, such as the CPCM solvation model.

Pseudopotentials: There is no significant change in computational expenses for using PP basis compared to all electron basis description for TM complexes. However, for actinides, using all-electron basis description significantly increases the computational time and expenses, and may not include relativistic effects. To overcome this, different PPs were developed. For actinides, small-core PPs that replace 60 core electrons by the PP are recommended since the LC-PPs (78 e$^-$) have been found to produce large systematic errors in the prediction of redox potentials of actinyl aqua complexes in solution.

Spin-Orbit Coupling: Spin-orbit coupling interaction is not very significant for the 3d and 4d transition metal coordination complexes; however this effect has significant influence when the 5d elements and actinides are considered.

ZPE, FEc, and Standard State Corrections: Even though these corrections improve the calculated reduction potentials, more often these effects are ignored because the improvements do often not justify the computational expense.

Isodesmic vs. *Direct Methods*: In the literature, both of these models have been successfully applied for the redox potential predictions; however, the isodesmic method often works better because of error cancellations and the lack of need to include the reference electrode potential. In order to minimize systematic errors further, the isodesmic reference complex has to be chosen carefully, it should belong from the same period of the periodic table, and it has to have a similar coordination and electronic environment as the redox-active complex. For example, one of the aqua complex of the TM 3d series can be used as the isodesmic reference to predict the redox potential of other M (3+/2+) aqueous redox complexes in the series. If the direct method is used for the redox potential prediction, care should be taken to choose the absolute value of the reference electrode potential. For instance, there are different values available for the SHE; however, we suggest that the absolute value recommended by the IUPAC is being used.

We hope that the methods described in this review serve as a helpful tool to choose appropriate computational methods for redox-potential predictions in areas of chemistry, biology, and mineralogy. These techniques may also help explore wherever redox chemistry is playing an important role, for instance, in geochemistry and biochemistry, in which the transport of electrons is crucial for many life processes. In addition, there are areas of cross-fertilization for example in designing battery materials with more efficiency and cost effectiveness than currently used materials or in understanding the corrosion of materials.

Moreover, this study has profound implications, in particular to the environmental science where a deeper understanding of redox reactions taking place on semiconducting mineral surfaces would help us design new remediation methods for radioactive actinide materials. For example, reductive immobilization processes slow the transport of these elements into the geo- and hydrosphere down and prevent long-term damages.

Rather than applying conventional approaches for the materials synthesis without any prior knowledge about the materials is often a black-box approach, in which success rate is serendipitous. Utilizing the gained knowledge from credible *ab initio* methods to predict the materials properties may result in promising materials for tailored applications.

Acknowledgments

This study was funded by the U.S. Department of Energy, Office of Basic Energy Sciences, Heavy Element Chemistry program under Grant DE-FG02-06ER15783. In addition, we would like to express our sincere thanks to the two anonymous reviewers for their constructive comments and suggestions for the improvement of this manuscript.

Author Contributions

Both Krishnamoorthy Arumugam and Udo Becker were involved in writing and revising all parts of the manuscript. While Krishnamoorthy Arumugam did the bulk of the calculations, the latter were inspired and revised by Udo Becker.

Conflicts of Interest

The authors declare no conflict of interest.

References

1. Levina, A.; Lay, P.A. Mechanistic studies of relevance to the biological activities of chromium. *Coord. Chem. Rev.* **2005**, *249*, 281–298.

2. Sundararajan, M.; Campbell, A.J.; Hillier, I.H. How do enzymes reduce metals? The mechanism of the reduction of Cr(VI) in chromate by cytochrome c_7 proteins proposed from DFT calculations. *Faraday Discuss.* **2011**, *148*, 195–205.

3. Mullet, M.; Demoisson, F.; Humbert, B.; Michot, L.J.; Vantelon, D. Aqueous Cr(VI) reduction by pyrite: Speciation and characterisation of the solid phases by X-ray photoelectron, Raman and X-ray absorption spectroscopies. *Geochim. Cosmochim. Acta* **2007**, *71*, 3257–3271.

4. Fendorf, S.; Berkeley, L. Reduction of hexavalent chromium by amorphous iron sulfide. *Environ. Sci. Technol.* **1997**, *31*, 2039–2044.

5. Mullet, M.; Boursiquot, S.; Ehrhardt, J.-J. Removal of hexavalent chromium from solutions by mackinawite, tetragonal FeS. *Colloids Surf. A Physicochem. Eng. Asp.* **2004**, *244*, 77–85.

6. Hughes, M.F. Arsenic toxicity and potential mechanisms of action. *Toxicol. Lett.* **2002**, *133*, 1–16.

7. Sun, F.; Dempsey, B.A.; Osseo-Asare, K.A. As(V) and As(III) reactions on pristine pyrite and on surface-oxidized pyrite. *J. Colloid Interface Sci.* **2012**, *388*, 170–175.

8. Wolthers, M.; Charlet, L.; van Der Weijden, C.H.; van der Linde, P.R.; Rickard, D. Arsenic mobility in the ambient sulfidic environment: Sorption of arsenic(V) and arsenic(III) onto disordered mackinawite. *Geochim. Cosmochim. Acta* **2005**, *69*, 3483–3492.

9. Renock, D.; Mueller, M.; Yuan, K.; Ewing, R.C.; Becker, U. The energetics and kinetics of uranyl reduction on pyrite, hematite, and magnetite surfaces: A powder microelectrode study. *Geochim. Cosmochim. Acta* **2013**, *118*, 56–71.

10. Scott, T.B.; Riba Tort, O.; Allen, G.C. Aqueous uptake of uranium onto pyrite surfaces; Reactivity of fresh *versus* weathered material. *Geochim. Cosmochim. Acta* **2007**, *71*, 5044–5053.

11. Moyes, L.N.; Jones, M.J.; Pattrick, R.A.D. An X-ray absorption spectroscopy study of neptunium (V) reactions with mackinawite (FeS). *Environ. Sci. Technol.* **2002**, *36*, 179–183.

12. Kirsch, R.; Fellhauer, D.; Altmaier, M.; Neck, V.; Rossberg, A.; Fangh, T.; Charlet, L.; Scheinost, A.C. Oxidation state and local structure of plutonium reacted with magnetite, mackinawite, and chukanovite. *Environ. Sci. Technol.* **2011**, *45*, 7267–7274.

13. Bruggeman, C.; Maes, N. Uptake of uranium(VI) by pyrite under boom clay conditions: Influence of dissolved organic carbon. *Environ. Sci. Technol.* **2010**, *44*, 4210–4216.

14. Latta, D.E.; Pearce, C.I.; Rosso, K.M.; Kemner, K.M.; Boyanov, M.I. Reaction of UVI with titanium-substituted magnetite: Influence of Ti on UIV speciation. *Environ. Sci. Technol.* **2013**, *47*, 4121–4130.

15. Morris, D.E. Redox energetics and kinetics of uranyl coordination complexes in aqueous solution. *Inorg. Chem.* **2002**, *41*, 3542–3547.

16. Austin, J.P.; Sundararajan, M.; Vincent, M.A.; Hillier, I.H. The geometric structures, vibrational frequencies and redox properties of the actinyl coordination complexes ($[AnO_2(L)_n]^m$; An = U, Pu, Np; L = H_2O, Cl^-, CO_3^{2-}, $CH_3CO_2^-$, OH^-) in aqueous solution, studied by density functional theory methods. *Dalt. Trans.* **2009**, 5902–5909.

17. Steele, H.; Taylor, R.J. A theoretical study of the inner-sphere disproportionation reaction mechanism of the pentavalent actinyl ions. *Inorg. Chem.* **2007**, *46*, 6311–6318.

18. Alessi, D.S.; Uster, B.; Veeramani, H.; Suvorova, E.I.; Lezama-Pacheco, J.S.; Stubbs, J.E.; Bargar, J.R.; Bernier-Latmani, R. Quantitative separation of monomeric U(IV) from UO_2 in products of U(VI) reduction. *Environ. Sci. Technol.* **2012**, *46*, 6150–6157.

19. Wall, J.D.; Krumholz, L.R. Uranium reduction. *Annu. Rev. Microbiol.* **2006**, *60*, 149–166.

20. Renshaw, J.C.; Butchins, L.J.C.; Livens, F.R.; May, I.; Charnock, J.M.; Lloyd, J.R. Bioreduction of uranium: Environmental implications of a pentavalent intermediate. *Environ. Sci. Technol.* **2005**, *39*, 5657–5660.

21. Lovley, D.R.; Phillips, E.J.P.; Survey, U.S.G. Bioremediation of uranium contamination with enzymatic uranium reduction. *Envirn. Sci. Technol.* **1992**, *26*, 2228–2234.

22. Luo, W.; Wu, W.-M.; Yan, T.; Criddle, C.S.; Jardine, P.M.; Zhou, J.; Gu, B. Influence of bicarbonate, sulfate, and electron donors on biological reduction of uranium and microbial community composition. *Appl. Microbiol. Biotechnol.* **2007**, *77*, 713–721.

23. Sundararajan, M.; Campbell, A.J.; Hillier, I.H. Catalytic cycles for the reduction of $[UO_2]^{2+}$ by cytochrome c_7 proteins proposed from DFT calculations. *J. Phys. Chem. A* **2008**, *112*, 4451–4457.

24. Sundararajan, M.; Assary, R.S.; Hillier, I.H.; Vaughan, D.J. The mechanism of the reduction of $[AnO_2]^{2+}$ (An = U, Np, Pu) in aqueous solution, and by Fe(II) containing proteins and mineral surfaces, probed by DFT calculations. *Dalt. Trans.* **2011**, *40*, 11156–11163.

25. Kelly, C.P.; Cramer, C.J.; Truhlar, D.G. Single-ion solvation free energies and the normal hydrogen electrode potential in methanol , acetonitrile , and dimethyl sulfoxide. *J. Phys. Chem. B* **2007**, *111*, 408–422.

26. Trasatti, S. The absolute electrode potential: An explanatory note. *Pure Appl. Chem.* **1986**, *58*, 955–966.

27. Fawcett, W.R.; Acta, S.E. The ionic work function and its role in estimating absolute electrode potentials. *Langmuir* **2008**, *24*, 9868–9875.

28. Reiss, H. The absolute potential of the standard hydrogen electrode: A new estimate. *J. Phys. Chem.* **1985**, *89*, 4207–4213.

29. Hansen, W.N.; Kolb, D.M. The work function of emersed electrodes. *J. Electroanal. Chem. Interfacial Electrochem.* **1979**, *100*, 493–500.

30. Coriani, S.; Haaland, A.; Helgaker, T.; Jørgensen, P. The equilibrium structure of ferrocene. *Chemphyschem* **2006**, *7*, 245–249.

31. Gritzner, G.; Kuta, J. Recommendations on reporting electrode potenitals in nonaqueous solvents. *Pure Appl. Chem.* **1982**, *54*, 1527–1532.

32. Pavlishchuk, V.V.; Addison, A.W. Conversion constants for redox potentials measured *versus* different reference electrodes in acetonitrile solutions at 25 °C. *Inorganica Chim. Acta* **2000**, *298*, 97–102.

33. Isse, A.A.; Gennaro, A. Absolute potential of the standard hydrogen electrode and the problem of interconversion of potentials in different solvents. *J. Phys. Chem. B* **2010**, *114*, 7894–7899.

34. Konezny, S.J.; Doherty, M.D.; Luca, O.R.; Crabtree, R.H.; Soloveichik, G.L.; Batista, V.S. Reduction of systematic uncertainty in DFT redox potentials of transition-metal complexes. *J. Phys. Chem. C* **2012**, *116*, 6349–6356.

35. Bartmess, J.E. Thermodynamics of the electron and the proton. *J. Phys. Chem.* **1994**, *98*, 6420–6424.

36. Cramer, C.J. *Essentials of Computational Chemistry Theories and Models*, 2nd ed.; John Wiley and Sons: Hoboken, NJ, USA, 2004; pp. 378–379.

37. Kelly, C.P.; Cramer, C.J.; Truhlar, D.G. Aqueous solvation free energies of ions and ion—Water clusters based on an accurate value for the absolute aqueous solvation free energy of the proton. *J. Phys. Chem. B* **2006**, *110*, 16066–16081.

38. Namazian, M.; Almodarresieh, H.A.; Noorbala, M.R.; Zare, H.R. DFT calculation of electrode potentials for substituted quinones in aqueous solution. *Chem. Phys. Lett.* **2004**, *396*, 424–428.

39. Zhu, X.-Q.; Wang, C.-H.; Liang, H. Scales of oxidation potentials, pK(a), and BDE of various hydroquinones and catechols in DMSO. *J. Org. Chem.* **2010**, *75*, 7240–7257.

40. Alizadeh, K.; Shamsipur, M. Calculation of the two-step reduction potentials of some quinones in acetonitrile. *J. Mol. Struct. Theochem.* **2008**, *862*, 39–43.

41. Zhu, X.-Q.; Wang, C.-H. Accurate estimation of the one-electron reduction potentials of various substituted quinones in DMSO and CH_3CN. *J. Org. Chem.* **2010**, *75*, 5037–5047.

42. Fernandez, L.E.; Horvath, S.; Hammes-Schi, S. Theoretical analysis of the sequential proton-coupled electron transfer mechanisms for H_2 oxidation and production pathways catalyzed by nickel molecular electrocatalysts. *J. Phys. Chem. C* **2012**, 3171–3180.

43. Solis, B.H.; Hammes-Schiffer, S. Computational study of anomalous reduction potentials for hydrogen evolution catalyzed by cobalt dithiolene complexes. *J. Am. Chem. Soc.* **2012**, *134*, 15253–15256.

44. Solis, B.H.; Hammes-Schiffer, S. Substituent effects on cobalt diglyoxime catalysts for hydrogen evolution. *J. Am. Chem. Soc.* **2011**, *133*, 19036–19039.

45. Arumugam, K. Redox Chemistry of Actinyl Complexes in Solution: A DFT Study. Ph.D. Thesis, The University of Manchester, Manchester, UK, 2012.

46. Hohenberg, P.; Kohn, W. Inhomogeneous electron gas. *Phys. Rev.* **1964**, *136*, B864–B871.

47. Kohn, W.; Sham, L.J. Self-consistent equations including exchange and correlation effects. *Phys. Rev.* **1965**, *140*, A1133–A1138.

48. Cohen, A.J.; Mori-Sánchez, P.; Yang, W. Challenges for density functional theory. *Chem. Rev.* **2012**, *112*, 289–320.

49. Zhao, Y.; Truhlar, D.G. The M06 suite of density functionals for main group thermochemistry, thermochemical kinetics, noncovalent interactions, excited states, and transition elements: Two new functionals and systematic testing of four M06-class functionals and 12 other functionals. *Theor. Chem. Acc.* **2007**, *120*, 215–241.

50. Cramer, C.J.; Truhlar, D.G. Density functional theory for transition metals and transition metal chemistry. *Phys. Chem. Chem. Phys.* **2009**, *11*, 10757–10816.

51. Becke, A.D. Density-fnnctional exchange-energy approximation with correct asymptotic behavior. *Phys. Rev. A* **1988**, *38*, 3098–3100.

52. Becke, A.D. A new mixing of Hartree–Fock and local density-functional theories. *J. Chem. Phys.* **1993**, *98*, 1372–1377.

53. Lee, C.; Yang, W.; Parr, R.G. Development of the Colle-Salvetti correlation-energy formula into a functional of the electron density. *Phys. Rev. B* **1988**, *37*, 785–789.

54. Heyd, J.; Scuseria, G.E.; Ernzerhof, M. Hybrid functionals based on a screened Coulomb potential. *J. Chem. Phys.* **2003**, *118*, 8207–8215.

55. Castro, L. Calculations of one-electron redox potentials of oxoiron(IV) porphyrin complexes. *J. Chem. Theory Comput.* **2013**, *10*, 243–251.

56. Roy, L.E.; Jakubikova, E.; Guthrie, M.G.; Batista, E.R. Calculation of one-electron redox potentials revisited. Is it possible to calculate accurate potentials with density functional methods? *J. Phys. Chem. A* **2009**, *113*, 6745–6750.

57. Roy, L.E.; Batista, E.R.; Hay, P.J. Theoretical studies on the redox potentials of Fe dinuclear complexes as models for hydrogenase. *Inorg. Chem.* **2008**, *47*, 9228–9237.

58. Austin, J.P.; Burton, N.A.; Hillier, I.H.; Sundararajan, M.; Vincent, M.A. Which density functional should be used to study actinyl complexes? *Phys. Chem. Chem. Phys.* **2009**, *11*, 1143–1145.

59. Steele, H.M.; Guillaumont, D.; Moisy, P. Density functional theory calculations of the redox potentials of actinide(VI)/actinide(V) couple in water. *J. Phys. Chem. A* **2013**, *117*, 4500–4505.

60. Jaque, P.; Marenich, A.V.; Cramer, C.J.; Truhlar, D.G. Computational electrochemistry: The aqueous $Ru^{3+}|Ru^{2+}$ reduction potential. *J. Phys. Chem. C* **2007**, *111*, 5783–5799.

61. Baik, M.-H.; Friesner, R.A. Computing redox potentials in solution: Density functional theory as a tool for rational design of redox agents. *J. Phys. Chem. A* **2002**, *106*, 7407–7412.

62. Mennucci, B.; Tomasi, J.; Cammi, R.; Cheeseman, J.R.; Frisch, M.J.; Devlin, F.J.; Gabriel, S.; Stephens, P.J. Polarizable continuum model (PCM) Calculations of solvent effects on optical rotations of chiral molecules. *J. Phys. Chem. A* **2002**, *106*, 6102–6113.

63. Klamt, A.; Schuurmann, G. COSMO: A new approach to dielectric screening in solvents with explicit expressions for the screening energy and its gradient. *J. Chem. Soc. Perkin Trans. 2* **1993**, *1993*, 799–805.

64. Barone, V.; Cossi, M. Quantum calculation of molecular energies and energy gradients in soluion by a conductor solvent model. *J. Phys. Chem. A* **1998**, *102*, 1995–2001.

65. Cossi, M.; Rega, N.; Scalmani, G.; Barone, V. Energies, structures, and electronic propoerties of molecules in solution with the C-PCM solvation model. *J. Comput. Chem.* **2003**, *24*, 669–681.

66.   Mennucci, B.; Cancès, E.; Tomasi, J. Evaluation of solvent effects in isotropic and anisotropic dielectrics and in ionic solutions with a unified integral equation method: theoretical bases, computational implementation, and numerical applications. *J. Phys. Chem. B* **1997**, *101*, 10506–10517.

67.   Cancès, E.; Mennucci, B.; Tomasi, J. A new integral equation formalism for the polarizable continuum model: Theoretical background and applications to isotropic and anisotropic dielectrics. *J. Chem. Phys.* **1997**, *107*, 3032–3041.

68.   Marenich, A.V.; Cramer, C.J.; Truhlar, D.G. Universal solvation model based on solute electron density and on a continuum model of the solvent defined by the bulk dielectric constant and atomic surface tensions. *J. Phys. Chem. B* **2009**, *113*, 6378–6396.

69.   Klamt, A. Conductor-like screening model for real solvents: A new approach to the quantitative calculation of solvation phenomena. *J. Phys. Chem.* **1995**, *99*, 2224–2235.

70.   Klamt, A.; Jonas, V.; Burger, T.; Lohrenz, J.C.W. Refinement and parametrization of COSMO-RS. *J. Phys. Chem. A* **1998**, *102*, 5074–5085.

71.   Edinger, S.R.; Cortis, C.; Shenkin, P.S.; Friesner, R.A. Solvation free energies of peptides: Comparison of approximate continuum solvation models with accurate solution of the Poisson–Boltzmann equation. *J. Phys. Chem. B* **1997**, *101*, 1190–1197.

72.   Friedrichs, M.; Zhou, R.; Edinger, S.R.; Friesner, R.A. Poisson–Boltzmann analytical gradients for molecular modeling calculations. *J. Phys. Chem. B* **1999**, *103*, 3057–3061.

73.   Marten, B.; Kim, K.; Cortis, C.; Friesner, R.A.; Murphy, R.B.; Ringnalda, M.N.; Sitkoff, D.; Honig, B. New model for calculation of solvation free energies: Correction of self-consistent reaction field continuum dielectric theory for short-range hydrogen-bonding effects. *J. Phys. Chem.* **1996**, *100*, 11775–11788.

74.   Tomasi, J.; Mennucci, B.; Cammi, R. Quantum mechanical continuum solvation models. *Chem. Rev.* **2005**, *105*, 2999–3093.

75.   Frisch, M.J.; Trucks, G.W.; Schlegel, H.B.; Scuseria, G.E.; Robb, M.A.; Cheeseman, J.R.; Scalmani, G.; Barone, V.; Mennucci, B.; Petersson, G.A.; Nakatsuji, H.; Caricato, M.; Li, X.; Hratchian, H.P.; Izmaylov, A.F.; Bloino, J.; Zheng, G.; Sonnenberg, J.L.; Hada, M.; Ehara, M.; Toyota, K.; Fukuda, R.; Hasegawa, J.; Ishida, M.; Nakajima, T.; Honda, Y.; Kitao, O.; Nakai, H.; Vreven, T.; Montgomery, J.A., Jr.; Peralta, J.E.; Ogliaro, F.; Bearpark, M.; Heyd, J.J.; Brothers, E.; Kudin, K.N.; Staroverov, V.N.; Kobayashi, R.; Normand, J.; Raghavachari, K.; Rendell, A.; Burant, J.C.; Iyengar, S.S.; Tomasi, J.; Cossi, M.; Rega, N.; Millam, J.M.; Klene, M.; Knox, J.E.; Cross, J.B.; Bakken, V.; Adamo, C.; Jaramillo, J.; Gomperts, R.; Stratmann, R.E.; Yazyev, O.; Austin, A.J.; Cammi, R.; Pomelli, C.; Ochterski, J.W.; Martin, R.L.; Morokuma, K.; Zakrzewski, V.G.; Voth, G.A.; Salvador, P.; Dannenberg, J.J.; Dapprich, S.; Daniels, A.D.; Farkas, Ö.; Foresman, J.B.; Ortiz, J.V.; Cioslowski, J.; Fox, D.J. Gaussian 09, Revision A.02; Gaussian, Inc.: Wallingford, CT, USA, 2009.

76.   Takano, Y.; Houk, K.N.; Angeles, L. Benchmarking the conductor-like polarizable continuum model (CPCM) for aqueous solvation free energies of neutral and ionic organic molecules. *J. Chem. Theory Comput.* **2005**, *1*, 70–77.

77. Methods, I.S.; Chiorescu, I.; Deubel, D.V.; Arion, V.B.; Keppler, B.K. Computational electrochemistry of ruthenium anticancer agents. Unprecedented benchmarking of implicit solvation methods. *J. Chem. Theory Comput.* **2008**, *4*, 499–506.

78. Bondi, A. Van der Waals volumes and radii. *J. Phys. Chem.* **1964**, *68*, 441–451.

79. Uudsemaa, M.; Tamm, T. Density-functional theory calculations of aqueous redox potentials of fourth-period transition metals. *J. Phys. Chem. A* **2003**, *107*, 9997–10003.

80. Keith, T.A.; Frisch, M.J. Inclusion of explicit solvent molecules in a self-consistent-reaction field model of solvation. *ACS Symp. Ser.* **1994**, *569*, 22–35.

81. Hughes, T.F.; Friesner, R.A. Development of accurate DFT methods for computing redox potentials of transition metal complexes: Results for model complexes and application to cytochrome P450. *J. Chem. Theory Comput.* **2012**, *8*, 442–459.

82. Srnec, M.; Chalupský, J.; Fojta, M.; Zendlová, L.; Havran, L.; Hocek, M.; Kývala, M.; Rulísek, L. Effect of spin-orbit coupling on reduction potentials of octahedral ruthenium(II/III) and osmium(II/III) complexes. *J. Am. Chem. Soc.* **2008**, *130*, 10947–10954.

83. Hay, P.J.; Martin, R.L.; Schreckenbach, G. Theoretical studies of the properties and solution chemistry of AnO_2^{2+} and AnO_2^+ aquo complexes for An = U , Np , and Pu. *J. Phys. Chem. A* **2000**, *104*, 6259–6270.

84. Wang, L.-P.; Van Voorhis, T. A polarizable QM/MM explicit solvent model for computational electrochemistry in water. *J. Chem. Theory Comput.* **2012**, *8*, 610–617.

85. Li, G.; Zhang, X.; Cui, Q. Free energy perturbation calculations with combined QM/MM potentials complications, simplifications, and applications to redox potential calculations. *J. Phys. Chem. B* **2003**, *107*, 8643–8653.

86. Formaneck, M.S.; Li, G.; Zhang, X.; Cui, Q. Calculating accurate redox potentials in enzymes with a combined QM/MM free energy perturbation approach. *J. Theor. Comput. Chem.* **2002**, *1*, 53–67.

87. Zeng, X.; Hu, H.; Hu, X.; Yang, W. Calculating solution redox free energies with *ab initio* quantum mechanical/molecular mechanical minimum free energy path method. *J. Chem. Phys.* **2009**, *130*, 164111, doi:10.1063/1.3120605.

88. Costanzo, F.; Sulpizi, M.; Della Valle, R.G.; Sprik, M. The oxidation of tyrosine and tryptophan studied by a molecular dynamics normal hydrogen electrode. *J. Chem. Phys.* **2011**, *134*, 244508, doi:10.1063/1.3597603.

89. Rauschnot, J.C.; Yang, C.; Yang, V.; Bhattacharyya, S. Theoretical determination of the redox potentials of NRH:Quinone oxidoreductase 2 using quantum mechanical/molecular mechanical simulations. *J. Phys. Chem. B* **2009**, *113*, 8149–8157.

90. Oxidase, C.; Bhattacharyya, S.; Stankovich, M.T.; Truhlar, D.G.; Gao, J. Combined quantum mechanical and molecular mechanical simulations of one- and two-electron reduction potentials of flavin cofactor in water, medium-chain acyl-CoA dehydrogenase, and cholesterol oxidase. *J. Phys. Chem. A* **2007**, *111*, 5729–5742.

91. Van den Bosch, M.; Swart, M.; Snijders, J.G.; Berendsen, H.J.C.; Mark, A.E.; Oostenbrink, C.; van Gunsteren, W.F.; Canters, G.W. Calculation of the redox potential of the protein azurin and some mutants. *Chembiochem* **2005**, *6*, 738–746.

92. Blumberger, J.; Tateyama, Y.; Sprik, M. *Ab initio* molecular dynamics simulation of redox reactions in solution. *Comput. Phys. Commun.* **2005**, *169*, 256–261.

93. VandeVondele, J.; Ayala, R.; Sulpizi, M.; Sprik, M. Redox free energies and one-electron energy levels in density functional theory based *ab initio* molecular dynamics. *J. Electroanal. Chem.* **2007**, *607*, 113–120.

94. Adriaanse, C.; Cheng, J.; Chau, V.; Sulpizi, M.; VandeVondele, J.; Sprik, M. Aqueous redox chemistry and the electronic band structure of liquid water. *J. Phys. Chem. Lett.* **2012**, *3*, 3411–3415.

95. Blumberger, J.; Sprik, M. *Ab initio* molecular dynamics simulation of the aqueous Ru^{2+}/Ru^{3+} redox reaction: The Marcus perspective. *J. Phys. Chem. B* **2005**, *109*, 6793–6804.

96. Kamerlin, S.C.L.; Haranczyk, M.; Warshel, A. Progress in *ab initio* QM/MM free-energy simulations of electrostatic energies in proteins: accelerated QM/MM studies of pK_a, redox reactions and solvation free energies. *J. Phys. Chem. B* **2009**, *113*, 1253–1272.

97. Cheng, J.; Sulpizi, M.; Sprik, M. Redox potentials and pK_a for benzoquinone from density functional theory based molecular dynamics. *J. Chem. Phys.* **2009**, *131*, 154504, doi:10.1063/1.3250438.

98. Evans, D.H. One-electron and two-electron transfers in electrochemistry and homogeneous solution reactions. *Chem. Rev.* **2008**, *108*, 2113–2144.

99. Costentin, C. Electrochemical approach to the mechanistic study of proton-coupled electron transfer. *Chem. Rev.* **2008**, *108*, 2145–2179.

100. Savéant, J.-M. Molecular catalysis of electrochemical reactions. Mechanistic aspects. *Chem. Rev.* **2008**, *108*, 2348–2378.

101. Yoshida, J.; Kataoka, K.; Horcajada, R.; Nagaki, A. Modern strategies in electroorganic synthesis. *Chem. Rev.* **2008**, *108*, 2265–2299.

102. Keith, J.A.; Carter, E.A. Theoretical insights into pyridinium-based photoelectrocatalytic reduction of CO_2. *J. Am. Chem. Soc.* **2012**, *134*, 7580–7583.

103. Namazian, M.; Coote, M.L. Accurate calculation of absolute one-electron redox potentials of some *para*-quinone derivatives in acetonitrile. *J. Phys. Chem. A* **2007**, *3*, 7227–7232.

104. Gogoll, A.; Strømme, M.; Sjo, M. Investigation of the redox chemistry of isoindole-4,7-diones. *J. Phys. Chem. C* **2013**, *117*, 894–901.

105. Karlsson, C.; Ja, E.; Strømme, M.; Sjo, M. Computational electrochemistry study of 16 isoindole-4,7-diones as candidates for organic cathode materials. *J. Phys. Chem. C* **2012**, *116*, 3793–2801.

106. Francke, R.; Little, R.D. Optimizing Electron transfer mediators based on arylimidazoles by ring fusion: Synthesis, electrochemistry and computational analysis of 2-aryl-1-methylphenanthro[9,10-*d*]imidazoles. *J. Am. Chem. Soc.* **2013**, *136*, 427–435.

107. Fu, Y.; Liu, L.; Yu, H.-Z.; Wang, Y.-M.; Guo, Q.-X. Quantum-chemical predictions of absolute standard redox potentials of diverse organic molecules and free radicals in acetonitrile. *J. Am. Chem. Soc.* **2005**, *127*, 7227–7234.

108. Fu, Y.; Liu, L.; Wang, Y.-M.; Li, J.-N.; Yu, T.-Q.; Guo, Q.-X. Quantum-chemical predictions of redox potentials of organic anions in dimethyl sulfoxide and reevaluation of bond dissociation enthalpies measured by the electrochemical methods. *J. Phys. Chem. A* **2006**, *110*, 5874–5886.

109. Lynch, E.J.; Speelman, A.L.; Curry, B.A.; Murillo, C.S.; Gillmore, J.G. Expanding and testing a computational method for predicting the ground state reduction potentials of organic molecules on the basis of empirical correlation to experiment. *J. Org. Chem.* **2012**, *77*, 6423–6430.

110. Bogart, J.A.; Lee, H.B.; Boreen, M.A.; Jun, M.; Schelter, E.J. Fine-tuning the oxidative ability of persistent radicals: Electrochemical and computational studies of substituted 2-pyridylhydroxylamines. *J. Org. Chem.* **2013**, *78*, 6344–6349.

111. Davis, A.P.; Fry, A.J. Experimental and computed absolute redox potentials of polycyclic aromatic hydrocarbons are highly linearly correlated over a wide range of structures and potentials. *J. Phys. Chem. A* **2010**, *114*, 12299–12304.

112. Blinco, J.P.; Hodgson, J.L.; Morrow, B.J.; Walker, J.R.; Will, G.D.; Coote, M.L.; Bottle, S.E. Experimental and theoretical studies of the redox potentials of cyclic nitroxides. *J. Org. Chem.* **2008**, *73*, 6763–6771.

113. Hodgson, J.L.; Namazian, M.; Bottle, S.E.; Coote, M.L. One-electron oxidation and reduction potentials of nitroxide antioxidants: A theoretical study. *J. Phys. Chem. A* **2007**, *111*, 13595–13605.

114. Guerard, J.J.; Arey, J.S. Critical evaluation of implicit solvent models for predicting aqueous oxidation potentials of neutral organic compounds. *J. Chem. Theory Comput.* **2013**, *9*, 5046–5058.

115. Psciuk, B.T.; Lord, R.L.; Munk, B.H.; Schlegel, H.B. Theoretical determination of one-electron oxidation potentials for nucleic acid bases. *J. Chem. Theory Comput.* **2012**, *8*, 5107–5123.

116. Psciuk, B.T.; Schlegel, H.B. Computational prediction of one-electron reduction potentials and acid dissociation constants for guanine oxidation intermediates and products. *J. Phys. Chem. B* **2013**, *117*, 9518–9531.

117. Baik, M.-H.; Silverman, J.S.; Yang, I.V.; Ropp, P.A.; Szalai, V.A.; Yang, W.; Thorp, H.H. Using density functional theory to design DNA base analogues with low oxidation potentials. *J. Phys. Chem. B* **2001**, *105*, 6437–6444.

118. Crespo-Hernandez, C.E.; Arce, R.; Ishikawa, Y.; Gorb, L.; Leszczynski, J.; Close, D.M. *Ab initio* ionization energy thresholds of DNA and RNA bases in gas phase and in aqueous solution. *J. Phys. Chem. A* **2004**, *108*, 6373–6377.

119. Paukku, Y.; Hill, G. Theoretical determination of one-electron redox potentials for DNA bases, base pairs, and stacks. *J. Phys. Chem. A* **2011**, *115*, 4804–4810.

120. Lin, M.-J.; Liu, W.-X.; Peng, C.R.; Lu, W.-C. A First-principles method for predicting redox potentials of nucleobases and the metabolites in aqueous solution. *Acta Phys. Chim. Sin.* **2011**, *27*, 595–603.

121. Li, X.-L.; Fu, Y. Theoretical study of reduction potentials of substituted flavins. *J. Mol. Struct. Theochem* **2008**, *856*, 112–118.

122. North, M.A.; Bhattacharyya, S.; Truhlar, D.G. Improved density functional description of the electrochemistry and structure-property descriptors of substituted flavins. *J. Phys. Chem. B* **2010**, *114*, 14907–14915.

123. Walsh, J.D.; Miller, A. Flavin reduction potential tuning by substitution and bending. *J. Mol. Struct. Theochem* **2003**, *623*, 185–195.

124. Wade, K. The structural significance of the number of skeletal bonding electron-pairs in carboranes, the higher boranes and borane anions, and various transition-metal carbonyl cluster compounds. *J. Chem. Soc. D Chem. Commun.* **1971**, *1971*, 792–793.

125. Welch, A.J. The significance and impact of Wade's rules. *Chem. Commun.* **2013**, *49*, 3615–3616.

126. Wahab, A.; Stepp, B.; Douvris, C.; Valášek, M.; Štursa, J.; Klíma, J.; Piqueras, M.-C.; Crespo, R.; Ludvík, J.; Michl, J. Measured and calculated oxidation potentials of 1-X-12-Y-CB$_{11}$Me$_{10}$$^-$ anions. *Inorg. Chem.* **2012**, *51*, 5128–5137.

127. Boeré, R.T.; Bolli, C.; Finze, M.; Himmelspach, A.; Knapp, C.; Roemmele, T.L. Quantum-chemical and electrochemical investigation of the electrochemical windows of halogenated carborate anions. *Chemistry* **2013**, *19*, 1784–1795.

128. Lee, T.B.; Mckee, M.L. Redox energetics of *Hypercloso* boron hydrides B$_n$H$_n$ (n = 6–13) and B$_{12}$X$_{12}$ (X = F, Cl, OH, and CH$_3$). *Inorg. Chem.* **2012**, *51*, 4205–4214.

129. Moens, J.; Geerlings, P.; Roos, G. A conceptual DFT approach for the evaluation and interpretation of redox potentials. *Chem. Eur. J.* **2007**, *13*, 8174–8184.

130. Li, J.; Fisher, C.L.; Chen, J.L.; Bashford, D.; Noodleman, L. Calculation of redox potentials and pk_a values of hydrated transition metal cations by a combined density functional and continuum dielectric theory. *Inorg. Chem.* **1996**, *35*, 4694–4702.

131. Chen, S.; Wang, L.-W. Thermodynamic oxidation and reduction potentials of photocatalytic semiconductors in aqueous solution. *Chem. Mater.* **2012**, *24*, 3659–3666.

132. Matsui, T.; Kitagawa, Y.; Shigeta, Y.; Okumura, M. A density functional theory based protocol to compute the redox potential of transition metal complex with the correction of pseudo-counterion: General theory and applications. *J. Chem. Theory Comput.* **2013**, *9*, 2974–2980.

133. Moens, J.; De Proft, F.; Geerlings, P. A density functional theory study on ligand additive effects on redox potentials. *Phys. Chem. Chem. Phys.* **2010**, *12*, 13174–13181.

134. Haines, D.E.; O'Hanlon, D.C.; Manna, J.; Jones, M.K.; Shaner, S.E.; Sun, J.; Hopkins, M.D. Oxidation-potential tuning of tungsten-alkylidyne complexes over a 2 V range. *Inorg. Chem.* **2013**, *52*, 9650–9658.

135. Holland, J.P.; Green, J.C.; Dilworth, J.R. Probing the mechanism of hypoxia selectivity of copper bis(thiosemicarbazonato) complexes: DFT calculation of redox potentials and absolute acidities in solution. *Dalt. Trans.* **2006**, 783–794.

136. Berard, J.J.; Schreckenbach, G.; Arnold, P.L.; Patel, D.; Love, J.B. Computational density functional study of polypyrrolic macrocycles: analysis of actinyl-oxo to 3d transition metal bonding. *Inorg. Chem.* **2008**, *47*, 11583–11592.

137. Shamov, G.A.; Schreckenbach, G. Density functional studies of actinyl aquo complexes studied using small-core effective core potentials and a scalar four-component relativistic method. *J. Phys. Chem. A* **2005**, *109*, 10961–10974.

138. Shamov, G.A.; Schreckenbach, G. Relativistic density functional theory study of dioxoactinide(VI) and -(V) complexation with alaskaphyrin and related Schiff-base macrocyclic ligands. *J. Phys. Chem. A* **2006**, *110*, 9486–9499.

139. Horowitz, S.E.; Marston, J.B. Strong correlations in actinide redox reactions. *J. Chem. Phys.* **2011**, *134*, 064510, doi:10.1063/1.3549571.

140. Tsushima, S.; Wahlgren, U.; Grenthe, I. Quantum chemical calculations of reduction potentials of AnO$_2$$^{2+}$/AnO$_2$$^+$ (An = U, Np, Pu, Am) and Fe^{3+}/Fe^{2+} couples. *J. Phys. Chem. A* **2006**, 9175–9182.

141. Schreckenbach, G.; Shamov, G.A. Theoretical actinide molecular science. *Acc. Chem. Res.* **2010**, *43*, 19–29.

142. Rappe, A.K.; Casewit, C.J.; Colwell, K.S.; Goddard, W.A., III; Skiff, W.M. UFF, a full periodic table force field for molecular mechanics and molecular dynamics simulations. *J. Am. Chem. Soc.* **1992**, *114*, 10024–10035.

143. Rappe, A.K.; Colwell, K.S.; Casewit, C.J. Application of a universal force field to metal complexes. *Inorg. Chem.* **1993**, 3438–3450.

144. Elkechai, A.; Boucekkine, A.; Belkhiri, L.; Amarouche, M.; Clappe, C.; Hauchard, D.; Ephritikhine, M. A DFT and experimental investigation of the electron affinity of the triscyclopentadienyl uranium complexes Cp$_3$UX. *Dalt. Trans.* **2009**, *2009*, 2843–2849.

145. Elkechai, A.; Meskaldji, S.; Boucekkine, A.; Belkhiri, L.; Bouchet, D.; Amarouche, M.; Clappe, C.; Hauchard, D.; Ephritikhine, M. A relativistic DFT study of the electron affinity of the biscyclopentadienyl uranium complexes Cp$_{*2}$UX$_2$. *J. Mol. Struct. Theochem* **2010**, *954*, 115–123.

146. Elkechai, A.; Boucekkine, A.; Belkhiri, L.; Hauchard, D.; Clappe, C.; Ephritikhine, M. Electron affinities of biscyclopentadienyl and phospholyl uranium(IV) borohydride complexes: Experimental and DFT studies. *Comptes Rendus Chim.* **2010**, *13*, 860–869.

147. Elkechai, A.; Mani, Y.; Boucekkine, A.; Ephritikhine, M. Density functional theory investigation of the redox properties of tricyclopentadienyl- and phospholyluranium(IV) chloride complexes. *Inorg. Chem.* **2012**, *51*, 6943–6952.

148. Küchle, W.; Dolg, M.; Stoll, H.; Preuss, H. Energy-adjusted pseudopotentials for the actinides. Parameter sets and test calculations for thorium and thorium monoxide. *J. Chem. Phys.* **1994**, *100*, 7535–7542.

149. Cao, Z.; Balasubramanian, K. Theoretical studies of UO$_2$(H$_2$O)$_n^{2+}$, NpO$_2$(H$_2$O)$_n^+$, and PuO$_2$(H$_2$O)$_n^{2+}$ complexes (*n* = 4–6) in aqueous solution and gas phase. *J. Chem. Phys.* **2005**, *123*, 114309, doi:10.1063/1.2018754.

150. Goodenough, J.B.; Kim, Y. Challenges for rechargeable Li batteries. *Chem. Mater.* **2010**, *22*, 587–603.

151. Meng, Y.S.; Arroyo-de Dompablo, M.E. First principles computational materials design for energy storage materials in lithium ion batteries. *Energy Environ. Sci.* **2009**, *2*, 589–609.

152. Wang, L.; Zhou, F.; Meng, Y.; Ceder, G. First-principles study of surface properties of LiFePO$_4$: Surface energy, structure, Wulff shape, and surface redox potential. *Phys. Rev. B* **2007**, *76*, 165435, doi:10.1103/PhysRevB.76.165435.

153. Zhou, F.; Cococcioni, M.; Marianetti, C.; Morgan, D.; Ceder, G. First-principles prediction of redox potentials in transition-metal compounds with LDA+U. *Phys. Rev. B* **2004**, *70*, 235121, doi:10.1103/PhysRevB.70.235121.

154. Chevrier, V.L.; Ong, S.P.; Armiento, R.; Chan, M.K.Y.; Ceder, G. Hybrid density functional calculations of redox potentials and formation energies of transition metal compounds. *Phys. Rev. B* **2010**, *82*, 075122, doi:10.1103/PhysRevB.82.075122.

155. Hautier, G.; Jain, A.; Ong, S.P.; Kang, B.; Moore, C.; Doe, R.; Ceder, G. Phosphates as lithium-ion battery cathodes: An evaluation based on high-throughput *ab initio* calculations. *Chem. Mater.* **2011**, *23*, 3495–3508.

156. Mueller, T.; Hautier, G.; Jain, A.; Ceder, G. Evaluation of tavorite-structured cathode materials for lithium-ion batteries using high-throughput computing. *Chem. Mater.* **2011**, *23*, 3854–3862.

157. Balducci, G.; Kas, J.; Fornasiero, P.; Graziani, M.; Islam, M.S. Surface and reduction energetics of the CeO$_2$–ZrO$_2$ catalysts. *J. Phys. Chem. B* **1998**, *102*, 557–561.

158. Balducci, G.; Kas, J.; Fornasiero, P.; Graziani, M.; Islam, M.S.; Gale, J.D. Computer simulation studies of bulk reduction and oxygen migration in CeO$_2$–ZrO$_2$ solid solutions. *J. Phys. Chem. B* **1997**, *101*, 1750–1753.

159. Balducci, G.; Islam, M.S.; Kašpar, J.; Fornasiero, P.; Graziani, M. Bulk reduction and oxygen migration in the ceria-based oxides. *Chem. Mater.* **2000**, *12*, 677–681.

160. Balducci, G.; Islam, M.S.; Kas, J.; Fornasiero, P. Reduction process in CeO$_2$–MO and CeO$_2$–M$_2$O$_2$ mixed oxides: A computer simulation study. *Chem. Mater.* **2003**, *15*, 3781–3785.

161. Speight, J.G. *Lange's Handbook of Chemistry*, 16th ed.; McGraw-Hill: New York, NY, USA, 2005.

162. Lide, D.R. *CRC Handbook of Chemistry and Physics*, 84th ed.; CRC Press: Boca Raton, FL, USA, 2003.

163. Zanoni, R.; Cossi, M.; Iozzi, M.F.; Cattaruzza, F.; Dalchiele, E.A.; Decker, F.; Marrani, A.G.; Valori, M. Tuning the redox potential in molecular monolayers covalently bound to H–Si(100) electrodes via distinct C–C tethering arms. *Superlattices Microstruct.* **2008**, *44*, 542–549.

164. Boccia, A.; Decker, F.; Marrani, A.G.; Stranges, S.; Zanoni, R.; Cossi, M.; Iozzi, M.F. Role of the extent of -electron conjugation in visible-light assisted molecular anchoring on Si(111) surfaces. *Superlattices Microstruct.* **2009**, *46*, 30–33.

165. Pro, T.; Buckley, J.; Huang, K.; Calborean, A.; Marc, G.; Delapierre, G.; Member, S.; Duclairoir, F.; Marchon, J.; Jalaguier, E.; Maldivi, P.; De Salvo, B.; Deleonibus, S. Investigation of hybrid molecular/silicon memories with redox-active molecules acting as storage media. *IEEE Trans. Nanotechnol.* **2009**, *8*, 204–213.

166. Pro, T.; Buckley, J.; Barattin, R.; Calborean, A.; Aiello, V.; Nicotra, G.; Gély, M.; Delapierre, G.; Jalaguier, E.; Duclairoir, F.; Chevalier, N.; Lombardo, S.; Maldivi, P.; Ghibaudo, G.; De Salvo, B.; Deleonibus, S. From atomistic to device level investigation of hybrid redox molecular/silicon field-effect memory devices. *IEEE Trans. Nanotechnol.* **2011**, *10*, 275–283.

Precious Metals in Automotive Technology: An Unsolvable Depletion Problem?

Ugo Bardi [1,*] and Stefano Caporali [2,3]

[1] Dipartimento di Scienze della Terra, Università di Firenze, Via G. La Pira 4, 50121 Firenze, Italy

[2] Dipartimento di Chimica, Università di Firenze, Via della Lastruccia 3, 50019 Sesto Fiorentino, Italy; E-Mail: stefano.caporali@unifi.it

[3] Consorzio Interuniversitario Nazionale per la Scienza e Tecnologia dei Materiali, Via Giusti 9, 50123 Firenze, Italy

* Author to whom correspondence should be addressed; E-Mail: ugo.bardi@unifi.it

Abstract: Since the second half of the 20th century, various devices have been developed in order to reduce the emissions of harmful substances at the exhaust pipe of combustion engines. In the automotive field, the most diffuse and best known device of this kind is the "three way" catalytic converter for engines using the Otto cycle designed to abate the emissions of carbon monoxide, nitrogen oxides and unburnt hydrocarbons. These catalytic converters can function only by means of precious metals (mainly platinum, rhodium and palladium) which exist in a limited supply in economically exploitable ores. The recent increase in prices of all mineral commodities is already making these converters significantly expensive and it is not impossible that the progressive depletion of precious metals will make them too expensive for the market of private cars. The present paper examines how this potential scarcity could affect the technology of road transportation worldwide. We argue that the supply of precious metals for automotive converters is not at risk in the short term, but that in the future it will not be possible to continue using this technology as a result of increasing prices generated by progressive depletion. Mitigation methods such as reducing the amounts of precious metals in catalysts, or recycling them can help but cannot be considered as a definitive solution. We argue that precious metal scarcity is a critical factor that may determine the future development of road transportation in the world. As the problem is basically unsolvable in the long run, we must explore new technologies for road transportation and we conclude that it is likely that the clean engine of the future will be electric and powered by batteries.

Keywords: platinum; platinum group metals (PGM); automotive; catalytic converter

---

1. Introduction

Road transportation today is mostly based on vehicles powered by internal combustion engines. These engines need fuels which can be easily gasified and which can provide a large amount of energy per unit weight and volume. In practice, all the engines commonly available on the market use hydrocarbons as fuels. In particular, liquid hydrocarbons such as gasoline and diesel fuel are the most commonly used even though, in recent times, gas phase fuels such as methane and liquefied petroleum gas (LPG) have become popular due to their lower cost. The combustion of hydrocarbons in these engines creates a number of polluting substances, including unburnt hydrocarbons, particulate matter, and harmful chemicals such as carbon monoxide (CO) and nitrogen oxides (NO_x). In engines operating with the Otto cycle and using gasoline as fuel, these chemicals are removed using catalytic converters at the exhaust. These devices can substantially reduce the amount of toxic substances emitted, but they are also expensive because of the need of using platinum group metals (PGM) as active catalytic substrates. On average, an automotive catalytic converter can store $1–3 \times 10^{-3}$ kg of platinum and smaller amounts of rhodium and palladium. As a consequence, nowadays, automotive converters use more than half of the world's mineral production of platinum [1]. That raises the question of whether there exist sufficient PGM mineral resources extractable at reasonable prices in order to satisfy the future demand.

This subject has been studied in previous papers and a popularized discussion of the platinum depletion problem was reported by Cohen in 2007 in "The New Scientist" [2]. Several academic papers discussing the issue of PGM supply were published during the past few years, such as by Glaister *et al.* [3] and Mudd *et al.* [4], while a paper specifically dedicated to the problem of PGM depletion in view of the needs of the automotive industry was published by Yang in 2009 [5]. In the present paper, we update the previous results and we discuss the issue in view of what appears to be a "production peak" for PGMs observed in recent years. We discuss how the depletion of PGM may affect the world's road transportation system and we arrive to the conclusion that high costs of platinum group metals is a problem destined to get worse with time. That creates a critical problem for a large sector of the world's road transportation system which cannot run without PGM-based catalysts, unless we were to return to unacceptable levels of pollution. This situation is a strong incentive for developing radically different alternatives, in particular battery powered vehicles which are inherently cleaner and appear to suffer from less important depletion problems.

2. Pollution Removal from Combustion Engines by Means of Catalytic Converters

Practically all internal combustion engines available on the market today use hydrocarbons as fuels. In principle, non-hydrocarbon fuels, such as pure hydrogen or hydrogen-nitrogen compounds (e.g., ammonia), could be also be used, but at present they find no market applications. The combustion of hydrocarbons in internal combustion engines generates mainly water (H_2O) and carbon dioxide (CO_2). Neither is considered a harmful substance even though CO_2 is toxic for human beings at very high concentrations [6]. Both water and carbon dioxide are greenhouse gases, but only carbon dioxide creates global warming because, unlike water, it remains in the atmosphere for times of the order of tens of thousands of years [7]. Then, the untreated emissions of an internal combustion engine normally contain substances which are toxic for human beings even at low concentrations. The most important ones are: (1) unburnt hydrocarbons, especially if aromatic, (2) carbon monoxide (CO), (3) nitrogen oxides (NO_x) and (4) particulate matter, typically in the form of carbon micro and nano-particles (much debate is ongoing about the harmful effect of these particles but it is generally agreed that they are a major health problem [8]). Additives to fuels may also create dangerous materials at the exhaust and, until not long ago, tetra-ethyl lead and ethyl bromide were common additives to gasoline, fortunately today forbidden by law in most (although not all) countries of the world [9].

The removal of carbon dioxide from the exhaust gases of a mobile engine is normally considered impossible, although it can be contemplated in the case of large, stationary engines. However, most of the toxic substances emitted can be strongly reduced in concentration by a combination of suitable operating parameters and catalytic chemical filters at the exhaust. At present, there exist two main approaches in this field. For diesel engines, the "lean" mixture of fuel and air reduces the problem of carbon monoxide and hydrocarbons, so that the main problem is to eliminate particulate matter and nitrogen oxides. The abatement of these pollutants is normally obtained by means of selective catalytic reactions (SCR), that is by a combination of an oxidation catalyst based on cerium oxide (to remove particulate) and by reaction with ammonia to remove nitrogen oxides. Ammonia, in turn, is generated by the injection of urea into the exhaust gas. Exhaust gas recirculation can also be used to reduce NO_x production when starting a cold engine.

For gasoline engines, instead, the problem of particulate matter is less important and the exhaust filter must address the problem of eliminating three different harmful gases: CO, NO_x and unburnt hydrocarbons. This is accomplished by means of "three way" catalysts based on noble metals (Pt, Pd and Rh, collectively referred to as "PGM" or "platinum group metals". Of these three metals, rhodium catalyzes reduction while palladium catalyzes oxidation; platinum, is active for both. The task of the catalyst is complex because it must perform several tasks at the same time: oxidize CO and unburnt hydrocarbons, while reducing NO_x. In order to optimize the yield of these reactions, the exhaust gas must contain a specific fraction of oxygen. The correct gas composition is obtained by controlling the air/fuel mix by means of oxygen sensors at the exhaust. In general, when in good conditions and operated properly, the converter can remove up to about 90% of the three gases; as described, for instance, by Kummer [10].

Considerable efforts have been dedicated to developing non-PGM materials that can catalyze these three reactions, but the task has turned out to be very difficult and a practical solution has not been found [11,12]. A catalyst which does not use PGMs called "Noxicat™" has been recently developed, but it is designed mainly for the abatement of NO_x in diesel engines. Other solutions based on oxides such as perovskites [13] and boehmites [14] as catalysts have been proposed but they seem to be far from industrial applications. In the end, the electronic structure of the platinum group metals is unique and it generates chemical properties that are not matched by any other element of the periodic table nor by compounds which can remain stable for a long time in the conditions of high temperature of automotive catalytic converters. Therefore, although it is not possible to exclude an unexpected breakthrough, the present situation raises a serious problem of future availability of PGMs in sufficient amounts, as it will be discussed in the next section.

3. Platinum Group Metals Abundance and Production

PGMs are rare in the Earth's crust, with typical average abundances of the order of a few parts per billion (ppb) at most. Of the three PGMs used in automotive catalysts, the most abundant is palladium (average 15 ppb), followed by platinum with 5 ppb and rhodium with about 1 ppb (data from [15]). PGMs are often found in sulphide minerals [16] and are also known to be siderophilic (iron-loving). The latter property accounts for their scarcity on the Earth's crust, since they were efficiently extracted by the metallic Fe-Ni phases in the Earth's core during planetary accretion. PGMs may occur in native form associated with gold, iron, copper and chromium and, due to their high weight and chemical inertness, can also be found in placer deposits. The production of PGMs is concentrated in a few mines: the main ones are the Bushveld igneous complex (South Africa), the sulphide deposits of Norilsk in Russia, placer deposits in the Ural mountains (Russia), the Sudbury mine (Ontario, Canada), the Hartley mine (Zimbabwe), the Still-water complex (Montana, USA), Northern Territory (Australia) and the Zechstein copper deposit in Poland. South Africa produces about 85% of the total world PGM production has 82% of the world's resources [17].

According to the United States Geological Survey [18], the total reserves of platinum group metals (PGMs) amount to 66 million tonnes, to be compared to a total combined use of platinum and palladium in 2011 of 400,000 metric tons. Hence, the ratio of reserves to production (R/P, with production assumed to be constant and equal to the present value), is of about 130 years. This result may appear comforting but the question here is not for how long we can produce PGMs in the unlikely hypothesis of constant future production, but how and if it will be possible to keep a sufficiently large production at costs compatible with the needs of road vehicles—*i.e.*, at costs which would not destroy the demand for these elements.

This question is related to a well known effect in economics, that of "diminishing returns". As less and less concentrated ores are exploited, the energy needed for extraction increases. As a consequence, production costs increase and, ultimately, market prices must increase since, obviously, nobody can produce at a loss for a long time. This effect had been described for the first time for mineral resources by William Stanley Jevons in his "The Coal Question" of 1866 [19]. Today, it is known that in many cases, the mineral industry is forced to access resources from lower and lower grade ores, as it has been shown, for instance, for the case of copper in a recent paper by Mudd and Weng [20]. This effect is surely a factor in the observed increasing prices of most mineral resources worldwide (see, e.g., Valero [21]). However, the increasing prices of all mineral commodities are also directly related to the increased prices of fossil fuels which provide most of the energy needed for extraction. As shown by Hall *et al.* [22], the progressive depletion of fossil fuels causes a reduction of the energy return on energy invested (EROI) which in turn generates a rise in prices as the result of the decreasing economic returns on extraction. In the case of platinum group metals, all these factors are at play and

Mudd [4] clearly shows that the industry is progressively forced to exploit lower grade, more expensive PGE metal ores.

In Figure 1, we can observe how the production of both platinum and palladium appear to have peaked in 2006, maintaining a plateau at lower levels afterward. "Peaking" is a phenomenon which appears to be of general validity in mineral extraction and occurs as a result of growing extractive costs. However, it is not easy to predict when a certain mineral resource will peak. The standard method to analyze peaking trends is often termed the "Hubbert method" (so called because of the name of the author who first mentioned the concept for the case of crude oil [23]). It is based on an estimate of the amount of resources which are defined as "extractable" and this estimation involves assumptions on the future economy in terms of demand and prices. On the basis of this method, Mudd *et al*. [4], have estimated that the actual "production peak" for PGMs should not occur before approximately 2050 if the present level of demand is maintained. However, we cannot exclude that a weak economy would depress demand and make the 2006 peak as the ultimate production peak for these metals. In any case, the plateauing of the past few years clearly indicates the strain placed on the industry by a combination of high costs of extraction and high costs of energy. The result is the observed increase in the prices of PGMs. For instance, platinum prices have increased of a factor of about 5 from 1992 to 2012 [24] reaching today a level of about 1500 U.S. dollars per ounce, that is more than 50 dollars per gram. The historical maximum has been in 2008, when platinum reached levels of about 80 dollars per gram (see Figure 1). The other PGMs, palladium and rhodium, have shown similar increases in relative terms.

Figure 1. (**A**) Pd (red) and Pt (black) monthly average price in U.S. dollars per troy oz and (**B**) world production in thousands of tons. Data sources [25,26].

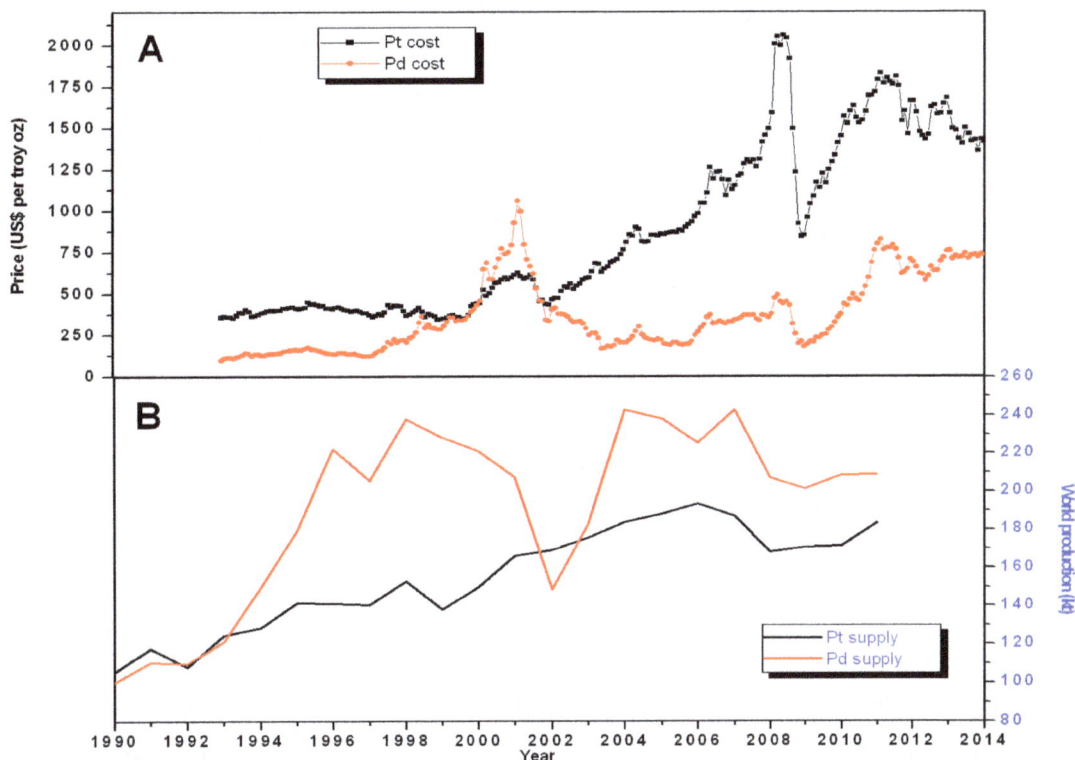

At these levels, the cost of the active metals in a three way catalytic converter can be 200–300 dollars. If the rising trend continues, the cost of the converter may become a major fraction of the cost of small car. We also need to take into account that catalytic converters are not the only technological applications which need PGMs. Fine chemical catalysts, anticancer drugs, as well as ORR (oxygen reduction reactions) are fields of great industrial relevance on which the use of PGMs is considered unavoidable [27].

What can be done to ease the high costs problems that derive from increasing PGM scarcity? As discussed in the previous section, developing non noble metal catalysts appears to be a very difficult option, hence—if we want to maintain the present technology of pollution abatement in combustion engines—we can at least mitigate the problem by (1) reducing the amount of catalyst in the converters and (2) recycling platinum group elements more efficiently.

Reducing the amount of PGMs in catalytic converters—and in particular of the expensive platinum—is possible, but there are limits to this approach. Often, it is possible to attain such a reduction by increasing the surface/volume ratio of the catalytic particles, that is making them smaller. However, below some dimensions, the particles become unstable, may move and coalesce with other particles with an overall loss of catalytic activity, or simply, they can be removed from the substrates and be carried away by the exhaust. It is also possible to vary the ratio of the different metals in the catalyst, for instance partly replacing platinum with palladium, which has a market price about one third lower. This is a route presently explored by catalyst manufacturers but, of course, it doesn't solve the problem at its roots.

Regarding recycling, there exist established procedures to recover platinum and the other noble metals from automotive converters with good efficiency [28]. The concentration of platinum in converters may be as high as 2 g/t in the ceramic catalyst brick, of the same order of magnitude as the gold content in primary ores (on average < 10 g/t). However, the end of life recycling rates of platinum from catalytic converters reach a global average of only 50%–60% [29].

This relatively low amount recycled is the result of two factors: one is the loss of noble metals during the life cycle of the catalyst, the other is that not all catalytic converters are actually recycled because cars may end their life in remote areas where there are no recycling facilities, or be lost in conditions where the catalyst cannot be conveniently recovered. While the recovery rate of old converters can surely be improved, we face a fundamental problem when considering the PGM loss at the exhaust. In an early study [30] the loss (or "attrition") of noble metals during operation has been estimated as 6% over 80,000 km of operation of the car. These metals are potentially dangerous pollutants and have generated serious concerns regarding their effects on the environment [31] and on human health [32,33]. Apart from this, these metals are dispersed in the environment at very low concentrations and are lost forever for all practical purposes. For this reason, recycling alone cannot solve the PGM depletion problem.

4. Consequences of PGM Scarcity: Moving to Electric Transportation

Given the inherent limitations of the previously discussed solutions for the limited availability of PGMs, it appears clear that the scarcity of platinum group metals is a critical factor in the future of road transportation. What alternatives can be conceived to solve the problem? As a first possibility, we may think of fuels not based on hydrocarbons: pure hydrogen (H_2) and compounds of hydrogen and nitrogen (such as ammonia, NH_3) can power internal combustion engines and the exhaust is not expected to contain unburnt hydrocarbons, particulate, or carbon monoxide. It may still contain nitrogen oxides, but these can be abated with catalysts which do not contain PGMs. The problem with this approach, however, is that these fuels cannot simply substitute presently used hydrocarbon fuels: shifting to hydrogen or ammonia would require completely different facilities for fuel distribution and transportation. Moreover, in the end, these fuels are still anchored to the inefficient internal combustion engine which is a major disadvantage in a period of high energy costs as the one in which we are living.

A more drastic and effective idea to solve the PGM problem would be move to a completely different category of engines: electric motors. These motors are lighter, more durable, more efficient and emitting no pollutants. Electric vehicles are already well known and diffused all over the world, although they remain, still today, a minority component of the transportation system. Their diffusion has been limited so far by a series of factors, including the poor performance in terms of power stored per unit weight of the current generation of batteries based on lead. The weight and the volume of batteries strongly limits the range of electric vehicles, except for the case of electric trains or other vehicles powered by aerial wires. As this limited range is considered to be a critical defect of road based electric vehicle, much research has been performed on ways to power electric motors by means of on-board power sources which could provide a significantly higher energy density than lead batteries. A much discussed possibility in this field is to use fuel cells operated using hydrogen as fuel. Fuel cells are efficient converters of the chemical energy stored in hydrogen, able to transform it directly into electrical energy. Because of this factor, fuel cell powered road vehicles can attain an acceptable range by avoiding the need of an inefficient thermal engine. Unfortunately, this approach raises an even worse platinum depletion problem than that encountered with exhaust catalysts. Low temperature fuel cells, usually using proton exchange membranes as electrolyte, need about 1–3×10^{-3} kg of platinum per kW of engine power as catalyst at the electrodes. Replacing the present world fleet of road vehicles with this kind of technology would simply not be possible with the limited platinum reserves available [34]. The industry is making a considerable effort in order to reduce the amount of platinum used in fuel cells, but it does not appear possible to eliminate it completely.

So, a better idea to provide power for road vehicles may be based on the new generation of lightweight batteries for automotive use. In the past, several new electrochemical systems were proposed and tested, such as nickel-cadmium, or nickel-metal hydride. However, at present the main effort in this field is directed toward batteries based on lithium compounds, which provide the best available values of energy density. The range of a road vehicle powered by lithium batteries is still lower than that obtained by traditional thermal engines, but it is often perceived acceptable by customers. The problem with lithium is that it may also suffer from depletion problems and this fact has generated a lively debate on the subject [35–39].

On this point, we remark that there are three main types of lithium sources: brines, minerals (e.g., pegmatites), and seawater. Brines formed by evaporation are commonly found in salt flats, such as those located in South America, China, and Tibet. Among these salt flats, the Salar de Atacama in Chile is at present the world's largest currently exploited lithium deposit, producing almost 40% of world lithium. At the current production rate (37,000 t per year), the known lithium reserves

(13 million tonnes) [18] would last for more than 300 years. If we could exploit all the land-based estimated resources then we would have about a millennium's supply, even without considering the other possible land sources. About seawater, lithium is one of the few minerals whose concentration is sufficiently high that extraction from the sea is an economically conceivable task [40], even though it is not industrially performed today.

However, just as it was discussed for PGMs, simply listing theoretically available resources is not a good way to understand how depletion will affect extraction costs and, hence, market prices. A detailed comparison of the relative depletion trends for PGMs and lithium is outside the scope of the present paper. However, we wish to remark that: (1) Unlike platinum and other PGMs, lithium production, so far, has shown no production peaks. (2) Lithium prices have increased during the past few years, following the general trend of mineral commodities, however—unlike the case of PGMs—the pure cost of lithium is still a negligible fraction of the total cost of an electric car. (3) Lithium recycling does not suffer of the dispersion problem that strongly limits the fraction of the PGMs which can be recycled from a catalytic converter. At present, lithium prices are still so low that recycling is not normally performed, but in the future that will be certainly possible. (4) Most of the negative views of lithium future availability, e.g., those expressed by Tahil [34] are the result of the assumption of a continued growth in the number of road vehicles for the foreseeable future. This assumption looks unrealistic in the present situation of economic constraints. The world sales of cars are still weakly increasing [41] but have stalled and are going down in many countries. This situation appears to be leading to a stasis and perhaps a contraction in the number of road vehicles that society will be able to afford in the future and that will surely ease the depletion problem with lithium, especially considering that, unlike the case of PGM, a very high recycling rate is possible with lithium batteries.

As a final note, we need to consider also that a radical shift to electric vehicles would also generate the problem of obtaining sufficient electric energy. This subject is beyond the scope of the present paper, but it is a very general problem that involves the transition from a fossil fuel based economy to a renewable (or nuclear) based one. In general terms, the transition is ongoing [42,43] and it is involving a shift from chemical energy obtained from fossil carbon to electric power directly obtained from non-carbon fueled sources. This transition is obviously favoring applications which can directly use this electric power, such as electric vehicles.

So, at least for those applications which do not demand long range transportation, the substitution of internal combustion engines with battery powered electric motors would greatly reduce pollution and lengthen the life span of the presently available mineral resources of platinum group metals which could therefore be saved for other purposes in catalysis as well as in other fields of the chemical industry.

5. Conclusions

The peaking observed in the production curve for platinum group metals indicates that the mining industry is already under heavy strain in maintaining a sufficient supply of PGMs at costs compatible with those of road transportation vehicles. This is a critical problem for the whole world's transportation system and it is not too early to start developing new technologies for road transportation which do not involve the use of extremely rare and precious materials where, even in the short term, supply disruption and price spikes could threaten the whole system. In the long run, we argue that the only definitive solution for the PGM depletion problem will be to replace vehicles powered by fossil hydrocarbons by battery powered electric vehicles.

Acknowledgments

The authors would like to thank the Club of Rome for providing a grant that made this study possible within the production of the 33rd report to the Club of Rome titled "Extracted" [44].

Author Contributions

Both the authors were involved in writing and revising all parts of the manuscript.

Conflicts of Interest

The authors declare no conflict of interest.

References

1. Alonso, E.; Field, F.R.; Kirchain, R.E. A Case Study of the Availability of Platinum Group Metals for Electronics Manufacturers. In Proceedings of the 2008 IEEE International Symposium on Electronics and the Environment, San Francisco, CA, USA, 19–22 May 2008.
2. Cohen, D. Earth's Natural Wealth: An Audit. Available online: http://www.sciencearchive.org.au/nova/newscientist/027ns_005.htm?q=nova/newscientist/027ns_005.htm (accessed on 25 April 2014).
3. Glaister, B.J.; Mudd, G.M. The environmental costs of platinum–PGM mining and sustainability: Is the glass half-full or half-empty? *Miner. Eng.* **2010**, *23*, 438–450.
4. Mudd, G.M. Key trends in the resource sustainability of platinum group elements. *Ore Geol. Rev.* **2012**, *46*, 106–117.
5. Yang, C.J. An impending platinum crisis and its implications for the future of the automobile. *Energy Policy* **2009**, *37*, 1805–1808.
6. Rice, S.A. Human Health Risk Assessment of CO_2 Survivors of Acute High Level Exposure and Populations Sensitive to Prolonged Low Level Exposure. In Proceedings of 3rd Annual Conference on Carbon Sequestration, Alexandria, VA, USA, 3–6 May 2004.
7. Archer, D. Fate of fossil fuel CO_2 in geologic time. *J. Geophys. Res. Ocean.* **1978**, *110*, C09S05, doi:10.1029/2004JC002625.
8. Pope, C.A., III; Burnett, R.T.; Thun, M.J.; Calle, E.E.; Krewski, D.; Ito, K.; Thurston, G.D. Lung cancer, cardiopulmonary mortality, and long-term exposure to fine particulate air pollution. *J. Am. Med. Assoc.* **2002**, *287*, 1132–1141.
9. Taylor, R.; Gethin-Damon, Z. Countries Where Leaded Petrol Is possibly still Sold for Road Use as at 17th June 2011. Available online: http://www.lead.org.au/fs/fst27.html (accessed on 10 April 2014).
10. Kummer, J.T. Use of noble metals in automobile exhaust catalysts. *J. Phys. Chem.* **1986**, *90*, 4747–4752.
11. Kapteijn, F.; Stegenga, S.; Dekker, N.J.J.; Bijsterbosch, J.W.; Moulijn J.A. Alternatives to noble metal catalyst for automotive exhaust purification. *Catal. Today* **1993**, *16*, 273–287.

12. Fechete, I.; Wang, Y.; Védrine, J.C. The past present and future of heterogeneous catalysis. *Catal. Today* **2012**, *189*, 2–27.

13. Qi, G.; Li, W. Pt-free LaMnO$_3$ based lean NO$_x$ trap catalyst. *Catal. Today* **2012**, *184*, 72–77.

14. Gálvez, M.E.; Ascaso, S.; Tobías, I.; Moliner, R.; Lázaro, M.J. Catalytic filters for the simultaneous removal of soot and NO$_x$: Influence of the alumina precursor on monolith washcoating and catalytic activity. *Catal. Today* **2012**, *191*, 96–105.

15. Linde, D.R. Section 14, Geophysics, Astronomy, and Acoustics; Abundance of Elements in the Earth's Crust and in the Sea. In *CRC Handbook of Chemistry and Physics*, 85th ed.; Lide, D.R., Ed.; CRC Press: Boca Raton, FL, USA, 2005.

16. Rankama, K.; Sahama, T.G. *Geochemistry*; The University of Chicago Press: Chicago, IL, USA, 1952.

17. Rao, C.R.M.; Reddi, G.S. Platinum group metals (PGM); Occurrence, use and recent trends in their determination. *Trends Anal. Chem.* **2000**, *19*, 565–586.

18. U.S. Department of the Interior; U.S. Geological Survey. *Mineral Commodity Summaries 2012*; U.S. Geological Survey: Reston, VA, USA; pp. 120–121. Available online: http://minerals.usgs.gov/ minerals/pubs/mcs/2012/mcs2012.pdf (accessed on 10 April 2014).

19. Jevons, W.S. *The Coal Question: An Enquiry Concerning the Progress of the Nation, and the Probable Exhaustion of Our Coal-Mines*, 2nd ed.; Macmillan and Co.: London, UK, 1866.

20. Mudd, G.M.; Weng, Z. Base Metals. In *Materials for a Sustainable Future*; Peter, L., Letcher, T.M., Scott, J.L., Salminen, J., Eds.; Royal Society of Chemistry: London, UK, 2012; pp. 11–55.

21. Valero, A. Mineral Resource Depletion Assessment. In *Eco-Efficient Construction and Building Materials*; Pacheco-Torgal, F., Cabeza, L., Labrincha J.; De Magalhaes, A., Eds.; Woodhead Publishing: Cambridge, UK, 2014; pp. 13–37.

22. Hall, C.A.S.; Lambert, J.G.; Balogh, S.B. EROI of different fuels and the implications for society. *Energy Policy* **2014**, *64*, 141–152.

23. Hubbert, M.K. Nuclear Energy and the Fossil Fuels. Presented at the Spring Meeting of the Southern District, American Petroleum Institute, San Antonio, TX, USA, 7–9 March 1956.

24. Kitco Web Page. Available online: http://www.kitco.com/charts (accessed on 10 April 2014).

25. Platinum Today, Johnson Matthey Market Data Tables. Available online: http://www.platinum.matthey.com/prices/price-charts (accessed on 4 April 2014).

26. U.S. Geological Survey Web Page. Platinum-Group Metals: Statistics and Information. Available online: http://minerals.usgs.gov/minerals/pubs/commodity/platinum/ (accessed on 10 April 2014).

27. Ungvary, F. Transition metals in organic synthesis: Hydroformylation, reduction, and oxidation: Annual survey covering the year 1993. *Coord. Chem. Rev.* **1995**, *141*, 371–493.

28. Hagelüken, C. Recycling the platinum group metals: A European perspective. *Platinum Metals Rev.* **2012**, *56*, 29–35.

29. Graedel, T.E.; Allwood, J.; Birat, J.-P.; Reck, B.K.; Sibley, S.F.; Sonnemann, G.; Buchert, M.; Hagelüken, C. *Recycling Rates of Metals—A Status Report*; United Nations Environment Programme: Paris, France, 2011.

30. Hill, R.F.; Mayer, W.J. Radiometric determination of platinum and palladium attrition from automotive catalysts. *IEEE Trans. Nucl. Sci.* **1977**, *24*, 2549–2554.

31. Moldovan, M.; Palacios, M.A.; Gómez, M.M.; Morrison, G.; Rauch, S.; McLeod, C.; Ma, R.; Caroli, S.; Alimonti, A.; Petrucci, F.; Bocca, B.; Schramel, P.; Zischka, M.; Petterson, C.; Wass, U.; Luna, M.; Saenz, J.C.; Santamaria, J. Environmental risk of particulate and soluble platinum group elements released from gasoline and diesel engine catalytic converters. *Sci. Total Environ.* **2002**, *296*, 199–208.

32. Caroli, S.; Alimonti, A.; Petrucci, F.; Bocca, B.; Krachler, M.; Forastiere, F.; Sacerdote, M.T.; Mallone, S. Assesment of exposure to platinum-group metals in urban children. *Spectrochim. Acta B* **2001**, *56*, 1241–1248.

33. Bocca, B.; Alimonti, A.; Cristaudo, A.; Cristallini, A.; Petrucci, F.; Caroli, S. Monitoring of the exposure to platinum group elements for two Italian population groups through urine analysis. *Anal. Chim. Acta* **2004**, *512*, 19–25.

34. Bossel, U. Does a hydrogen economy make sense? *Proc. IEEE* **2006**, 94, 1826–1836.

35. Gruber, P.W.; Medina, P.A.; Keoleian, G.A.; Kesler, S.E.; Everson, M.P.; Wallington, T.J. Global lithium availability. *J. Ind. Ecol.* **2011**, *5*, 760–775.

36. Tahil, W. *The Trouble with Lithium 2: Under the Microscope*; Meridian International Research: Martainville, France, 2008.

37. Yaksic, A; Tilton, J.E. Using the cumulative availability curve to assess the threat of mineral depletion: The case of lithium. *Resour. Policy* **2009**, *34*, 185–194.

38. Kushnir, D.; Sandén, B.A. The time dimension and lithium resource constraints for electric vehicles. *Resour. Policy.* **2012**, *37*, 93–103.

39. Kesler, S.E.; Gruber, P.W.; Medina, P.A.; Keoleian, G.A.; Everson, M.P.; Wallington, T.J. Global lithium resources: Relative importance of pegmatite, brine and other deposits. *Ore Geol. Rev.* **2012**, *48*, 55–69.

40. Bardi, U. Extracting minerals from seawater: An energy analysis. *Sustainability* **2010**, *2*, 980–992.

41. Gomes, C. *Global Auto Report*; Scotiabank Economics: Toronto, ON, Canada, 2014. Available online: http://www.globalstrategicmetalsnl.com/_content/documents/405.pdf (accessed on 10 April 2014).

42. Bardi, U. The grand challenge of the energy transition. *Front. Energy Res.* **2013**, *1*, doi:10.3389/fenrg.2013.00002.

43. Jacobson, M.Z.; Delucchi, M.A. Providing all global energy with wind, water, and solar power, Part I: Technologies, energy resources, quantities and areas of infrastructure, and materials. *Energy Policy* **2011**, *39*, 1154–1169.

44. Bardi, U. *Extracted: How the Quest for Mineral Wealth Is Plundering the Planet*; Chelsea Green: Hartford, VT, USA, 2014.

4

Characterization of Green Liquor Dregs, Potentially Useful for Prevention of the Formation of Acid Rock Drainage

Maria Mäkitalo [1,*], Christian Maurice [1,2], Yu Jia [1] and Björn Öhlander [1]

[1] Department of Civil, Environmental and Natural Resources Engineering, Luleå University of Technology, SE-97187 Luleå, Sweden; E-Mails: christian.maurice@ltu.se (C.M.); yu.jia@ltu.se (Y.J.); bjorn.ohlander@ltu.se (B.Ö.)

[2] Ramböll Sverige AB, Kyrkogatan 2, Box 850, SE-97126 Luleå, Sweden

* Author to whom correspondence should be addressed; E-Mail: maria.makitalo@ltu.se

Abstract: Using alternative materials such as residual products from other industries to mitigate the negative effects of acid rock drainage would simultaneously solve two environmental problems. The main residual product still landfilled by sulphate paper mills is the alkaline material green liquor dregs (GLD). A physical, mineralogical and chemical characterization of four batches of GLD was carried out to evaluate the potential to use it as a sealing layer in the construction of dry covers on sulphide-bearing mine waste. GLD has relatively low hydraulic conductivity (10^{-8} to 10^{-9} m/s), a high water retention capacity (WRC) and small particle size. Whilst the chemical and mineralogical composition varied between the different batches, these variations were not reflected in properties such as hydraulic conductivity and WRC. Due to relatively low trace element concentrations, leaching of contaminants from the GLD is not a concern for the environment. However, GLD is a sticky material, difficult to apply on mine waste deposits and the shear strength is insufficient for engineering applications. Therefore, improving the mechanical properties is necessary. In addition, GLD has a high buffering capacity indicating that it could act as an alkaline barrier. Once engineering technicalities have been overcome, the long-term effectiveness of GLD should be studied, especially the effect of aging and how the sealing layer would be engineered in respect to topography and climatic conditions.

Keywords: green liquor dregs; acid rock drainage; sulphidic mine waste; sealing layer

1. Introduction

The major potential long-term environmental effect of mining is the formation of acid rock drainage (ARD) in sulphide-bearing mine waste, which can last for hundreds of years in a specific deposit [1,2]. The traditional strategy to reduce ARD generation is to prevent oxygen from reaching the waste to decelerate oxidation. This can be achieved by applying a dry cover or a water cover [1,2]. A composite dry cover usually consists of a protective layer and a sealing layer. The function of the protective layer is to protect the integrity of the sealing layer (e.g., root penetration, freezing/thawing, erosion) [3]. The sealing layer material in a dry cover should have the ability to minimize oxygen diffusion and reduce water percolation into the waste. Using a material with low hydraulic conductivity minimizes water percolation, oxygen transport and the risk of ARD formation. A high degree of water saturation is important to reduce the oxygen diffusion and to avoid desiccation cracks through which water and oxygen may reach the waste. The oxygen diffusion rate is 10^4 times lower in water than in air. This, together with the fact that the O_2 concentration in water is ~20,000 times lower than in air, explains why oxygen transport is so strongly reduced in saturated sealing layers or through a water cover [1,2]. Moreover, a sealing layer material must have mechanical strength to ensure slope and surface stability. To avoid high maintenance costs the material should be durable and not be prone to degradation. The composite cover is usually constructed by using natural soils, in glaciated terrains often till.

Using alternative materials such as industrial waste to mitigate the negative effects of ARD would solve two environmental problems at the same time. However, there is an urgent need for detailed research on the function of alternative materials before they can be used on an industrial scale. In particular, reliable predictions of the long-term efficiency are important [2].

Pulp and paper mills generate large amounts of waste such as fiber sludge, mesa lime (ML), green liquor dregs (GLD) and fly ash (FA). The alkaline waste materials ML, GLD and FA have properties suggesting that they could be useful for remediation of reactive mine waste [4–8] or neutralization of ARD [9–12]. Most paper mills use the sulphate production process where wood chips are treated with sodium hydroxide and sodium sulphide to liberate the cellulose. The method enables the inorganic chemical to be recycled and reused. However, large amounts of waste are released where GLD makes up the largest fraction retrieved in the chemical recovery cycle. Reuse and recycling of the GLD is limited and most of the GLD produced in Sweden, ~240,000 t per year, is still landfilled, mainly at the pulp and paper mills own deposits. Even though the waste is not currently subjected to a tax for landfilling, the pulp and paper mills are facing disposal problems. GLD comprises one of the streams out of the pulping process to reduce the concentration of non-process elements (NPE), such as Mg, Si and Al originating from the wood [13]. The GLD retrieving process, using a pre-coat lime mud filter (the pre-coat is a mixture of $CaCO_3$, CaO and $Ca(OH)_2$), leads to various amounts of lime mud mixed with the green liquor, which strongly influences the final composition of the dregs. GLD is classified as a non-hazardous chemical waste by the Swedish EPA [14]. It has a high content of calcite, is alkaline and has been shown to have a low hydraulic conductivity [15]. This opens up the possibility of using it for construction of sealing layers in covers on sulphidic mine waste.

Since landfilling is costly, reusing the material would be beneficial for both the pulp and paper mills and the mining industry. Although GLD has been used in several sanitary landfill cover applications [16,17], no extensive characterization of the material has been performed. The objectives of the present study were, on the basis of laboratory investigations, (i) to characterize the GLD physically, mineralogically and chemically to define variations in the material properties and (ii) to evaluate the feasibility of using GLD in sealing layers.

2. Materials and Methods

2.1. Materials

GLD were obtained from the Billerud Karlsborg sulphate pulp and paper mill in northern Sweden. The sampling took place on four occasions and ~30 kg were collected each time and stored in sealed cans, to prevent the samples from drying, at room temperature. The four batches are referred to as A, B, C and D herein.

2.2. Methods

2.2.1. Mineralogical Characterization

X-ray diffraction analysis (XRD) was performed with a Siemens D5000 diffractometer (Siemens, Munich, Germany) using CuKα radiation generated at 40 kV and 40 mA. The scanning range was measured in Bragg-Brentano geometry from 5° to 90°.

2.2.2. Chemical Characterization

Concentrations of 31 elements were determined for all four batches of GLD with duplicate samples. The modified EPA method 200.7 (ICP-AES) and 200.8 (ICP-SMS) [18] were used and performed by an accredited laboratory (ALS Scandinavia, Luleå, Sweden). Dry matter content, was determined drying the sample at 105 °C for 24 h according to Swedish standard SS 028113-1 [19]. Ca concentrations were analyzed using ICP-OES (Varian Vista MPX, Varian, Palo Alto, CA, USA) at the Billerud paper mill for 14 samples of GLD.

The buffering capacity was determined for GLD batches A, B, C, and D with the batch titration method. The procedure was modified from the method described by Wyatt [20]. 0.75 g of dried GLD was placed in 50 mL plastic containers and mixed with 0–16 mL of 1 M HCl with 0.5 mL increments. Water was added to obtain a final volume of 20 mL. The samples were placed on an orbital shaker (IKA KS 260 basic, IKA-Werke GmbH & Co., Staufen, Germany) and agitated for 24 h to reach equilibrium. The pH was measured with a pH meter (Metrohm Ltd, 704 pH Meter, Herisau, Switzerland).

A leaching test was performed in triplicate of GLD. The procedure was modified from the Swedish standard SS-EN 12457-4 [21]. 70 g dry mass of GLD was added to 700 mL of high purity Milli-QTM water (Millipore, Bedford, MA, USA) with pH 7 (taking the water content of the sample into consideration) to achieve a liquid-to-solid (L/S) ratio of 10 mL/g. The samples were agitated once a day for 6 days and centrifuged to collect the supernatant. The eluates were filtered with a 45 μm nylon membrane and analyzed by an accredited laboratory (ALS Scandinavia, Luleå, Sweden) according to the previously described method. The L/S ratio was selected to simulate the long-term leaching. Assuming we have a 0.3 m cover with a weight of 400 kg/m² and the infiltration through the saturated cover is 50 L/m^2 per year gives a L/S of 0.125 L/kg per year. This means that it is expected to take 80 years to achieve L/S 10 in the field by percolation through a traditional cover [22].

2.2.3. Physical Characterization

Dry matter content was analyzed at the Billerud paper mill on 178 samples over one year. Paste pH was analyzed in duplicate with a pH meter (Metrohm Ltd., 704 pH Meter, Herisau, Switzerland) for all four batches of GLD according to Weber *et al.* [23].

Laser diffraction analysis was conducted on duplicate samples for all four batches of GLD using a CILAS Granulometer 1064 (CILAS, Orléans, France) to obtain the particle size distribution, which was calculated using the CILAS software.

To determine the surface properties of the particles, the pH effect on the electrophoretic mobility was analyzed with a ZetaPhoremeter IV (ZetaCOMPACT, CAD instrumentation, Les Essarts le Roi, France) with attached charge-couple device (CCD) video camera. The analysis was done after suspending GLD batch B ($n = 3$) in 1 mM KNO_3 and adjusting the pH with either HNO_3 or KOH to values between 2 and 12. The analysis was performed when the solution maintained the desired pH for 30 min. The results were analyzed with the zeta4 software (Parker Hannifin Corporation, Rohnert Park, CA, USA).

The surface area was analyzed in duplicate for all four batches of GLD. The samples were dried with a freeze dryer system (Freezone 4.5, Labconco Corp., Kansas City, MO, USA). To determine the surface area, the samples were analyzed with a DeSorb 2300A/FlowSorb instrument (Micromeritics, Norcross, GA, USA).

Constant Rate of Strain (CRS) tests were performed on GLD batches B and D according to a standard procedure [24]. Cylinders measuring 15 cm × 7 cm diameter were filled with GLD. The samples were set under a 30 kPa load for 28 days resembling a 1.5 m protective cover. The cylinders were placed in an oedometer and an increase of induced stress was applied on the samples. Drainage was only allowed from the top. The hydraulic conductivity was calculated based on the deformation and the pore pressure from the lower surface.

The water retention capacity (WRC) was measured on GLD batches B and D in duplicate. The samples were packed into cylinders and saturated from below. The cylinders were placed on a ceramic plate and pressurized from below using a pressure plate apparatus (Soilmoisture Corp., Goleta, CA, USA). The applied tensions were: 0.1, 0.5, 0.8, 1.0, 2.5 and 5.0 m water column (mwc). 50 and 150 mwc were determined on separate bulk samples. The volume of the loose samples was calculated using the bulk density, which was determined from the weight of the dried samples (105 °C for 24 h) divided by the cylinder volume. The total porosity was calculated for GLD batches B and D according to the equation: Total porosity = (Particle density – Bulk density)/Particle density. Density was determined from the WRC.

A multivolume helium pycnometer (Pycnometer 1305, Micromeritics, Nercross, GA, USA) was used to determine the density of GLD batch B ($n = 3$). The bulk density and compact density were calculated for GLD samples from batches B and D.

Surface and crystal structures were studied on GLD batch A using an FEI Magellan 400 XHR Scanning Electron Microscope.

Replicate samples of GLD were packed into cylinders measuring 15 cm × 7 cm diameter and cured for one month at 5 °C (soil temperature in field conditions) under a load of 20 kPa, equivalent to a 1 m top cover. The samples were tested for uniaxial compressive strengths. The deformation rate was set at 1.5%/min.

3. Results

Dry matter content of the GLD was measured on 178 occasions over one year at the Billerud Karlsborg paper mill. Average dry content was 46% ± 7% with 5th and 90th percentiles of 35% and 54%, respectively. For 14 of the samples, concentrations of Ca were measured at Billerud. Dry matter content and Ca concentration were correlated with a correlation coefficient of 0.75.

The XRD analyses showed that the crystalline phase of the GLD mainly is composed of calcite ($CaCO_3$). Brucite ($Mg(OH)_2$), a common form of Mg in ash, could also be detected. However, the detection limit is ~2% of sample and amorphous phases or poorly crystalline solids cannot be detected with the XRD. Particle size distribution according to laser diffraction analysis is shown in Figure 1. GLD appears to be a relatively fine-grained material with average d10, d50 and d90 values for the four batches of 1.9 ± 0.5 μm, 11.9 ± 3.9 and 26.4 ± 0.5 μm, respectively. The scanning electron microscope (SEM) measurements detected calcite crystals.

Table 1 shows the chemical composition of the different batches of GLD. Since calcite is the dominant mineral, the Ca concentration is high. Lime mud added in the process is the Ca source. The concentration of Mg is high, supporting the presence of brucite. Concentrations of Mn, Si, Al, K and P are also relatively high, indicating the presence of amorphous phases not detected by XRD. In the context of mine waste applications it is worth mentioning that the S content is high, ranging from 8330 to 18,650 mg/kg. Zn is the only trace metal that has a relatively high concentration, ranging from 733 to 1820 mg/kg. Table 2 shows the leached concentrations of elements, which are compared with the regulatory levels of important contaminants [25]. Most of the contaminants were below the level of leachates from inert waste. Of the trace metals, Hg and Pb exceeded this limit but not the limit for non-hazardous waste. K was readily leached where 97% was released. The average S concentration was 7990 mg/kg, which corresponds to a sulphate concentration of 23,970 mg/kg if it is assumed that S occurs as sulphate and corresponds to ~86% of the total sulphate in GLD. This is slightly higher than the leaching limit for non-hazardous waste. It should be mentioned that the leaching experiments performed when obtaining regulation limits were pursued using a one-step (one day) leaching test, whilst the leaching test used for GLD was performed with the batch leaching method that was ongoing for 6 days to simulate the long-term effect of leaching.

The zeta (ζ) potential for GLD batch B is negative over the entire pH range. Thus, cations could adsorb to the surface of the GLD particles. The isoelectric point (the pH where the net electric charge of the particles is 0) was not attained.

In Table 3, the results of the measurements of surface area, paste pH, porosity, hydraulic conductivity, compact density and bulk density are summarized. There are differences between the batches, but not so large that they change the character of the material. Paste pH is high (10.0 to 11.0), the hydraulic conductivity rather low (10^{-8} to 10^{-9} m/s), porosity relatively high (73.2% to 82.2%), bulk density ranges between 0.44 and 0.67 g/m^3 and compact density ranges from 2.47 to 2.60 g/cm^3.

Figure 1. Particle size distribution of green liquor dregs (GLD) batches A, B, C and D. For comparison, data for typical sealing layers consisting of clayey till is included [1].

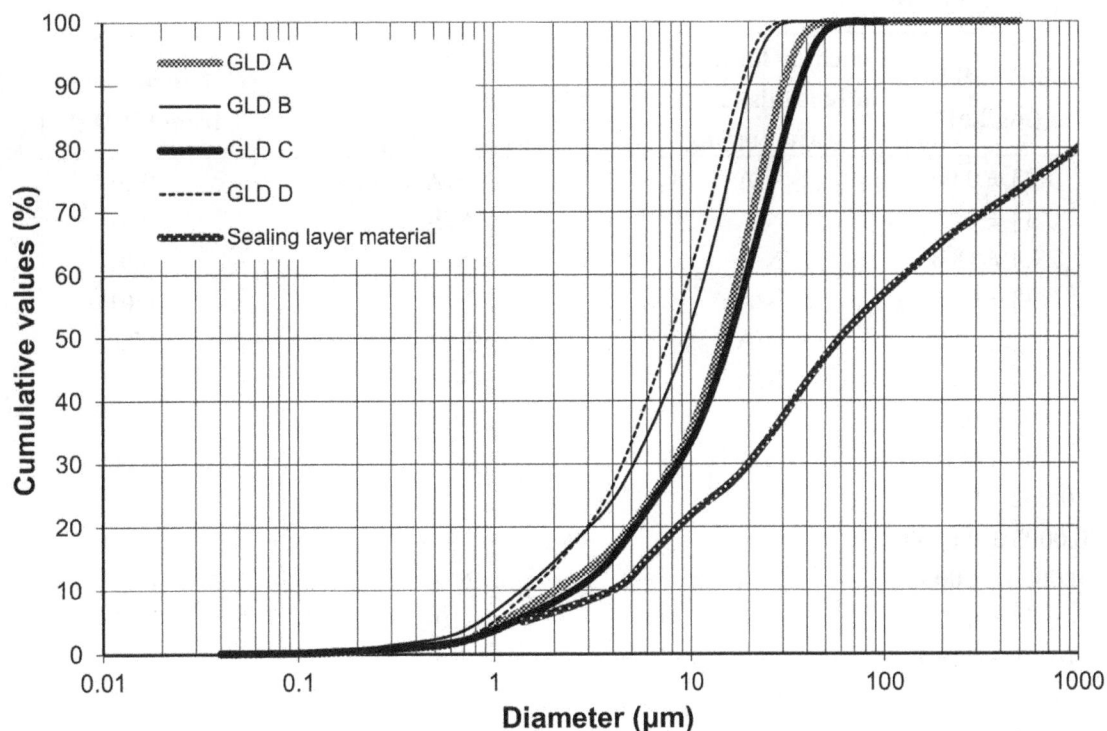

Table 1. Metal concentrations in GLD duplicate batches A–D expressed on a dry matter basis.

Element	Unit	Sample batch							
		A1	A2	B1	B2	C1	C2	D1	D2
Dry matter (DM)	%	51.4	51.6	47.8	47.2	37.9	38.0	35.7	35.5
Si	mg/kg	4,960	5,020	6,630	6,850	9,720	10,500	12,300	12,400
Al	mg/kg	2,700	2,720	2,900	3,090	5,280	5,180	9,820	9,890
Ca	mg/kg	317,000	317,000	295,000	296,500	252,000	247,000	219,000	217,000
Fe	mg/kg	3,190	3,170	4,540	3,360	5,700	6,520	9,650	7,260
K	mg/kg	1,910	2,050	2,910	2,970	3,010	3,060	3,360	3,570
Mg	mg/kg	37,800	38,300	43,400	43,900	50,600	49,600	72,400	73,300
Mn	mg/kg	8,350	8,430	10,800	10,900	12,900	12,700	17,300	17,500
P	mg/kg	4,000	4,030	3,590	3,610	1,270	1,250	922	911
LOI 1000 °C	% DM	41.3	41.4	40.7	40.8	43.1	43.1	39.9	40.0
As	mg/kg	0.305	0.225	0.327	0.172	0.332	0.383	0.474	0.431
Cd	mg/kg	3.54	3.69	5.80	5.53	4.98	4.87	8.29	8.65
Co	mg/kg	3.34	3.04	3.96	3.67	4.83	4.72	5.63	5.77
Cr	mg/kg	71.9	80.2	93.2	94.3	92.0	97.7	126	127
Cu	mg/kg	101	109	112	108	179	159	224	202
Hg	mg/kg	0.04	0.04	<0.04	<0.04	<0.04	<0.04	<0.04	<0.04
Ni	mg/kg	29.3	31.6	53.6	46.7	51.7	40.7	60.8	49.5
Pb	mg/kg	2.27	2.56	2.86	2.96	3.28	3.26	5.88	6.13
S	mg/kg	7,880	8,780	10,500	9,150	13,500	13,200	18,600	18,700
Zn	mg/kg	701	764	1,120	1,060	1,230	1,200	1,840	1,800
CaO	% DM	44.3	44.3	41.0	41.0	35.3	34.6	30.6	30.4

Table 2. Leaching of GLD (average of triplicate) with standard deviations at L/S 10 compared with regulatory limits for inert and non-hazardous leached with the same L/S ratio [25]. N/A = no data available.

Element	GLD L/S 10 (mg/kg)	Leaching limit values at landfills for inert waste (mg/kg)	Leaching limit values at landfills for non-hazardous waste (mg/kg)	Relative mass released from GLD (%)
Si	3.03 ± 0.00	N/A	N/A	0.01
Al	0.34 ± 0.19	N/A	N/A	0.01
Ca	29.4 ± 18.0	N/A	N/A	0.01
Fe	0.23 ± 0.00	N/A	N/A	0.01
K	1433 ± 12	N/A	N/A	97.0
Mg	8.3 ± 3.0	N/A	N/A	0.09
Mn	0.97 ± 0.76	N/A	N/A	0.01
P	3.07 ± 0.37	N/A	N/A	0.08
As	0.002 ± 0.002	0.5	2	0.32
Cd	0.0003 ± 0.0000	0.04	1	0.01
Co	0.0006± 0.0000	N/A	N/A	0.01
Cr	0.44 ± 0.07	0.5	10	0.7
Cu	0.045 ± 0.015	2	50	0.04
Hg	0.22 ± 0.00	0.01	0.2	0.08
Ni	7.01 ± 3.00	0.4	10	0.02
Pb	1.89 ± 0.00	0.5	10	0.06
S	7,990 ± 104	Sulphate: 1,000	Sulphate: 20,000	85.6
Zn	0.088 ± 0.051	4	10	0.01

Table 3. Surface area, pH, hydraulic conductivity, bulk density, compact density and porosity of GLD batches A, B, C and D. The analyses were performed on duplicate samples, except for hydraulic conductivity and pH which were measured in triplicate.

GLD batch	Surface area (m²/g)	pH	Hydraulic conductivity (m/s)	Bulk density (g/cm³)	Compact density (calculated) (g/cm³)	Compact density (by pycnometer) (g/cm³)	Porosity (%)
A	12.1 ± 0.1	11.0 ± 0.1	N/A	N/A	N/A	N/A	N/A
B	16.8 ± 0.1	10.0 ± 0.0	$1 \times 10^{-8} \pm 7 \times 10^{-9}$	0.67 ± 0.00	2.60 ± 0.00	2.57 ± 0.00	73.2 ± 0.4
C	21.4 ± 0.0	10.6 ± 0.1	N/A	N/A	N/A	N/A	N/A
D	20.4 ± 0.2	10.1 ± 0.0	$1 \times 10^{-8} \pm 7 \times 10^{-9}$	0.44 ± 0.00	2.47 ± 0.00	N/A	82.2 ± 0.1

The WRC curves for batches B and D are presented in Figure 2. Compared with five sealing layers made of clayey till, commonly used in glaciated terrains [1], the WRC is significantly higher for GLD. The acid neutralizing capacity pattern was similar between pH 7.8 and 11, for the four batches of GLD with a large buffering at pH 8. Small differences were found between pH 2 and 7.8. To maintain pH > 6, the average neutralization capacity was high with the average value 18.5 mmol H^+/g dry weight for the four batches of GLD.

The shear strength of GLD was assessed at 11.8 ± 1.6 kPa after one month curing time. This is a low value compared with other materials used for construction of sealing layers. Clayey till and Ca-bentonite have 4 times higher values and Na-bentonite 6 times higher values [26].

Figure 2. Water retention characteristics of GLD batches B and D. The results for two types of silt are also included for comparison [27]. The area between the dashed lines indicates the water retention capacity (WRC) of five sealing layers made of clayey till reported by Höglund *et al.* [1].

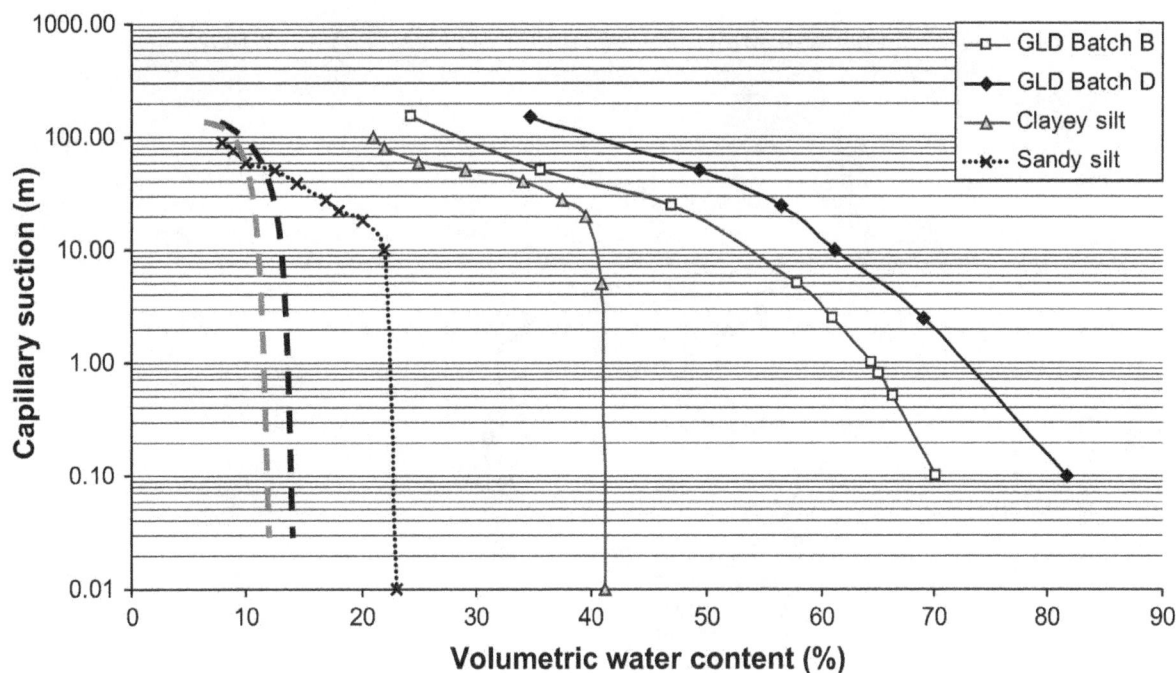

4. Discussion

4.1. Material Variation

To enable the reuse of GLD, the effect of varying material compositions on the desired function should be understood to identify possible limitations. The dry content of GLD monitored at the Billerud mill ranges between 29% and 72%, commonly within the range 36% and 54%. Increasing dry content is a consequence of larger amounts of lime mud scraped off from the filter coating. This decreases the relative concentration of non-process elements.

The XRD analysis indicated the presence of calcite and small amounts of brucite, which is in agreement with Martins *et al.* [28], while no other buffering minerals were detected. Calcite crystals were detected by SEM analysis. The chemical analysis showed large amounts of sulphur but no crystalline forms of this element was detected. However, Martins *et al.* [28] found gipsite in GLD, not detectible by XRD. Ettringite and gypsum are other solid forms of sulphate that may be present in small amounts and therefore not picked up by XRD due to the instruments detection limit. It is likely that mainly amorphous and liquid forms of sulphate are present in GLD and can therefore not be detected with XRD. Based on the ANC, up to 75% of the GLD content is $CaCO_3$, assuming that the source of alkalinity that is buffering around pH 8.3 is only $CaCO_3$. The higher content of Ca in batches A and B (Table 1) explains the higher acid buffering capacity observed compared to the two other batches (Figure 3). The observed variation in the chemical and mineralogical composition of the GLD can be caused both by the origin of the wood and the efficiency of the GLD retrieval. The variation between batches had no strong effect regarding the hydraulic conductivity, buffering capacity, pH and particle size of the studied materials compared to the natural variation observed in the replicates. However, the surface area, the density and the porosity varied significantly between the batches. The observed variation may complicate the use of the residual product on a large scale. Therefore potential negative effects of the quality variation on the intended application of the material should be addressed.

Figure 3. Acid neutralization capacity for GLD batch A, B, C and D.

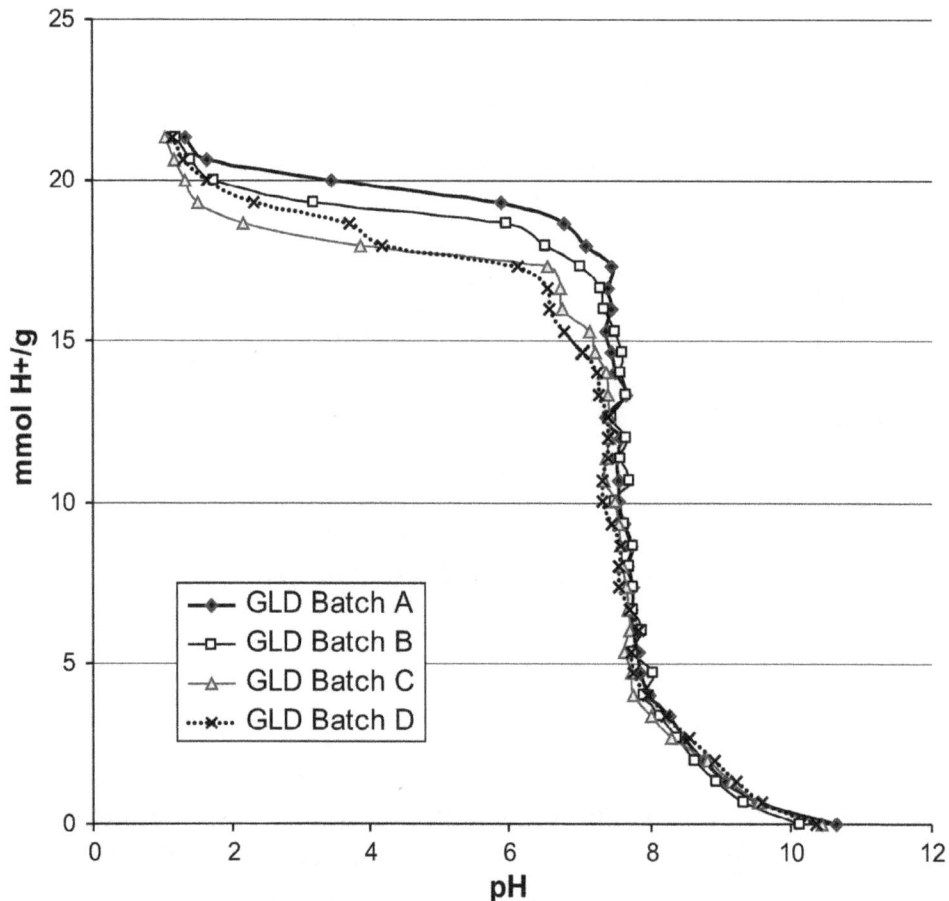

4.2. Use of GLD in Sealing Layers

A sealing layer applied on reactive mine waste should function as a barrier for oxygen transport. Pore size and degree of saturation affect the effective diffusion coefficient, D_e, which in turn affects the oxygen flux through the material. The porosity forms channels for oxygen diffusion if the pores are not saturated. D_e increases with increasing pore size and quantity [29] and the oxygen penetration decreases when the particles are irregularly shaped compared to a more regular structure [30]. Decreasing particle size is correlated to increasing total porosity [29]. GLD is a fine-grained material. Comparing the particle size distribution of GLD with sealing layers made of clayey till [1] shows that GLD particles are significantly smaller with a d50 of 12 μm compared to 60 μm for the till. A sealing layer should be homogeneous to achieve uniform water saturation and GLD has a small particle size throughout the material. As the porosity of GLD is high, >70%, oxygen diffusion may occur if it dries. Oxygen diffusion decreases sharply as the degree of saturation increases above 85% [1]. WRC is a hydraulic property affecting the transport of fluids in the material. GLD has very high water retention potential, much higher compared to other cover materials (Figure 2) such as clayey till. The high WRC is assumed to be due to the grain structure together with the ionic charge of the particles. Although the particle size falls into the range of silt, the WRC is more comparable to clayey materials than silty materials, especially at high under-pressures. Due to the high WRC, GLD has good potential to reduce the oxygen flux through a sealing layer by maintaining saturated or close to saturated conditions. The risk of shrinkage and cracking due to desiccation is reduced, which is otherwise a common problem in clay and may have a significant impact on the sealing layer performance.

A sealing layer should consist of a material that decreases water infiltration. Decreasing the particle size of a material is usually correlated to decreasing its hydraulic conductivity [31,32]. Hydraulic conductivity is negatively correlated to porosity in the case of fine-grained material [33,34] and the specific surface area [32]. A high porosity enables water to enter the material, but the large surface area also enhances chemical interactions binding water to particles, thereby limiting water transport, resulting in high grade of saturation. Our results showed that GLD has adequate characteristics, such as a small particle size, high porosity and average surface area of 18 m^2/g, to limit the water flow through the material. The tests confirmed a relatively low hydraulic conductivity, between 1×10^{-8} and 7×10^{-9} m/s. However, it did not reach the recommended minimum hydraulic conductivity [1] of 1×10^{-9} m/s.

There are several advantages of using GLD over virgin material such as till and other natural soils for construction of sealing layers. Opening quarries is not only costly but also has a significant environmental impact. A shortage of fine-grained cover material close to the mines is often an issue [2]. However, because the quality of GLD varies, there is a risk that not all of the GLD produced would have the appropriate characteristics for use in sealing layer applications. Based on the results, GLD with low hydraulic conductivity would be most suitable to use for construction of sealing layers. However, the small variations observed between the batches of GLD from Billerud are not expected to lead to complications if used for barrier applications. GLD has previously been used in applications such as a stabilization agent for road construction [15].

Ideally, the material used as a sealing layer should both decrease oxygen penetration, water infiltration and simultaneously function as a chemically reactive barrier retaining heavy metal ions originating from the mine waste. A major concern when using residual products is assessing their potentially negative impact on the environment. Both undesired alteration of the material, which may alter its sealing function, and side effects on underlying mine waste caused by GLD leachates are potential issues to be addressed. The GLD from Billerud, however, has low content of most potentially harmful metals. The leaching data indicated that a large amount of the K and S content was leached. It is possible that the sulphate leached from the GLD may increase the chance of gypsum precipitation and could decrease the chance of jarosite formation in the underlying sulphidic waste [35]. It can be concluded that calcite was not dissolved despite of the repeated leaching test. The pH remained at a high level (pH 11.6). The leaching behavior is likely to change if the material is acidified resulting in a lower pH. Due to the production process of GLD (burned at high temperatures) it does not contain organic matter meaning that GLD is not susceptible to organic matter degradation that may harm the integrity of the sealing layer [28]. The material also lacks elements that are likely to interact and transform into harmful substances or oxidize easily, thereby changing its initial structure and properties. The ζ-potential is similar to those of kaolinite and montmorillonite at lower pH, but the clay minerals show a larger negative charge at high pH [36]. The ζ-potential had a negative correlation with pH which can be related to its high content of $CaCO_3$ and is consistent with other electrokinetic studies on calcite summarized by Guichet et al. [37]. For pH below 8, the presence of $CaCO_3$ decreases leading to more free Ca^{2+} in solution [37]. The measured ζ-potential should therefore be less negative for pH below 8, which is consistent with the result. The electrical properties of calcite are a controversial topic. Electrokinetic measurements have shown to be dependent on CO_2 partial pressure and have yielded different results on calcites of various origins [37]. The calcite in the GLD has a negative ζ-potential at all pH levels, which is not the case with kaolinite. This indicates that the GLD particles are relatively stable and resist the formation of aggregates [38]. This, together with the high WRC, suggests that crack formation due to aggregation, providing a pathway for water and oxygen, would be limited. The function can be maintained independent of the pH.

The buffering capacity of GLD is high, which is consistent with its high carbonate content. To maintain pH > 6, the average neutralization capacity was 18.5 mmol H^+/g dry weight for the four batches of GLD. The high buffering capacity indicates that GLD could act as an alkaline barrier. The high grade of saturation will minimize oxygen intrusion into the underlying sulphidic waste, slowing down oxidation. Only small amounts of alkalinity would therefore be consumed by the mine waste, resulting in a long lasting sealing layer. This is in agreement with Jia et al. [22] who showed that it is feasible to use GLD as a cover material to immobilize trace metals in tailings. Leaching experiments have shown that the amount of elements released was below the leaching limits for non-hazardous waste regulated by the European Union [25], making it a good candidate for sealing layer applications. In addition, it has been shown that addition of 10% green liquor dregs to reactive, sulphide-rich tailings reduced leaching of metals such as Cu, Co, Cd and Ni [39]. A probable explanation is that GLD increases pH, thereby immobilizing metal ions by precipitation of secondary minerals and absorption to mineral surfaces [1].

4.3. Limitations

The results indicate that the application of GLD may be effective in the Swedish subarctic climate since the neutralization capacity is expected to last for a long time because the dissolution of buffering compounds such as CaO, $Ca(OH)_2$ and $CaCO_3$ will be slow due to the low hydraulic conductivity of the material and low precipitation. However, the applicability of a GLD sealing layer may not be suitable in all environments. Waste rock piles have to be carefully engineered diverting the water away from the pyritic waste. In climate zones with very high annual precipitation, there is a risk of water flow on top of the sealing layer. This may harm the integrity of the sealing layer or slowly dissolve the chemical compounds compromising its long-term function. The function may also be compromised if the material dries, which is a risk arid regions.

Although GLD has shown desirable properties to function as a sealing layer, some practical issues have to be addressed beforehand. The material is very sticky, making it difficult to handle on a large scale. In addition, the total amount of GLD annually produced is small in comparison with the amount of mine waste. In Sweden, as an example, the annual production of GLD is ca. 240,000 t produced in many mills spread over the country, many situated far away from the mining areas. Logistics and transport economics must be considered. During 2010, 89 million tons of mine waste was produced, most of it from mines where sulphide-bearing ores were mined [40].

The shear strength of GLD was assessed at 11.8 ± 1.6 kPa after one month curing time, which is low compared to the shear strength of bentonite and clayey till. Such low shear strength is insufficient for engineering applications [26]. It is important that the material exhibit sufficient compressive strength to support the weight of a protective layer but also exhibit plasticity to resist cracking that can be caused by differential settlements in the landfill. If GLD is applied as a sealing layer in a slope, there is a risk of sliding due to the low shear strength and the high water content. Improving the mechanical properties is therefore necessary. A mixture with the proportions 7:2:1 of tailings: GLD: fly ash was found to be geotechnically satisfactory but its function as oxygen barrier remains to be proven [41]. GLD may also be frost sensitive, which is a common problem with silty soils, causing damage to the cover (*i.e.*, frost heave) when used at shallow depths in cold regions.

5. Conclusions

GLD has the potential to be used as sealing layer in dry covers on mine waste since it has relatively low hydraulic conductivity, a high WRC and small particle size. Although the chemical and mineralogical composition varied between the different batches, these variations were not reflected in properties such as hydraulic conductivity and WRC. Leaching of contaminants from the GLD is not a concern for the environment. However, GLD is a sticky material, difficult to apply on mine waste deposits and the shear strength is insufficient for engineering applications. Improving the mechanical properties is therefore necessary. Mixing the GLD with tailings and additives such as fly ash is a promising methodology to increase the shear strength and even lower the hydraulic conductivity, but further research is necessary before using such mixtures for construction of sealing layers on an industrial scale.

GLD shows a high buffering capacity indicating that it could act as an alkaline barrier. Due to the low hydraulic conductivity, the buffering capacity will not be consumed in a very long time in subarctic regions. A practical evaluation of the use of GLD taking the climate/geographical region into consideration needs to be pursued. Covers on mine waste should be efficient in a very long-term perspective. Therefore, the effects of aging on GLD properties should be studied.

Acknowledgments

Financial support from Swedish Research Council for Environment, Agricultural Sciences and Spatial Planning (FORMAS), the European Union Structural Funds, the Northern Sweden Soil Remediation Centre, Sveriges Ingenjörer environmental fund and the Center of Advanced Mining and Metallurgy (CAMM) at Luleå University of Technology are gratefully acknowledged. The authors wish to thank Dr. Bertil Pålsson at Luleå University of Technology, for providing equipment for particle analysis, Ulf Nordström for providing technical support and David Stenman for his assistance. The reviewers are also thanked for helpful comments to improve the article.

Authors Contributions

Experimental design was conducted by Maria Mäkitalo. Leaching experiments were conducted by Yu Jia. Maria Mäkitalo, Christian Maurice and Björn Öhlander contributed to the preparation and writing of the manuscript.

Conflicts of Interest

The authors declare no conflict of interests.

References and Notes

1. Höglund, L.-O.; Herbert, R.; Lövgren, L.; Öhlander, B.; Neretniks, I.; Moreno, L.; Malmström, M.; Elander, P.; Lindvall, M.; Lindström, B. *MiMi—Performance Assessment, Main Report*; MiMi-Report 2003:3; MiMi Print: Stockholm, Sweden, 2004.
2. International Network for Acid Drainage (INAP). The Global Acid Rock Drainage Guide (GARD Guide). Available online: http://www.gardguide.com (accessed on 12 December 2013).
3. Carlsson, E. Sulphide-Rich Tailings Remediated by Soil Cover: Evaluation of Cover Efficiency and Tailings Geochemistry, Kristineberg, Northern Sweden. Ph.D. Thesis, Department of Environmental Engineering, Luleå University of Technology, 2002; p. 44.
4. Hallberg, R.O.; Granhagen, A.; Liljemark, A. A fly ash/biosludge dry cover for the mitigation of AMD at the Falun mine. *Chem. Erde Geochem.* **2005**, *65*, 43–63.
5. Cabral, A.; Racine, I.; Burnotte, F.; Lefebvre, G. Diffusion of oxygen through a pulp and paper residue barrier. *Can. Geotech. J.* **2000**, 201–207.
6. Catalan, L.J.J.; Kumari, A. Efficency of lime mud residues from kraft mills to amend oxidized mine tailings before permanent flooding. *J. Environ. Eng. Sci.* **2005**, *4*, 241–256.
7. Bäckström, M.; Karlsson, S.; Sartz, L. *Utvärdering och Demonstration av Efterbehandlingsalternativ för Historiskt Gruvavfall Med Aska och Alkaliska Restprodukter*; Report 1099; Värmeforsk Service AB: Stockholm, Sweden, 2009. (In Swedish)
8. Cousins, C.; Penner, G.H.; Liu, B.; Beckett, G.; Spiers, G. Organic matter degradation in paper sludge amendments over gold mine tailings. *Appl. Geochem.* **2009**, *24*, 2293–2300.
9. Doye, I.; Duchesne, J. Neutralisation of acid mine drainage with alkaline industrial residues: Laboratory investigation using batch-leaching tests. *Appl. Geochem.* **2003**, *18*, 1197–1213.
10. Bellaloui, A.; Chtaini, A.; Ballivy, G.; Narasiah, S. Laboratory investigation of the control of acid mine drainage using alkaline paper mill waste. *Water Air Soil Pollut.* **1999**, *111*, 57–73.
11. Chtaini, A.; Bellaloui, A.; Ballivy, G.; Narasiah, S. Field investigation of controlling acid mine drainage using alkaline paper mill waste. *Water Air Soil Pollut.* **2001**, 125, 357–374.
12. Sartz, L.; Bäckström, M.; Karlsson, S.; Allard, B. Stabilization of Acid-Generating Waste Rock with Alkaline By-Products: Results from a Meso-Scale Experiment. In Proceedings of Securing the Future and 8th International Conference on Acid Rock Drainage (ICARD), Skellefteå, Sweden, 22–26 June 2009.
13. Lundqvist, F.; Brelid, H.; Saltberg, A.; Gellerstedt, G.; Tomani, P. Removal of non-process elements from hardwood chips prior to kraft cooking. *Appita* **2006**, *59*, 493–499.
14. *SFS, 2001:1063, Swedish Ordinance of Waste*; Sweden Ministry of Environment: Stockholm, Sweden, 2001. (In Swedish)
15. Toikka, M. *Minimization and Utilization of Green Liquor Dregs and Ashes in Pulp and Paper*; Regional Environment Publication No. 82; Southeast Finland Regional Environment Centre, Ed.; Oy Edita Ab: Helsinki, Finland, 1998; pp. 1–82.

16. Pousette, K.; Mácsik, J. *Kartläggning av Restprodukter från Assi Domän AB: Research Report*; Luleå University of Technology: Luleå, Sweden, 2000. (In Swedish)

17. Hargelius, K. *Pilotyta med Tätskikt på Ätrans Deponi, Fältförsök-Värö-FAVRAB-Hylte*; Ramböll Sverige AB, Region Väst: Gothenburg, Sweden, 2008. (In Swedish)

18. U.S. Environmental Protection Agency (EPA). *Methods for the Determination of Metals in Environmental Samples*; EPA/600/4-91/010; Office of Research and Development, U.S. Environmental Protection Agency: Washington, DC, USA, 1991.

19. Swedish Standards Institute (SIS). *Determination of Dry Matter and Ignition Residue in Water, Sludge and Sediment*; SS 028113; SIS Förlag AB: Stockholm, Sweden, 1981. (In Swedish)

20. Wyatt, P.H. *A Technique for Determining the Acid Neutralizing Capacity of Till and Other Surficial Sediments*; Paper No. 84-1A; Geological Survey of Canada: Ottawa, ON, Canada, 1984; pp. 597–600.

21. Swedish Standards Institute (SIS). *Characterisation of Waste, Leaching, Compliance Test for Leaching of Granular Waste Materials and Sludges*; SS-EN 12457-4; SIS Förlag AB: Stockholm, Sweden, 2003. (In Swedish)

22. Jia, Y.; Maurice, C.; Öhlander, B. Effect of the alkaline industrial residues fly ash, green liquor dregs, and lime mud on mine tailings oxidation when used as covering material. *Environ. Earth Sci.* **2013**, doi:10.1007/s12665-013-2953-3.

23. Weber, P.A.; Hughes, J.B.; Conner, L.B.; Lindsay, P.; Smart, R. Short-Term Acid Rock Drainage Characteristics Determined by Paste pH and Kinetic NAG Testing: Cypress, Prospect, New Zealand. In Proceedings of 7th International Conference on Acid Rock Drainage (ICARD), St. Louis, MO, USA, 26–30 March 2006.

24. Swedish Standards Institute (SIS). *Geotechnical Tests, Compression Properties, Oedometer Test, CRS-Test, Cohesive Soil*; SS 02 71 26; SIS förlag AB: Stockholm, Sweden, 1991. (In Swedish)

25. European Council. Council decision of 19 December 2002 establishing criteria and procedures for the acceptance of waste at landfills pursuant to Article 16 of and Annex II to Directive 1999/31/EC. *Off. J. Eur. Commun.* **2002**, *L11*, 27–49.

26. Swedish Geotechnical Institute (SIG). *Deponiers Stabilitet, Vägledning för Beräkning*; SGI Information 19; Swedish Geotechnical Institute: Linköping, Sweden, 2007. (In Swedish)

27. Bussière, B.; Chapuis, R.P.; Aubertin, M. Unsaturated Flow Modeling for Exposed and Covered Tailings Dams. In Proceedings of the International Symposium on Major Challenges in Tailings Dams, Montreal, QC, Canada, 15 June 2003.

28. Martins, F.M.; Martins, J.M.; Ferracin, L.C.; da Cunha, C.J. Mineral phases of green liquor dregs, slaker grits, lime mud and wood ash of a Kraft pulp and paper mill. *J. Hazard. Mater.* **2007**, *147*, 610–617.

29. Mbonimpa, M.; Aubertin, M.; Aachib, M.; Bussière, B. Diffusion and consumption of oxygen in unsaturated cover materials. *Can. Geotech. J.* **2003**, *40*, 916–932.

30. Erickson, J.; Tyler, E.J. Soil Oxygen Delivery to Wastewater Infiltration Surfaces. In Proceedings of 2000 National Onsite Wastewater Recycle Association (NOWRA), Laurel, MD, USA, 5 August 2000.

31. Sivapullaiah, P.V.; Sridharan, A.; Stalin, V.K. Hydraulic conductivity of bentonite-sand mixtures. *Can. Geotech. J.* **2000**, *37*, 406–413.

32. Benson, C.H.; Trast, J.M. Hydraulic conductivity of thirteen compacted clays. *Clays Clay Miner.* **1995**, *43*, 669–681.

33. Marion, D.; Nur, A.; Yin, H.; Han, D. Compressional velocity and porosity in sand-clay mixtures. *Geophysics* **1992**, *57*, 554–563.

34. Koltermann, C.E.; Gorelick, S.M. Fractional packing model for hydraulic conductivity derived from sediment mixtures. *Water Resour. Res.* **1995**, *31*, 3283–3297.

35. Jia, Y. Luleå University of Technology, Luleå, Sweden. Unpublished Modeling Data, 2014.

36. Kosmulski, M.; Dahlsten, P. High ionic strength electrokinetics of clay minerals. *Colloids Surf. Physicochem. Eng. Aspects* **2006**, *291*, 212–218.

37. Guichet, X.; Jouniaux, L.; Catel, N. Modification of streaming potential by precipitation of calcite in a sand-water system: Laboratory measurements in the pH range from 4 to 12. *Geophys. J. Int.* **2006**, *166*, 445–460.

38. Hunter, R.J. *Zeta Potential in Colloid Science*; Academic Press: London, UK, 1981.

39. Maurice, C.; Villain, L.; Hargelius, K. Green Liquor Dregs for the Remediation of Abandoned Mine Tailings—Opportunities and Limitations. In Proceedings of Securing the Future and 8th International Conference on Acid Rock Drainage (ICARD), Skellefteå, Sweden, 22–26 June 2009.

40. Naturvårdverket (Swedish Environmental Protection Agency). *Avfall i Sverige 2010*; Report 6520; CM Gruppen AB: Bromma, Sweden, 2012. (In Swedish)

41. Jia, Y.; Stenman, D.; Mäkitalo, M.; Maurice, C.; Öhlander, B. Use of amended tailings as mine waste cover. *Waste Biomass Valoriz.* **2013**, *4*, 709–718.

5

Distribution and Substitution Mechanism of Ge in a Ge-(Fe)-Bearing Sphalerite

Nigel J. Cook [1], **Barbara Etschmann** [1,2,3], **Cristiana L. Ciobanu** [1], **Kalotina Geraki** [4], **Daryl L. Howard** [5], **Timothy Williams** [6], **Nick Rae** [2,5], **Allan Pring** [3,7], **Guorong Chen** [8], **Bernt Johannessen** [5] **and Joël Brugger** [2,3,*]

[1] School of Chemical Engineering, University of Adelaide, Adelaide, SA 5005, Australia;
E-Mails: nigel.cook@adelaide.edu.au (N.J.C.); barbara.etschmann@monash.edu (B.E.);
cristiana.ciobanu@adelaide.edu.au (C.L.C.)

[2] School of Geosciences, Monash University, Clayton, VIC 3800, Australia;
E-Mail: nicholas.rae@synchrotron.org.au

[3] South Australian Museum, North Terrace, Adelaide, SA 5000, Australia;
E-Mail: allan.pring@flinders.edu.au

[4] Diamond Light Source, Harwell Science and Innovation Campus, Didcot, Oxon OX11 0QX, UK;
E-Mail: Tina.Geraki@diamond.ac.uk

[5] Australian Synchrotron, 800 Blackburn Rd., Clayton, VIC 3168, Australia;
E-Mails: Daryl.Howard@synchrotron.org.au (D.L.H.); bernt.j@synchrotron.org.au (B.J.)

[6] The Monash Centre for Electron Microscopy, Monash University, Clayton, VIC 3800, Australia;
E-Mail: Timothy.williams@monash.edu (T.W.)

[7] School of Chemical and Physical Sciences, Flinders University, GPO Box 2100, Adelaide,
SA 5000, Australia

[8] Key Laboratory for Ultrafine Materials of Ministry of Education, School of Materials Science and
Engineering, East China University of Science and Technology, Shanghai 200237, China;
E-Mail: grchen@ecust.edu.cn (G.C.)

* Author to whom correspondence should be addressed; E-Mail: joel.brugger@monash.edu

Academic Editor: Mostafa Fayek

Abstract: The distribution and substitution mechanism of Ge in the Ge-rich sphalerite from the Tres Marias Zn deposit, Mexico, was studied using a combination of techniques at μm- to atomic scales. Trace element mapping by Laser Ablation Inductively Coupled

Mass Spectrometry shows that Ge is enriched in the same bands as Fe, and that Ge-rich sphalerite also contains measurable levels of several other minor elements, including As, Pb and Tl. Micron- to nanoscale heterogeneity in the sample, both textural and compositional, is revealed by investigation using Focused Ion Beam-Scanning Electron Microscopy (FIB-SEM) combined with Synchrotron X-ray Fluorescence mapping and High-Resolution Transmission Electron Microscopy imaging of FIB-prepared samples. Results show that Ge is preferentially incorporated within Fe-rich sphalerite with textural complexity finer than that of the microbeam used for the X-ray Absorption Near Edge Structure (XANES) measurements. Such heterogeneity, expressed as intergrowths between 3C sphalerite and 2H wurtzite on $[1\bar{1}0]$ zones, could be the result of either a primary growth process, or alternatively, polystage crystallization, in which early Fe-Ge-rich sphalerite is partially replaced by Fe-Ge-poor wurtzite. FIB-SEM imaging shows evidence for replacement supporting the latter. Transformation of sphalerite into wurtzite is promoted by (111)* twinning or lattice-scale defects, leading to a heterogeneous ZnS sample, in which the dominant component, sphalerite, can host up to ~20% wurtzite. Ge K-edge XANES spectra for this sphalerite are identical to those of the germanite and argyrodite standards and the synthetic chalcogenide glasses GeS_2 and $GeSe_2$, indicating the Ge formally exists in the tetravalent form in this sphalerite. Fe K-edge XANES spectra for the same sample indicate that Fe is present mainly as Fe^{2+}, and Cu K-edge XANES spectra are characteristic for Cu^+. Since there is no evidence for coupled substitution involving a monovalent element, we propose that Ge^{4+} substitutes for (Zn^{2+}, Fe^{2+}) with vacancies in the structure to compensate for charge balance. This study shows the utility of synchrotron radiation combined with electron beam micro-analysis in investigating low-level concentrations of minor metals in common sulfides.

Keywords: synchrotron radiation; XANES spectroscopy (Ge; Fe; Cu K-edges); sphalerite; germanium; oxidation state

---

1. Introduction

Consumption of germanium for use in light-emitting diodes, fiber-optic systems, and satellite and terrestrial solar cells, has significantly increased in recent years, highlighting the need to ensure an adequate future supply of germanium. Currently, germanium is extracted commercially from some Ge-rich coal seams and from zinc concentrates from some Zn-Pb mining operations, in which Ge is hosted within the common sulfide mineral sphalerite (ZnS) [1,2]. Germanium is particularly enriched in relatively Fe-poor sphalerite from Mississippi Valley-type (MVT) deposits formed at relatively low temperatures [2].

The mechanism by which Ge is substituted into the sphalerite crystal lattice has long been the subject of debate. Some authors (e.g., [3–5]) have favoured incorporation of Ge^{4+} into the sphalerite structure, implying either coupled substitution or vacancies to achieve charge compensation. For example, [5,6] invoke the $3Zn^{2+} \leftrightarrow Ge^{4+} + 2Ag^+$ substitution for the incorporation of Ge (up to

1200 ppm) in Ag-rich (max 1000 ppm) sphalerite from a French vein-type deposit, based on a coarse correlation between Ge and Ag. In contrast, [6,7] found no significant correlation between Ge and other elements in sphalerite from different base metal mineral deposits containing tens to hundreds of ppm Ge, and suggested that Ge might be directly substituted as Ge^{2+} for Zn^{2+} or, alternatively, the substitution mechanism involves Ge^{4+}, but with vacancies to maintain charge balance.

Understanding the oxidation state of Ge in natural sphalerite is critical for modeling element substitution mechanisms. We used X-ray Absorption Near-Edge Structure (XANES) micro-spectroscopy to investigate the Ge oxidation state in a relatively Ge-rich sphalerite (~1000 ppm); higher concentrations up to ~3000 ppm are only known from a limited number of occurrences worldwide [8]. We studied cm-sized pieces of sphalerite ore from the Tres Marias Zn deposit, Mexico [9]. In this material, Ge concentrations correlate positively with the Fe contents. This presents us with the opportunity to also investigate the oxidation state of Fe in substituted sphalerite. Although Fe is normally considered to occur only as Fe^{2+} in sphalerite (e.g., [10,11]), we aim to establish whether this assumption is also true for Ge-rich sphalerite, or whether there is any evidence for the presence of Fe^{3+}.

2. Sample Description

2.1. Macro- to μm-Scale

The sample comprises massive sphalerite with a characteristic bladed appearance, possibly suggesting the co-presence of wurtzite at the smallest scale, as in other specimens considered to consist only of sphalerite (Pring et al. unpublished results). Powder X-ray diffraction studies showed that the sample consists of a fine-scale intergrowth of sphalerite and wurtzite-2H, with the ratio of the two minerals ranging from 10:1 down to 4:1. Two compositionally distinct areas (Fe-rich and Fe-poor) are recognized on the surface of our polished mount. A ragged boundary separates the two areas (Figure 1f in [6]). The Fe-rich sphalerite consists of aligned blades, typically 50–200 μm in length, and with irregular to lamellar grate-like features within some blades that are distinct in reflected light or back-scattered electron images. Figure 3a in [9] shows a back-scattered electron image with "delicate lamellar to dendritic" textures identical to those described here. The Fe-poor sphalerite has the same general appearance but is distinct by having a markedly greater porosity along the individual blades.

Published Electron Probe (EPMA) and Laser Ablation Inductively-Coupled Plasma Mass Spectrometry (LA-ICP-MS) spot analyses of the same hand specimen [6] gave Fe contents of 3.14 wt% [$(Zn_{0.95}Fe_{0.05})S$] and 8.72 wt% [$(Zn_{0.85}Fe_{0.15})S$] in the low- and high-Fe sphalerite, respectively. Germanium is enriched in the Fe-rich area of the polished mount. Mean Ge concentrations, as measured by LA-ICP-MS, are 252 and 1081 ppm in the Fe-poor and Fe-rich areas, respectively. Cadmium concentrations are ~5000 ppm in both areas. Other elements present at significant concentrations (LA-ICP-MS data) are As (means of 572 and 434 ppm in Ge-Fe-rich and Ge-Fe-poor areas, respectively), Pb (1349 and 3090 ppm) and Tl (158 and 53 ppm). Silver and Cu concentrations are a few ppm and a few tens of ppm, respectively. Gallium concentration is ~25 ppm in the Fe-poor sphalerite but an order of magnitude lower in the Fe-rich area.

LA-ICP-MS trace element maps across the boundary between the two compositionally distinct areas were obtained to further characterize the material. Analytical procedures and operating conditions follow [12], using the sulfide-matrix Mass-1 [13] as a reference material, and Zn as the internal standard. The maps (Figure 1) show a striking compositional heterogeneity, highlighting the differences between the Fe-rich areas, and that Ge enrichment closely follows Fe and is also associated with enrichment in Ag, Hg, Mn, Tl and As, and that there is an inverse relationship between (Ge,Fe) and Zn. Silver, In and Sb display a subtle zoning relative to the boundary between the two areas. The sum of cations likely present in the monovalent state, (Cu + Ag + Tl) which were observed to play a role in maintaining charge balance in galena [14], is only a fraction of that of Ge.

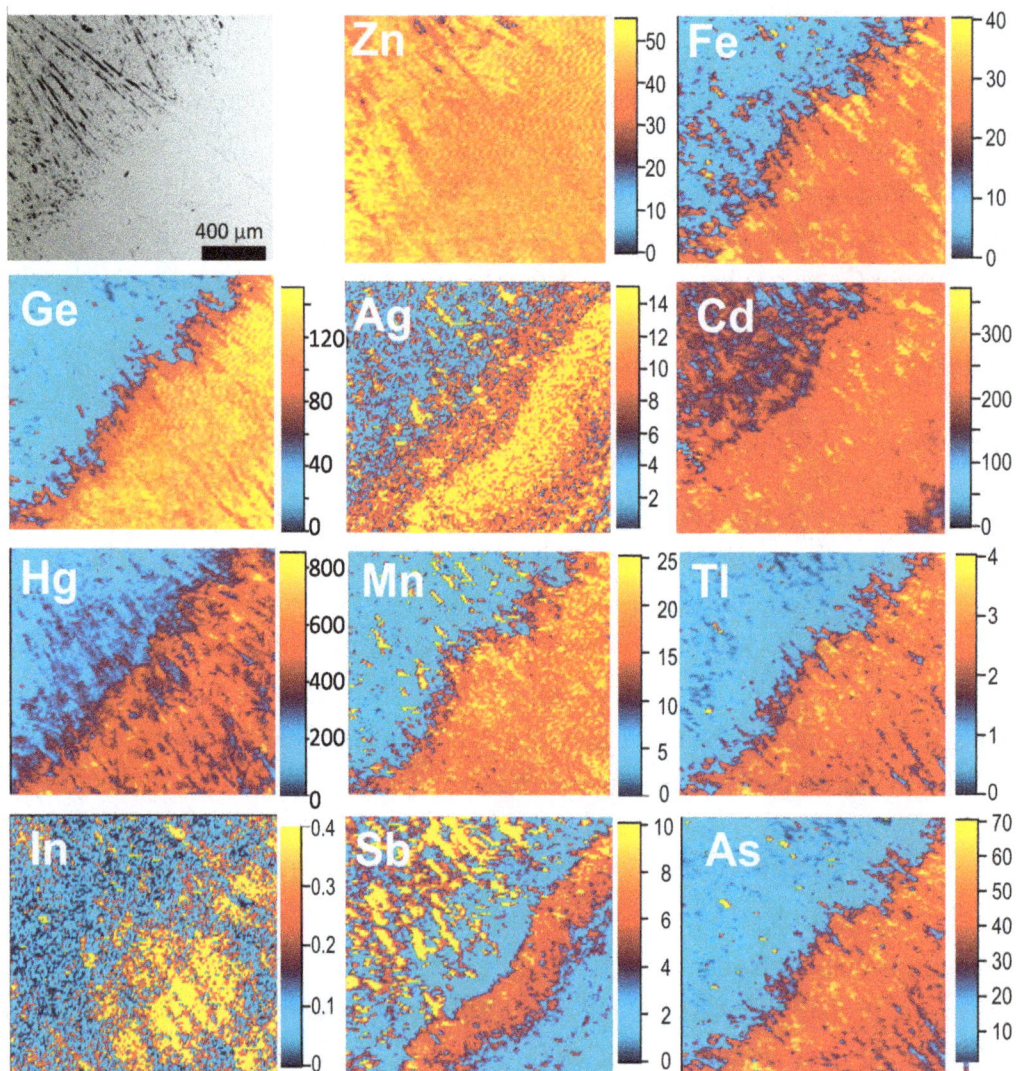

Figure 1. Reflected light image of sphalerite from the Tres Marias Zn deposit, Mexico (top left) and LA-ICP-MS element maps across the boundary between (Ge-Fe)-rich and (Ge-Fe)-poor areas. Scales in counts-per-second $\times 10^3$ (except Zn and Fe: $\times 10^6$).

2.2. Nanoscale Sample Characterization

FIB-SEM, synchrotron microbeam X-ray fluorescence microscopy and TEM techniques were employed, following [15], to assess (i) if the micron-scale textural and compositional heterogeneity observed in the Tres Marias material (Figure 1f in [6]; Figure 1) extend down to the nanoscale; and (ii) to understand the relationships between the ZnS polytypes at the nanoscale, in view of the co-existence of variable mixtures between sphalerite (3C) and wurtzite (2H) indicated by the X-ray powder diffraction data obtained on bulk material.

FIB-SEM cross-sectioning and imaging were performed in each of the two areas, (Ge-Fe)-rich and -poor (Figures 2 and 3). The FIB cut in the (Ge-Fe)-rich area was done across the boundaries of blades with or without the characteristic, intricate sub-structure defined by perpendicular grating (subsets of short lamellae; Figure 2a). This type of pattern extends in depth, in particular in the middle part of the wall exposed by FIB cross-sectioning (Figure 2b). Further complexity, however, is shown by the presence of coarse grains (equant domains) on the cross-section margins (arrowed on Figure 2b). The darkest and most homogenous parts on the Secondary Electron (SE) images are present either in these grains or in the adjacent vertical structures. In detail, both vertical and horizontal lamellar sets show sub-μm heterogeneity on SE images (Figure 2c). Moreover, sub-μm pores $\pm$ inclusions are present at the boundary and within the horizontal lamellar sets (circled on Figure 2c). The fine, sub-μm zoning is present in all types of structures (Figure 2d–f). Boundary relationships between darker and brighter structures (Figure 2b,d) show a degree of corrosion, *i.e.*, the brighter grain intrudes the outline of the adjacent darker grain. The most complex patterns are represented by sub-micron-scale banding in the brighter portions, both granular and vertical (Figure 2e,f). Furthermore, the brighter structures are typified by numerous pores $\pm$ inclusions (circled on Figure 2e), unlike the darker ones which are clean. The corrosion relationship, as well as the presence of pores $\pm$ inclusions within the brighter domains, suggests replacement of the darker, homogeneous domains (relict) by the brighter sub-structures.

The cut in the (Ge-Fe)-poor area was also performed perpendicular to the elongation highlighted in this case by sets of microfractures, trails of pores and secondary mineral inclusions (Figure 3a). FIB cross-sectioning reveals filled fractures and cavities at depth, as well as the presence of irregularly shaped darker areas within a brighter matrix (Figure 3b). In detail, the relict character of the darker portions is further highlighted by marginal cracks and relationships with larger cavities (Figure 3c). Compared to the (Ge-Fe)-rich area, the replacement character is well developed, and the brighter matrix features patchiness in shades rather than fine zoning.

Synchrotron X-ray Fluorescence Microscopy using a Vortex silicon drift detector [16] at the Australian Synchrotron was used to obtain element maps of the FIB-prepared slice from the (Ge-Fe)-rich sphalerite (Figure 4) following the same sample preparation approach as [17]. The incident X-ray beam was focused with a Fresnel zone plate to a spot size of ~200 nm. The maps show that the μm-scale substructures identified by FIB cross-section imaging are also expressed by compositional variation, *i.e.*, the darker (on FIB image; Figure 2b,c), relict (?) parts are Zn- and As-poor but clearly Fe- and Ge-rich relative to the brighter parts.

High-resolution TEM imaging of FIB-prepared foils from the (Ge-Fe)-rich area shows irregular boundaries separating grains of different orientation (Figure 5a). In detail, the nanoscale textural complexity is further highlighted by sets of irregular [1$\bar{1}$0] twins. Such twinning is one of the main mechanisms for polytype transformation between the 3C cubic and nH hexagonal ZnS polytypes (e.g., [15,18,19]. However, the dominant component of the sample is 3C sphalerite featuring lattice-scale defects of which most abundant are stacking faults along (hkl)* directions (Figure 5b). Indexing of the image in Figure 5b is shown in the Fast Fourier Transform (FFT) diffraction in Figure 5e. Similar irregular twin domains and defects have been described for Fe-rich Toyoha sphalerite [15].

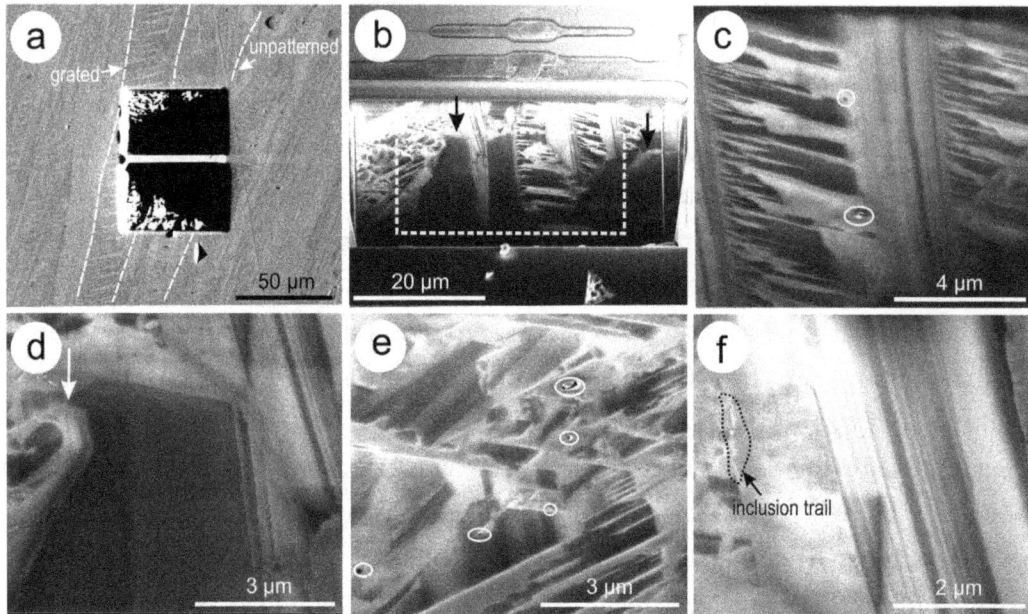

Figure 2. Secondary Electron (SE) images of (Ge-Fe)-rich areas showing: (**a**) location of FIB cut across the intricate sub-structure combining elongate blades and perpendicular subsets of short lamellae (marked); (**b**) extension of surface structures in depth exposed by FIB cross-sectioning in the middle part of the wall. Note presence of equant domains (arrowed) on the cross-section margins; the rectangle shows the area mapped in Figure 4; (**c**) detail of lamellar structures showing sub-μm scale heterogeneity; pores ± inclusions are circled; (**d**) detail of a coarse grain with darkest appearance showing boundary relationships with brighter grain; note partial corrosion on one side (arrowed); (**e,f**) details of sub-μm-scale heterogeneity expressed as fine banding in the brighter domains exposed in different orientations; pores ± inclusions are circled. FIB-SEM work performed on a Dual Beam FEI Helios Nanolab FIB-SEM platform at Adelaide Microscopy following methods outlined by [15].

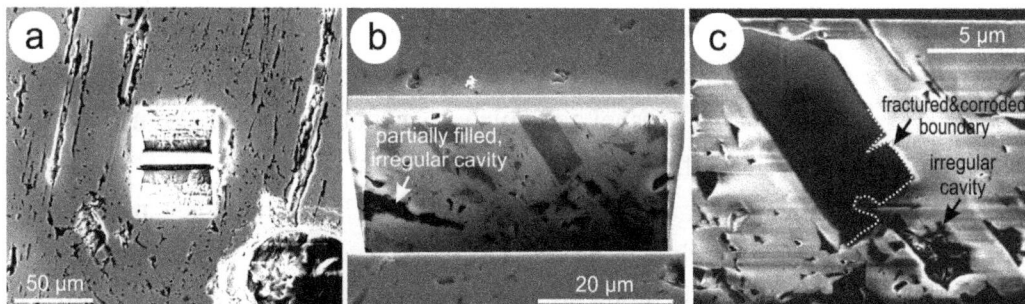

Figure 3. Secondary Electron (SE) images of Ge-Fe-poor area showing: (**a**) location of FIB cut roughly perpendicular to the elongation highlighted by sets of microfractures, trails of pores and secondary mineral inclusions; (**b**) extension of surface structures in depth exposed by FIB cross-sectioning. Note filled fractures and cavities (marked) at depth, as well as the presence of irregularly-shaped darker areas within a brighter matrix; (**c**) marginal cracks and relationships with larger pores lending the darker domains a relict character. The experimental methodology is the same as for Figure 2.

Relevant crystallographic relationships between the two polytypes are illustrated in Figure 5a, *i.e.*, lattice fringes of the upper and lower grains, with orientations shown by indexed FFT diffractions (Figure 5c,d computed from the image in Figure 5a). Relationships across the grain boundary can also be clearly seen from the orientation of dominant lattice fringes (75° to one-another). Transformation into wurtzite takes place by twinning along the (111)* direction of initial sphalerite (3C polytype) which corresponds to the c* axis of wurtzite or other hexagonal ZnS polytypes (c*$_{nH}$). This means that the stacking direction of horizontal sets of lamellae in Figure 2 are along the (111)* axis of sphalerite epitaxial with c* axis in wurtzite.

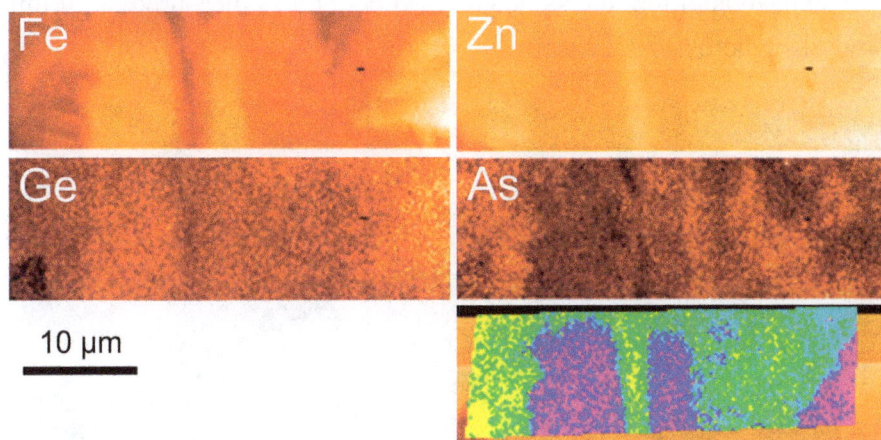

Figure 4. Microbeam X-ray Fluorescence element maps for Fe, Zn, Ge and As in a slice cut from Ge-Fe-rich area in the Tres Marias sphalerite (rectangle on Figure 2b). A 200 nm step size was used; which corresponded to the approximate beam size (Australian Synchrotron) and was also nearly identical to the smallest possible beam size. The maps show that the μm-scale structures identified by FIB cross-section imaging are also expressed by compositional variation, *i.e.*, the darker, relict domains are Zn and As-poor but clearly Fe- and Ge-rich relative to the brighter domains. The map lower right is a composite of the Fe (red), Zn (green) and Ge (blue) maps.

3. μ–XANES Data

Germanium, Cu and Fe K-edge XANES spectra were collected at the Microfocus Spectroscopy Beamline (I18), Diamond Light Source Synchrotron facility, UK I18 uses a Si(111) monochromator, the incident X-ray beam was focused to ~2 × 2 μm^2 using Kirkpatrick-Baez (KB) mirrors. The fluorescence data were collected using a four-element Si-Drift detector and transmission data with ion chambers. Ge and Cu XANES data were collected in fluorescence mode from the Ge-Fe-rich area of the polished mount. Al and Ni filters were used to cut down fluorescence from Zn at the Ge edge. Fe Kα data were collected in transmission from pressed pellets made from the powdered sample diluted with boron nitride (BN). XANES/EXAFS (Extended X-ray Absorption Fine Structure) were measured on the reference materials, which included GeO$_2$ (99.999%, Strem Chemicals; tetragonal modification with quartz-like structure [20]); germanite Cu$_{13}$Fe$_2$Ge$_2$S$_{16}$ (South Australian Museum (SAM) G4777, Tsumeb Mine, Namibia; Ge, Cu, Fe); partially oxidized powdered Ge metal (British Drug House (BDH)) Chemicals; 99.99% Ge); stannite (Cu$_2$FeSnS$_4$, SAM G19206 Yaogangxian Mine, Hunan,

China; Cu, Fe); and chalcopyrite $CuFeS_2$ (SAM G22621, Moonta Mines, South Australia; Cu, Fe). In addition, argyrodite Ag_8GeS_6 (Museum Victoria (M), Melbourne, Australia, sample number M3071; [21]); renierite $Cu_{11}ZnGeFe_4S_{16}$ (M47647); synthetic GeS_2 glass (synthesized by heating high-purity (5N) Ge and S that had been vacuum sealed in quartz ampules to 900 °C, rocked for 10 h and then quenched in air [22–24]); synthetic $GeSe_2$ glass (synthesized via the melt-quenching method [25,26]); and synthetic GeS (orthorhombic; Strem Chemicals, 99.999%) were measured in transmission mode on pellets either at the X-ray Absorption Spectroscopy (XAS) beamline of the Australian Synchrotron or at BM18, the Core EXAFS beamline at the Diamond Light Source. Beamline energies were calibrated using Fe and Cu foils as well as Ge-metal; in addition, the same $GeO_2(s)$ pellet was measured at all three beamlines to provide a direct comparison across the datasets.

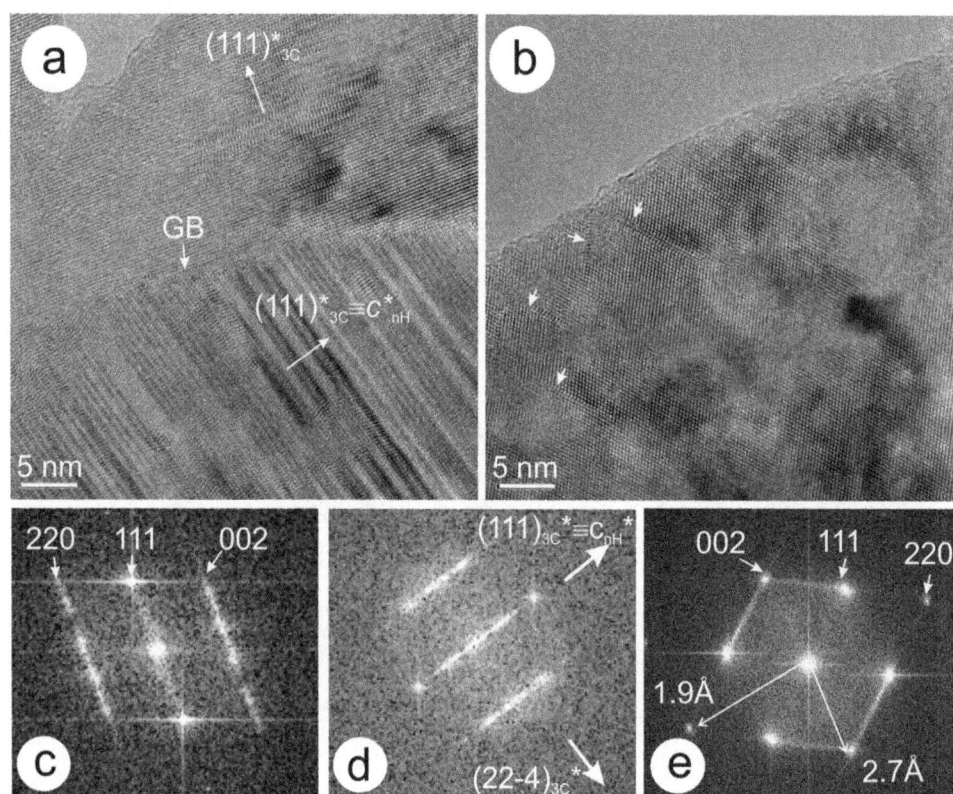

Figure 5. High Resolution Transmission Electron Microscopy (HRTEM) images of Ge-Fe-rich sphalerite showing (**a**) an irregular boundary (GB) separating grains of different orientation, *i.e.*, $(111)^*_{3C}$ at 75° to one another. Note sets of irregular [1–10] twins expressed by contrast differences on the lower grain; (**b**) 3C sphalerite down the same zone axis showing abundant stacking faults along (hkl)* directions (arrowed). Images recorded using a JEOL 2011 instrument (JEOL Co. Ltd., Akishima, Japan) at the Monash Centre for Electron Microscopy, operated at 200 kV. (**c**–**e**) Fast Fourier Transform diffractions computed from images in (**a**) and (**b**), showing orientations of upper and lower grains in (**a**) as (**c**) and (**d**), and of the grain from (**b**) in (**e**).

The assignment of the oxidation state of Ge in sulfide minerals is poorly constrained, as these minerals are generally structurally and compositionally complex. Ge is assumed to exist in tetravalent form in known Ge-sulfide minerals, and is present in tetrahedral coordination with Ge-S distances in the range of 2.19–2.35 Å (renierite, $Cu_{11}ZnGeFe_4S_{16}$, 2.27 Å [27]; Ag_2PbGeS_4, 2.2212 Å [28]; putzite, $(Cu_{4.7}Ag_{3.3})GeS_6$, 2.192 Å [29]; argyrodite, Ag_8GeS_6, 2.212 Å [21]; Cu_8GeS_6, 2.235 Å [30]). In germanite, nominally $Cu^+_{16}Cu^{2+}_{10}Fe^{3+}_4Ge^{4+}_4S_{32}$ assuming a Ge^{4+} oxidation state, Fe and Ge share a tetrahedral site, with (Ge,Fe)-S distances of 1 × 2.18 Å and 3 × 2.35 Å (mean 2.31 Å) [31]. Only in the GeS_2 glass can the oxidation state be formally assigned as Ge^{4+} and here the Ge is tetrahedral with Ge-S distances of 2.224 Å [32]. The Ge-S bond length in the GeS_2 glass was refined from our data to be 2.231(9) Å with a Debye-Waller of 0.0023(6) $Å^2$. Divalent germanium typically occurs in triangular pyramidal coordination in oxide and halogenide compounds, with a stereochemically active lone electron pair oriented opposite to the triangle of anions, similar to As^{3+} [33]. In GeS(orthorhombic), Ge^{2+} exists in 3 + 2 coordination, with Ge-bonding distances significantly greater than in 4 + compounds (3 × 2.441; 2 × 3.270 Å [34]).

The low Ge and high Zn concentrations (Ge $K_{\alpha1}$ fluorescence line at 9.886 keV receiving some background from the Zn $K_{\beta1}$ line) resulted in rather noisy Ge spectra for the Tres Marias sphalerite; however, the spectra measured on different points on the sample were similar. The average of five Ge $K\alpha$ edge spectra for Ge-bearing sphalerite is shown in Figure 6a, along with reference materials. The "white line" (peak) for the Tres Marias sample aligns with that of germanite, argyrodite, renierite, and the GeS_2 glass, indicating that the oxidation state and local environment (geometry, nature of ligands) of Ge in these samples are similar. Plots of the derivative for the XANES spectra (Figure 6b) show that: (i) the peak of the derivative (inflexion point) for GeO_2 is shifted by ~5 eV relative to the first peak of oxidized Ge metal; (ii) the second peak of the Ge metal aligns well with GeO_2, demonstrating that the metallic powder was partially oxidized into Ge^{+4}; and (iii) the peak for germanite lies between metallic Ge and GeO_2 (~3 eV below GeO_2), but 2 eV above that of GeS; the peak of GeS is at approximately the same position as metallic Ge. The alignment of the Tres Marias sphalerite, germanite, renierite and argydorite spectra with GeS_2 glass indicates that Ge is present as Ge^{4+} in all these minerals.

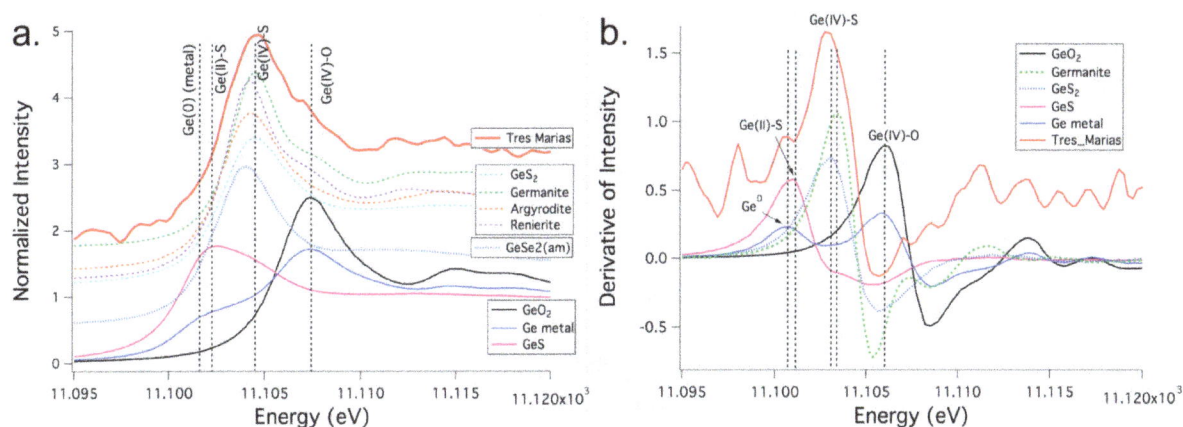

Figure 6. Ge K-edge XANES data for the Tres Marias sample compared to standards ((**a**) normalized; (**b**) first derivative).

In general, the edge position shifts towards higher energy with increasing valence state of the resonant atom, as the energy shift of a core state is directly related to the variation of the atomic electronic occupancy [35]. This correlation is strong for semi-metals such as As (e.g., [36]) and Te (e.g., [37]), for which the energy of the first peak in the derivative of the XANES spectrum shifts nearly monotonically with formal oxidation state, regardless of the associated ligand. Exceptions to this rule exist, notably chromium [38], where the shift in edge energy for Cr^{3+} with various ligands is comparable to the shift between Cr^0 and Cr^{6+}. After examining the density of states, [38] concluded that geometry and not electronegativity was the driving factor behind these large energy shifts. In the case of Ge, the large shift between GeO_2 and GeS_2, both of which contain tetrahedral-coordinated Ge^{4+} is surprising and cannot be explained by geometry. It must be predominantly related to the effect of the ligand. Note that both S and Se have a similar effect relative to O (Figure 6). The electronegativity of Ge favors covalent bonding with many ligands; Ge–O bonds have about 31% ionic character and Ge–S bonds only 7% [39].

Fe K-edge XANES were collected on the Tres Marias sphalerite, and Fe and Cu K-edge XANES were collected on germanite, a sphalerite-derivative [31]. The edge position in Fe and Cu K-edge XANES spectra is commonly obscured [40], and most determinations of oxidation states rely on empirical calibrations, which depend, in turn, on the material investigated [41]. For the Fe data, most calibrations rely on the position of the pre-edge ($1s \rightarrow 3d$) at around 7113 eV. In all the minerals shown in Figure 7a,b, Fe is present in tetrahedral coordination. Stannite contains Fe^{2+} [42]; the overall spectra and the pre-edge positions (7112.54 and 7112.56 eV, respectively) are similar in stannite and Tres Marias sphalerite, suggesting similar prevalent coordination and oxidation states. The formal oxidation state of Fe in chalcopyrite is likely to be Fe^{3+}, and the position of the pre-edge is shifted by ~0.8 eV (7113.37 eV). The Fe-K-edge XANES of germanite is intermediate between stannite and chalcopyrite (pre-edge at 7112.94 eV), suggesting a mixed oxidation state of Fe.

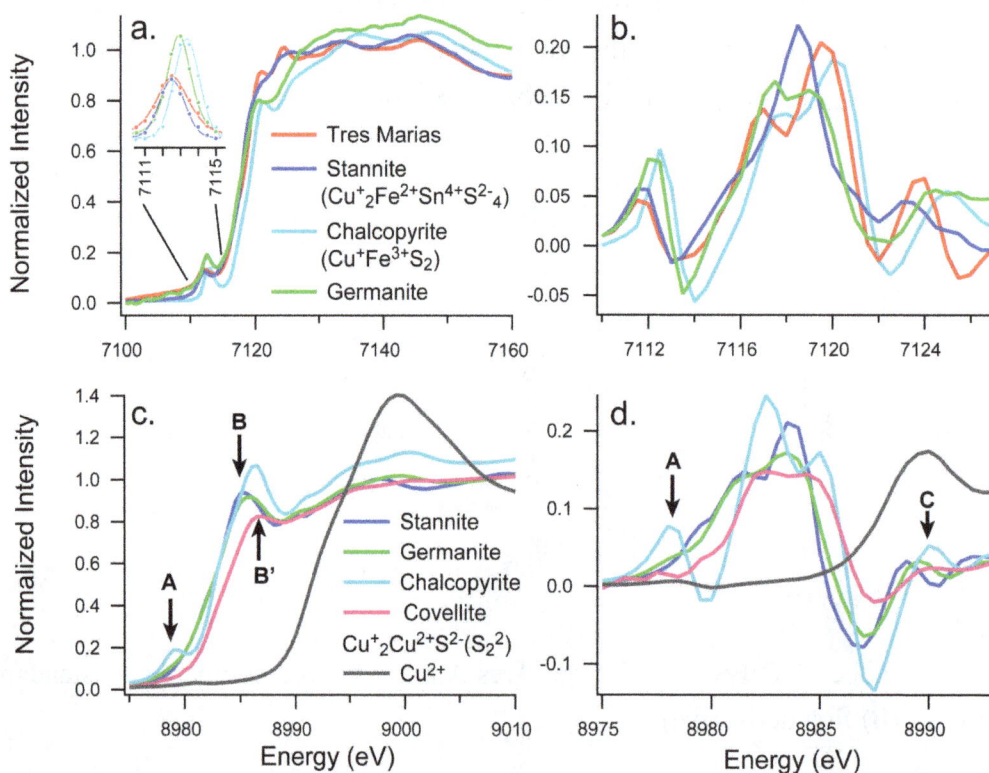

Figure 7. Fe and Cu K-edge XANES data for the Tres Marias sphalerite compared to standards ((**a**), (**c**), normalized; (**b**), (**d**), first derivative).

Charge balance considerations require that Cu exists in mixed oxidation state (Cu^+/Cu^{2+}) in germanite. Germanite contains four tetrahedrally-coordinated Cu sites [31]. The Cu-XANES spectrum of germanite is close to that of stannite, in which the formal oxidation state of copper is +1. The lack of pre-edge at ~8980 keV (arrow A in Figure 7c) and the distinct pre-edge appearing as a shoulder at 8983 keV (arrow B in Figure 7c) are key features of Cu^+ compounds, suggesting the strong monovalent character for some of the Cu in germanite ([40], Figure 7c,d). This situation is analogous to covellite (CuS), which based on the structural formula $Cu^+_2Cu^{2+}(S_2)^{2-}S^{2-}$ contains both Cu(I) and Cu(II) in 2:1 ratio [43]. However, XANES and X-ray photoelectron spectroscopy (XPS) data show little evidence for Cu(II) in covellite [44,45], reflecting the covalent nature of the Cu-S bonds and the mainly monovalent nature of Cu in this mineral.

4. Discussion

According to the XANES data, germanium is present as Ge^{4+} in the natural Ge-sulfides germanite, argyrodite, and renierite as well as in the Tres Marias sphalerite. Iron in the sphalerite is predominantly in divalent form, while Fe in germanite is mixed valence. Copper in germanite has a strong monovalent affinity and XANES does not confirm the presence of Cu(II) in germanite. This study also demonstrates the versatility of synchrotron radiation for the determination of chemical state of trace elements in natural minerals [46,47].

The stable oxidation states of Ge in the solid state are Ge^{2+} and Ge^{4+}. Germanium is an element of low crustal abundance; most Ge is found in small amounts (few ppm) in silicate minerals, due to isomorphous substitution of Ge^{4+} for the Si^{4+} [8]. Germanium forms discrete Ge-minerals in a limited number of ore deposits, notably in the Tsumeb Mine, Namibia [48] and the Apex Mine, Utah, USA [49]. Ge^{4+} is present in all known minerals where the germanium oxidation state can be established unambiguously. Ge^{2+} was proposed by [50] to occur in the Ge-spinel, brunogeierite (Fe_2GeO_4), but a recent re-investigation has shown that ideal, end-member brunogeierite is $(Fe^{2+})_2Ge^{4+}O_4$ [33]. [51] suggested that Ge^{2+} substitutes for Pb^{2+} in anglesite and cerussite from Tsumeb (50 to 500 ppm Ge). Germanium is assumed to be transported in the tetravalent state (e.g., germanic acid, H_4GeO_4(aq), and its dissociation products, and possibly chloride and fluoride complexes) in hydrothermal fluids [52]. However, we note that Ge^{2+} is stable in aqueous solutions even at room temperature (standard potential of 0 V for the $Ge^{4+} + 2e^- = Ge^{2+}$ reaction [53]). Complexing and increase in temperature will increase the relative stability of the Ge^{2+} oxidation state in hydrothermal solutions (e.g., compare with tellurium; [54]). The first unambiguous occurrence of Ge^{2+} in a natural mineral may have been observed by [55]. On the basis of XANES and EXAFS data, [55] suggest the presence of both Ge^{2+} and Ge^{4+} in Ge-bearing MVT-type sphalerite from Tennessee. The material studied by Bonnet et al. [55] is clearly different from that studied here. These differences are emphasized by the additional observation of (argutite-like) Ge^{4+} surrounded by oxygen atoms, and an inverse correlation between Ge and Fe in the Tennessee material.

Combined with the LA-ICP-MS data that show no correlation between Ge and monovalent cations substituting in the Tres Marias sphalerite, our XANES data suggest that the substitution of Ge^{4+} + (vacancy) for (Zn^{2+}, Fe^{2+}) is the main mechanism of Ge incorporation in the studied sphalerite. Why the substitution of $Zn^{2+} \leftrightarrow Ge^{4+}$ + (vacancy) is closely associated with Fe-rich zones is not clear.

[11] used autocorrelation analysis of infrared spectra of Fe-bearing sphalerites to show that there is little strain introduced into the structure associated with Fe substitution of Zn. Thus the concentration of Ge^{4+} in iron-rich domains appears to be a distinct characteristic of the Tres Marias sphalerite-dominated ZnS.

Compositional-textural heterogeneity in the Tres Marias sample is observed at a scale finer than that of the microbeam used for the XANES measurements ($\sim2 \times 2$ μm^2). This intrinsic micron- to nanoscale heterogeneity is expressed both texturally and compositionally in the Tres Marias material. This may be the result of a "two-stage deposition process: early Ge-Fe-rich Zn sulfide precipitated, surrounded by later Ge-Fe-poor Zn sulfide", the interpretation given by [9]. The data here indicate that early Fe-rich sphalerite (~8 wt% Fe; occurring as darkest structures on the SE images on Figures 2 and 3), is the main Ge-carrier (up to ~1000 ppm Ge; [6]), as shown by the correlation between compositional patterns (Figure 4) and textures revealed by FIB cross-sectioning (Figure 2). However, the sample also shows intergrowths between the dominant sphalerite and a lesser wurtzite component that could be part of a primary growth process rather than the product of two-stage crystallization. Considering the evidence for corrosion and pores ± inclusions in the (Ge-Fe)-rich area (Figure 2) it is more logical to assume a polystage formation. Moreover, from the more advanced porosity and cavities, partially filled by secondary minerals, in the (Ge-Fe)-poor areas (Figure 3), fluid-driven replacement can be inferred.

Transformation of sphalerite into wurtzite is promoted by (111)* twinning or lattice-scale defects (Figure 5) and leads to a heterogeneous ZnS sample, in which the dominant component, sphalerite, can host up to ~20% wurtzite (X-ray powder diffraction data). Whereas crystal-structural control of transformation between the two ZnS polytypes is a common feature in many Zn-ores, Ge-enrichment is consistent with formation of Fe-rich sphalerite at Tres Marias. We note the similarity to textures described by [56], in which Fe-rich, trace-element-bearing acicular sphalerite was considered to have formed as wurtzite and subsequently underwent transformation to sphalerite, even if the scenario favoured in this paper indicates a partial transformation of initial sphalerite into zones that contain intergrown wurtzite.

Acknowledgements

We gratefully acknowledge the Diamond Light Source Synchrotron facility for beamline access and excellent collaboration during our visit to UK (experiment sp7563). Part of this research was undertaken on the XAS and X-ray fluorescence microscopy (XFM) beamlines at the Australian Synchrotron, Victoria, Australia. We sincerely thank Bernhardt Saini-Eidukat and Frank Melcher for making the sample available to us. Rongping Wang (Australian National University (ANU), Canberra, Australia) generously provided the $GeSe_2$ glass sample. The authors acknowledge the use of facilities within the Monash Centre for Electron Microscopy and the use of equipment funded by Australian Research Council (ARC) grant RIEFP99. The manuscript benefitted from the comments from three anonymous reviewers.

Author contributions

N.J.C. led the grant application for synchrotron access and co-ordinated this work together with J.B. B.E. conducted the XAS data analysis, C.L.C. performed the nanoscale characterization; D.H., T.W., N.R., B.J. and K.G. assisted with experiments and expertise. A.P. provided samples and contributed to interpretations; G.C. contributed standards and expertise.

Conflicts of Interest

The authors declare no conflict of interest.

References

1. Seredin, V.V. From coal science to metal production and environmental protection: A new story of success Commentary. *Int. J. Coal Geol.* **2012**, *90*, 1–3.
2. Frenzel, M.; Ketris, M.P.; Gutzmer, J. On the geological availability of germanium. *Miner. Depos.* **2014**, *49*, 471–486.
3. Moh, G.H.; Jäger, A. Phasengleichgewichte des Systems Ge–Pb–Zn–S in Relation zu Germanium-Gehalten alpiner Pb–Zn-Lagerstätten. *Verh. Geol. Bundesanst. Wien* **1978**, *1978*, 437–440. (In German)
4. Johan, Z. Indium and germanium in the structure of sphalerite: An example of coupled substitution with copper. *Mineral. Petrol.* **1988**, *39*, 211–229.
5. Belissont, R.; Boiron, M.-C.; Luais, B.; Cathelineau, M. LA-ICP-MS analyses of minor and trace elements and bulk Ge isotopes in zoned Ge-rich sphalerites from the Noailhac–Saint-Salvy deposit (France): Insights into incorporation mechanisms and ore deposition processes. *Geochim. Cosmochim. Acta* **2014**, *126*, 518–540.
6. Cook, N.J.; Ciobanu, C.L.; Pring, A.; Skinner, W.; Shimizu, M.; Danyushevsky, L.; Saini-Eidukat, B.; Melcher, F. Trace and minor elements in sphalerite: A LA-ICPMS study. *Geochim. Cosmochim. Acta* **2009**, *73*, 4761–4791.
7. Lin, Y.; Cook, N.J.; Ciobanu, C.L.; Liu, Y.P.; Zhang, Q.; Liu, T.G.; Gao, W.; Yang, Y.L.; Danyushevskiy, L. Trace and minor elements in sphalerite from base metal deposits in South China: A LA-ICPMS study. *Ore Geol. Rev.* **2011**, *39*, 188–217.
8. Höll, R.; Kling, M.; Schroll, E. Metallogenesis of germanium—A review. *Ore Geol. Rev.* **2007**, *30*, 145–180.
9. Saini-Eidukat, B.; Melcher, F.; Lodziak, J. Zinc-germanium ores of the Tres Marias Mine, Chihuahua, Mexico. *Miner. Depos.* **2009**, *44*, 363–370.
10. Di Benedetto, F.; Andreozzi, G.B.; Bernardini, G.P.; Borgheresi, M.; Caneschi, A.; Cipriani, C.; Gatteschi, D.; Romanelli, M. Short-range order of Fe^{2+} in sphalerite by ^{57}Fe Mössbauer spectroscopy and magnetic susceptibility. *Phys. Chem. Miner.* **2005**, *32*, 339–348.
11. Pring, A.; Tarantino, S.C.; Tenailleau, C.; Etschmann, B.; Carpentep, M.A.; Zhang, M.; Lin, Y.; Withers, R.L. The crystal chemistry of Fe-bearing sphalerites: An infrared spectroscopic study. *Am. Mineral.* **2008**, *93*, 591–597.

12. Cook, N.J.; Ciobanu, C.L.; Meria, D.; Silcock, D.; Wade, B. Arsenopyrite-pyrite association in an orogenic gold ore: tracing mineralization history from textures and trace Elements. *Econ. Geol.* **2013**, *108*, 1273–1283.

13. Wilson, S.A.; Ridley, W.I.; Koenig, A.E. Development of sulfide calibration standards for the laser ablation inductively-coupled plasma mass spectrometry technique. *J. Anal. At. Spectrom.* **2002**, *17*, 406–409.

14. George, L.; Cook, N.J.; Ciobanu, C.L.; Wade, B.P. Trace and minor elements in galena: A reconnaissance LA-ICP-MS study. *Am. Mineral.* **2015**, *100*, 548–569.

15. Ciobanu, C.; Cook, N.J.; Utsunomiya, S.; Pring, A.; Green, L. Focussed ion beam-transmission electron microscopy applications in ore mineralogy: Bridging micro- and nanoscale observations. *Ore Geol. Rev.* **2011**, *42*, 6–31.

16. Paterson, D.; de Jonge, M.D.; Howard, D.L.; Lewis, W.; McKinlay, J.; Starritt, A.; Kusel, M.; Ryan, C.G.; Kirkham, R.; Moorhead, G., *et al.* The X-ray Fluorescence Microscopy Beamline at the Australian Synchrotron. In *10th International Conference on X-Ray Microscopy*; McNulty, I., Eyberger, C., Lai, B., Eds.; American Institute of Physics: College Park, MD, USA, 2011; pp. 219–222.

17. Cook, N.J.; Ciobanu, C.L.; Brugger, J.; Etschmann, B.; Howard, D.L.; de Jonge, M.D.; Ryan, C.; Paterson, D. Determination of the oxidation state of Cu in substituted Cu-In-Fe-bearing sphalerite via mu-XANES spectroscopy. *Am. Mineral.* **2012**, *97*, 476–479.

18. Fleet, M.E. Structural Transformations in Natural Zns. *Am. Mineral.* **1977**, *62*, 540–546.

19. Pósfai, M.; Buseck, P.R. Modular structures in sulphides: sphalerite/wurtzite-, pyrite/marcasite-, and pyrrhotite-type minerals. *EMU Notes in Mineral.* **1997**, *1*, 193–235.

20. Smith, G.S.; Isaacs, P.B. Crystal structure of quartz-like GeO_2. *Acta Crystallogr.* **1964**, *17*, 842–846.

21. Eulenberger, G. Die Kristallstruktur der Tieftemperaturmodifikation von Ag_8GeS_6, synthetischer Argyrodit. *Chem. Mon.* **1977**, *108*, 901–913. (In German)

22. Hilton, A.R.; Jones, C.E.; Brau, M. Non-Oxide IVA-VA-VIA Chalcogenide Glasses. I. Glass-Forming Regions and Variations in Physical Properties. *Phys. Chem. Glasses* **1966**, *7*, 105–112.

23. Zakery, A.; Elliott, S.R. Optical properties and applications of chalcogenide glasses: A review. *J. Non-Cryst. Solids* **2003**, *330*, 1–12.

24. Hilton, A.R.; Kemp, S. *Chalcogenide Glasses for Infrared Optics*; McGraw-Hill Companies Inc.: New York, NY, USA, 2010.

25. Wang, R.P.; Smith, A.; Luther-Davies, B.; Kokkonen, H.; Jackson, I. Observation of two elastic thresholds in $Ge_xAs_ySe_{1-x-y}$ glasses. *J. Appl. Phys.* **2009**, *105*, doi:10.1063/1.3079806.

26. Wei, W.H.; Wang, R.P.; Shen, X.; Fang, L.; Luther-Davies, B. Correlation between structural and physical properties in Ge-Sb-Se glasses. *J. Phys. Chem. C* **2013**, *117*, 16571–16576.

27. Bernstein, L.R.; Reichel, D.G.; Merlino, S. Renierite crystal-structure refined from Rietveld analysis of powder neutron-diffraction data. *Am. Mineral.* **1989**, *74*, 1177–1181.

28. Kogut, Y.; Fedorchuk, A.; Zhbankov, O.; Romanyuk, Y.; Kityk, I.; Piskach, L.; Parasyuk, O. Isothermal section of the Ag_2S-PbS-GeS_2 system at 300 K and the crystal structure of Ag_2PbGeS_4. *J. Alloy. Compd.* **2011**, *509*, 4264–4267.

29. Paar, W.H.; Roberts, A.C.; Berlepsch, P.; Armbruster, T.; Topa, D.; Zagler, G. Putzite, $(Cu_{4.7}Ag_{3.3})_{\Sigma 8}GeS_6$, a new mineral species from capillitas, Catamarca, Argentina: Description and crystal structure. *Can. Mineral.* **2004**, *42*, 1757–1769.

30. Ishii, M.; Onoda, M.; Shibata, K. Structure and vibrational spectra of argyrodite family compounds Cu_8SiX_6 (X = S, Se) and Cu_8GeS_6. *Solid State Ion.* **1999**, *121*, 11–18.

31. Tettenhorst, R.T.; Corbato, C.E. Crystal-structure of germanite, $Cu_{26}Fe_4Fe_4S_{32}$, determined by powder X-ray-diffraction. *Am. Mineral.* **1984**, *69*, 943–947.

32. Dittmar, G.; Schafer, H. Crystal-structure of low-temperature GeS_2. *Acta Cryst. B* **1976**, *32*, 1188–1192.

33. Cempirek, J.; Groat, L.A. Note on the formula of brunogeierite and the first bond-valence parameters for Ge^{2+}. *J. Geosci.* **2013**, *58*, 71–74.

34. Bissert, G.; Hesse, K.F. Refinement of structure of germanium(II) sulfide, GeS. *Acta Cryst. B* **1978**, *34*, 1322–1323.

35. Joly, Y.; Bunău, O.; Lorenzo, J.E.; Galera, R.M.; Grenier, S.; Thompson, B. Self-consistency, spin-orbit and other advances in the FDMNES code to simulate XANES and RXD experiments. *J. Phys. Conf. Ser.* **2009**, *190*, doi:10.1088/1742-6596/190/1/012007.

36. James-Smith, J.; Cauzid, J.; Testemale, D.; Liu, W.; Hazemann, J.; Proux, O.; Etschmann, B.; Philippot, P.; Banks, D.; Williams, P.; *et al.* Arsenic speciation in fluid inclusions using micro-beam X-ray absorption spectroscopy. *Am. Mineral.* **2010**, *95*, 921–932.

37. Grundler, P.; Brugger, J.; Meisser, N.; Ansermet, S.; Borg, S.; Etschmann, B.; Testemale, D.; Bolin, T. Xocolatlite, $Ca_2Mn_2^{4+}Te_2O_{12} \cdot H_2O$, a new tellurate related to kuranakhite: Description and measurement of Te oxidation state by XANES spectroscopy. *Am. Mineral.* **2008**, *93*, 1911–1920.

38. Tromp, M.; Moulin, J.; Reid, G.; Evans, J. Cr K-edge XANES spectroscopy: Ligand and oxidation state dependence—What is oxidation state? *X-Ray Absorpt. Fine Struct.* **2007**, *882*, 699–701.

39. Bernstein, L.R. Germanium Geochemistry and Mineralogy. *Geochim. Cosmochim. Acta* **1985**, *49*, 2409–2422.

40. Kau, L.-S.; Spira-Solomon, D.J.; Penner-Hahn, J.E.; Hodgson, K.O.; Solomon, E.I. X-ray absorption edge determination of the oxidation state and coordination number of copper: Application to the type 3 site in Rhus vernicifera laccase and its reaction with oxygen. *J. Am. Chem. Soc.* **1987**, *109*, 6433–6442.

41. Berry, A.J.; Yaxley, G.M.; Woodland, A.B.; Foran, G.J. A XANES calibration for determining the oxidation state of iron in mantle garnet. *Chem. Geol.* **2010**, *278*, 31–37.

42. Zalewski, W.; Bacewicz, R.; Antonowicz, J.; Pietnoczka, A.; Evstigneeva, T.L.; Schorr, S. XAFS study of kesterite, kuramite and stannite type alloys. *J. Alloy. Compd.* **2010**, *492*, 35–38.

43. Evans, H.T.; Konnert, J.A. Crystal-structure refinement of covellite. *Am. Mineral.* **1976**, *61*, 996–1000.

44. Pattrick, R.A.D.; Mosselmans, J.F.W.; Charnock, J.M.; England, K.E.R.; Helz, G.R.; Garner, C.D.; Vaughan, D.J. The structure of amorphous copper sulfide precipitates: An X-ray absorption study. *Geochim. Cosmochim. Acta* **1997**, *61*, 2023–2036.

45. Di Benedetto, F.; Borgheresi, M.; Caneschi, A.; Chastanet, G.; Cipriani, C.; Gatteschi, D.; Pratesi, G.; Romanelli, M.; Sessoli, R. First evidence of natural superconductivity: Covellite. *Eur. J. Mineral.* **2006**, *18*, 283–287.

46. Brugger, J.; Pring, A.; Reith, F.; Ryan, C.; Etschmann, B.; Liu, W.; O'Neill, B.; Ngothai, Y. Probing ore deposits formation: New insights and challenges from synchrotron and neutron studies. *Radiat. Phys. Chem.* **2010**, *79*, 151–161.

47. Brugger, J.; Etschmann, B.; Pownceby, M.; Liu, W.; Grundler, P.; Brewe, D. Oxidation state of europium in scheelite: Tracking fluid-rock interaction in gold deposits. *Chem. Geol.* **2008**, *257*, 26–33.

48. Melcher, F. The Otavi mountain land in Namibia—Tsumeb, germanium and snowball Earth. *Mitt. Österr. Mineral. Ges.* **2003**, *148*, 413–435.

49. Dutrizac, J.E.; Jambor, J.L.; Chen, T.T. Host minerals for the gallium-germanium ores of the Apex Mine, Utah. *Econ. Geol.* **1986**, *81*, 946–950.

50. Welch, M.D.; Cooper, M.A.; Hawthorne, F.C. The crystal structure of brunogeierite, Fe_2GeO_4 spinel. *Mineral. Mag.* **2001**, *65*, 441–444.

51. Frondel, C.; Ito, J. Geochemistry of germanium in the oxidized zone of the Tsumeb mine, South-West Africa. *Am. Mineral.* **1957**, *42*, 743–753.

52. Wood, S.A.; Samson, I.M. The aqueous geochemistry of gallium, germanium, indium and scandium. *Ore Geol. Rev.* **2006**, *28*, 57–102.

53. Vanýsek, P. Electrochemical Series. In *Handbook of Chemistry and Physics*, 92nd ed.; Chemical Rubber Company: Boca Raton, FL, USA, 2011.

54. Grundler, P.V.; Brugger, J.; Etschmann, B.E.; Helm, L.; Liu, W.H.; Spry, P.G.; Tian, Y.; Testemale, D.; Pring, A. Speciation of aqueous tellurium(IV) in hydrothermal solutions and vapors, and the role of oxidized tellurium species in Te transport and gold deposition. *Geochim. Cosmochim. Acta* **2013**, *120*, 298–325.

55. Bonnet, J.; Mösser-Ruck, R.; Cauzid, J.; Bailly, L.; André, A. Crystallographic control of trace element (Cu-Ga-Ge-Fe-Cd) distribution in MVT sphalerites, Tennessee, USA. In Proceedings of the 21st IMA Meeting, Johannesburg, South Africa, 1–5 September 2014.

56. Beaudoin, G. Acicular sphalerite enriched in Ag, Sb, and Cu embedded within color-banded sphalerite from the Kokanee Range, British Columbia, Canada. *Can. Mineral.* **2000**, *38*, 1387–1398.

Evolution of Acid Mine Drainage Formation in Sulphidic Mine Tailings

Bernhard Dold

SUMIRCO (Sustainable Mining Research & Consult EIRL), Casilla 28, San Pedro de la Paz 4130000, Chile; E-Mail: bernhard.dold@gmail.com

Abstract: Sulphidic mine tailings are among the largest mining wastes on Earth and are prone to produce acid mine drainage (AMD). The formation of AMD is a sequence of complex biogeochemical and mineral dissolution processes. It can be classified in three main steps occurring from the operational phase of a tailings impoundment until the final appearance of AMD after operations ceased: (1) During the operational phase of a tailings impoundment the pH-Eh regime is normally alkaline to neutral and reducing (water-saturated). Associated environmental problems include the presence of high sulphate concentrations due to dissolution of gypsum-anhydrite, and/or effluents enriched in elements such as Mo and As, which desorbed from primary ferric hydroxides during the alkaline flotation process. (2) Once mining-related operations of the tailings impoundment has ceased, sulphide oxidation starts, resulting in the formation of an acidic oxidation zone and a ferrous iron-rich plume below the oxidation front, that re-oxidises once it surfaces, producing the first visible sign of AMD, *i.e.*, the precipitation of ferrihydrite and concomitant acidification. (3) Consumption of the (reactive) neutralization potential of the gangue minerals and subsequent outflow of acidic, heavy metal-rich leachates from the tailings is the final step in the evolution of an AMD system. The formation of multi-colour efflorescent salts can be a visible sign of this stage.

Keywords: mining; metal; tailings; oxidation; acid mine drainage; waste management; pollution; solubility; reductive dissolution; sulphide; ore deposit; sustainability

1. Introduction

Mine tailings are among the largest mining wastes on Earth and can reach surface areas of up to 52 km^2 [1] and be several hundred meters high. As this waste type results mainly from the flotation process of sulphide mineral ores they are very likely to produce acid mine drainage (AMD), the main environmental problem of contemporary mining activity. The on-land deposition has many environmental, socio-economic, and geotechnical stability problems, which can make them a limiting factor to production in the mining industry. Tailings require large land areas and they have a great potential to produce ground and surface water contamination due to mineral dissolution in the operative and post-operative stage. Leaching from tailings results in an increase of oxyanions in solution (e.g., sulphate, arsenate, molybdate) during operation, and AMD formation after operation. Additionally it also represents a threat downstream in case of catastrophic dam failures, as has happened in the past [2]. The public becomes concerned and the mining operations have to compete with alternative land uses like agriculture, fisheries, or tourism. As a result, the mining industry is re-evaluating the option of submarine tailings disposal (STD), a heavily disputed practice used in some locations over the last few decades primarily resulting in negative impacts on the environment (reviewed in an other paper of this special issue on submarine tailings disposal (STD) [3]).

This review summarizes the work of 20 years of research on AMD in order to understand the evolution and the controlling parameters of AMD formation in this type of mine waste.

The review starts with an introduction into the biogeochemical processes occurring during sulphide oxidation and then focuses on the very beginning of the process in the transport channels of the tailings onto the tailings impoundments and the processes occurring in active operations. Then follows the evolution of AMD formation after the operation of the tailings impoundment has ceased, in relation to time, climate, deposition technique and flotation and finally ore deposit type will be analyzed. The biogeochemical processes involved are highlighted in multi-extreme environments. At the end of this review, problems of management, remediation, and prevention options are discussed in order to increase the sustainability of mining operations.

For this purpose, we use mainly studies from porphyry copper ore deposits as examples, but this knowledge on the mineralogy and the resulting geochemistry can be extended with due caution to other sulphide ore deposits. Terminology and technical descriptions in this article have been kept simple so as to provide a review that can be used by a wide audience. For more details please refer to the specific research articles.

2. Sulphide Oxidation

For the proper understanding of the formation of acid mine drainage, the biogeochemical interactions and the sequences in these processes have to be understood. This chapter is taken from Dold [4] for the convenience of the reader and more details on this issue can be found in this open access book chapter free of charge.

The problem of sulphide oxidation and the associated generation of acid mine drainage (AMD), or more generally acid rock drainage (ARD), as well as the dissolution and precipitation processes of metals and minerals, has been a major focus of investigation over the last 50 years [5–9]. The primary mineralogical composition has a strong influence on the oxidation processes. This has been well illustrated [10–12], showing that reaction rates display significant differences depending on which sulphides are being oxidized by Fe(III) and the potential Fe(III) hydroxide coating. Kinetic-type weathering experiments indicate the importance of trace element composition in the stability of individual sulphides. Where different sulphides are in contact with each other, electrochemical processes are likely to occur and influence the reactivity of sulphides [13].

Most mines are surrounded by piles, dumps, or impoundments containing pulverized material or waste from the benefaction process (Figure 1A), which are known as tailings, waste rock dumps, stockpiles, or leach dumps or pads. Waste rock dumps generally contain material with low ore grade, which is mined but not milled (Run of Mine; ROM). These materials can still contain large concentrations of sulphide minerals, which may undergo oxidation, producing a major source of metal and acid contamination [14]. In the following section the focus is on the acid producing sulphide minerals, mainly using pyrite as an example.

The most common sulphide mineral is pyrite (FeS_2). Oxidation of pyrite takes place in several steps including the formation of the meta-stable secondary products ferrihydrite ($5Fe_2O_3 \cdot 9H_2O$), schwertmannite (between $Fe_8O_8(OH)_6SO_4$ and $Fe_{16}O_{16}(OH)_{10}(SO_4)_3$), and goethite ($FeO(OH)$), as well the more stable secondary jarosite ($KFe_3(SO_4)_2(OH)_6$), and hematite (Fe_2O_3) depending on the geochemical conditions [6,9,11,15–18]. Oxidation of pyrite may be considered to take place in three major steps: (1) oxidation of sulphur (Equation (1)); (2) oxidation of ferrous iron (Equation (2)); and (3) hydrolysis and precipitation of ferric complexes and minerals (Equation (4)). The kinetics of each reaction is different and depends on the conditions prevalent in the tailings:

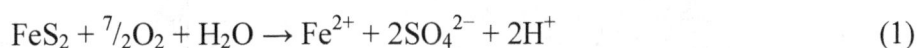

$$FeS_2 + \frac{7}{2}O_2 + H_2O \rightarrow Fe^{2+} + 2SO_4^{2-} + 2H^+ \tag{1}$$

$$Fe^{2+} + \frac{1}{4}O_2 + H^+ \rightarrow Fe^{3+} + \frac{1}{2}H_2O \tag{2}$$

Reaction rates are strongly increased by microbial activity (e.g., *Acidithiobacillus* spp. or *Leptospirillum* spp.):

$$FeS_2 + 14Fe^{3+} + 8H_2O \rightarrow 15Fe^{2+} + 2SO_4^{2-} + 16H^+ \tag{3}$$

Equation (1) describes the initial step of pyrite oxidation in the presence of atmospheric oxygen. The oxidation of ferrous iron to ferric iron, is strongly accelerated at low pH conditions by microbiological activity (Equation (2)), producing ferric iron as the primary oxidant of pyrite (Equation (3)) [7,19,20]. Under abiotic conditions the rate of oxidation of pyrite by ferric iron is controlled by the rate of oxidation of ferrous iron, which decreases rapidly with decreasing pH. Below about pH 3 the oxidation of pyrite by ferric iron is about ten to a hundred times faster than by oxygen [21].

It has been known for more than 50 years that microorganisms like *Acidithiobacillus ferrooxidans* or *Leptospirillum ferrooxidans* obtain energy by oxidizing Fe^{2+} to Fe^{3+} from sulphides by catalyzing this reaction [22] and this may increase the rate of Reaction (2) up to the factor of about 100 over abiotic oxidation [23]. More recent results show that a complex microbial community is responsible for sulphide oxidation [19,24–27]. Nordstrom and Southam [28] stated that the initiating step of pyrite oxidation does not require an elaborated sequence of different geochemical reactions that dominate at different pH ranges. *Acidithiobacillus* spp. forms nano-environments, which grow on sulphide mineral surfaces [29]. These nano-environments can develop thin layers of acidic water that do not affect the bulk pH of the water chemistry. With progressive oxidation, the nano-environments may change to microenvironments [30]. Evidence of acidic microenvironments in the presence of near neutral pH for the bulk water can be inferred from the presence of jarosite (this mineral forms at pH around 2) in certain soil horizons where the current water pH is neutral [31]. Barker *et al.* [32] observed microbial colonization of biotite and measured pH in microenvironments in the surroundings of living microcolonies. The solution pH decreased from near neutral at the mineral surface to pH 3–4 around micro-colonies living within confined spaces at interior colonized cleavage planes.

Figure 1. (**A**) Open pit mine surrounded by waste dumps and stock-piles. (**B**) Semi-Autogenous Grinding (SAG) mill. (**C**) Froth flotation of chalcopyrite concentrate. (**D**) Deposition point of a tailings impoundment. (**E**) Areal photograph of a valley dam tailings impoundment. Note the slight saturation of the tailings and the seepage in the dam (dark humid spots in the dam). And (**F**) areal view of a big tailings impoundment with near complete water saturation.

When mine water, rich in ferrous and ferric iron, reaches the surface it will fully oxidize and hydrolyze, resulting in the precipitation of ferrihydrite (Fh), schwertmannite (Sh), goethite (Gt), or jarosite (Jt) depending on the pH-Eh conditions, and availability of key elements such as potassium and sulphate (Figure 2). These secondary minerals like jarosite, schwertmannite and ferrihydrite are meta-stable and can transform into goethite [17].

The hydrolysis and precipitation of iron hydroxides (and to a lesser degree, jarosite) will produce most of the acid in this process. If the pH is less than about 2, ferric hydrolysis products like $Fe(OH)_3$ are not stable and Fe^{3+} remains in solution:

$$Fe^{3+} + 3H_2O \rightarrow Fe(OH)_{3(s)} + 3H^+ \tag{4}$$

Note that the net reaction of complete oxidation of pyrite, hydrolysis of Fe^{3+} and precipitation of iron hydroxide (sum of Reactions (1), (2) and (4) produces four moles of H^+ per mole of pyrite (in case of $Fe(OH)_3$ formation, see Reaction (5), i.e., pyrite oxidation is the most efficient producer of acid among the common sulphide minerals (net Reaction (5)). Nevertheless, it is important to be aware that the hydrolysis of $Fe(OH)_3$ is the main acid producer ($^3/_4$ of the moles of H^+ per mol pyrite).

$$FeS_2 + {}^{15}/_4O_2 + {}^7/_2H_2O \rightarrow Fe(OH)_3 + 2SO_4^{2-} + 4H^+ \tag{5}$$

The process of pyrite oxidation relates to all sulphide minerals once exposed to oxidizing conditions (e.g., chalcopyrite, bornite, molybdenite, arsenopyrite, enargite, galena, and sphalerite among others). In this process different amounts of protons are released [4] and the metals and other harmful elements or compounds are released to the environment.

3. From the Flotation Process to the Active Tailings Impoundment

The goal of the flotation process is to separate the economically valuable target minerals from the gangue minerals, which have no economic value at the time of exploitation [33]. In order to be able to do this, the rocks extracted from the mine (underground or open pit) as coarse ROM granulometry (including blocks of 1 m diameter down to rock powder), have to be broken, ground and milled (Figure 1B) to a very fine grain size, in order to be able to separate on the addition of chemical reagents, selectively the target minerals (i.e., to make it hydrophobic, which then enables it to attach to introduced air bubbles and so float towards the surface of the flotation cell (Figure 1C), were it can be harvested) [34,35]. Non-economic sulphide minerals, like pyrite can be suppressed from flotation as for example by pH adjustment (alkaline circuit), and end up in the waste materials, which are called tailings (Figure 1D). As the flotation process has a recovery of 80%–90%, between 10% and 20% of the target mineral ends up in the tailings together with the non-economic sulphides like pyrite or other accessory sulphides, which can contain other environmentally harmful elements. These tailings are then sent in suspension via tubes, channels or directly in riverbeds towards their final disposal sites (Figure 1D), i.e., a river, lake(s), or the sea, but mainly in mines today on-land in constructed, tailings impoundments or dams (Figure 1E,F). Depending on the geochemical conditions of this final disposal site, the mineral assemblage in the tailings can undergo geochemical oxidative processes, which can lead to the release of metals, toxic compounds, and acid. The geochemical and mineralogical effects of disposal of mine tailings in reducing environments is reviewed in another paper of this special issue concerning submarine tailings disposal (STD) [3]. The present review focuses on the processes resulting from the exposition of sulphidic mine tailings to oxidation in on-land tailings impoundments.

The whole flotation process is performed using a mineral suspension with a solids-water ratio of about 40%:60%. Thus, the flotation is a highly water consuming process, and therefore water is the limiting factor for mine development in many arid to semi-arid regions (e.g., Northern Chile and Southern Peru). Some mining operations have opted to use marine water for the flotation process [36,37]. Water recycling from the decantation pond of the tailings impoundment is also a common practice to recovery industrial water. New techniques like paste tailings and dry staking recover water before final deposition and increase geotechnical safety of the tailings deposit [38,39]. However, it should be noted that sulphide oxidation is enhanced by these new techniques, as the tailings are never completely water saturated, but humid, and oxygen can more easily reach the sulphides, compared to the traditional water-saturated tailings impoundments.

In the flotation process, tailings come in to contact with water and oxygen for the first time, leading to Reaction (1). However, at this stage the oxygen supply is limited, as only dissolved oxygen is available for the sulphide oxidation in the flotation process. As most flotation processes are maintained artificially at alkaline pH conditions in order to suppress the flotation of pyrite, sulphide oxidation during the flotation does not result in extensive acid generation. However, isotopic studies (δ^{34}S, δ^{18}O) of dissolved sulphate suggest along a 87 km long tailings channel that sulphide oxidation starts in the flotation process and during transport towards the final disposal site [40]. Additionally, if the ore has oxyanions associated with iron oxide minerals, for example when ore is slightly pre-oxidized by supergene processes in the upper part of the ore deposit, then, due to the alkaline flotation circuit, As and Mo can be desorbed during flotation and possible make it necessary to implement an abatement plant for these elements, as is the case for Mo in the El Teniente mine, Chile.

When the tailings reach the active tailings impoundment, they should then in a strict sense be maintained water saturated in order to minimize oxidation of the sulphide minerals (water contains a maximum of approximately 10 mg/L dissolved oxygen). This is not always the case or possible, for example due to high evaporation rates in dry climates, so that often parts of the tailings are exposed during summer time to a thin unsaturated zone to oxidation even in active tailings impoundments (Figure 1E). At this stage, the 21% of atmospheric oxygen will start to oxidize the sulphide mineral assemblage present in the tailings. This goes hand-in-hand with the increase of pore water concentration in metals and oxyanions like (Na, K, Cl, SO$_4$, Mg, Cu, Mo) towards the surface due to capillary transport, and the formation of efflorescent salts on the surface, like halite, gypsum, and Na-K-Ca-Mg sulphates like mirabilite $Na_2SO_4 \cdot 10H_2O$ and syngenite $K_2Ca(SO_4)_2 \cdot 4H_2O$ [1,40]. Due to neutral to alkaline pH at this stage, only major cations together with sulphate and chloride are mobile and the resulting efflorescent salts are mainly white in colour.

Another commonly observed geochemical process occurring in active tailings of porphyry copper deposits is a strong increase in sulphate concentrations, which typically range between 1500 and 2000 mg/L, with an annual trend to increase towards the end of summer (Figure 2) and sometimes a general increase with time can also be observed. The sulphate concentrations are controlled by the solubility of gypsum [40,41], often present in an ore deposit (gypsum or anhydrite), and the increase by the release of sulphate due to weathering processes associated with sulphide oxidation. Neutralization reactions, e.g., silicate weathering, liberates major cations into solution, which then form sulphate complexes, so that higher concentrations of sulphate can stay in solution, than can be explained by the solubility of gypsum alone.

Figure 2. Example of the evolution of dissolved sulphate concentrations (in mg/L) in the decantation pond of an active tailings impoundment during a five year period. A clear seasonal trend is observed, peaking end of summer due to evaporation effects.

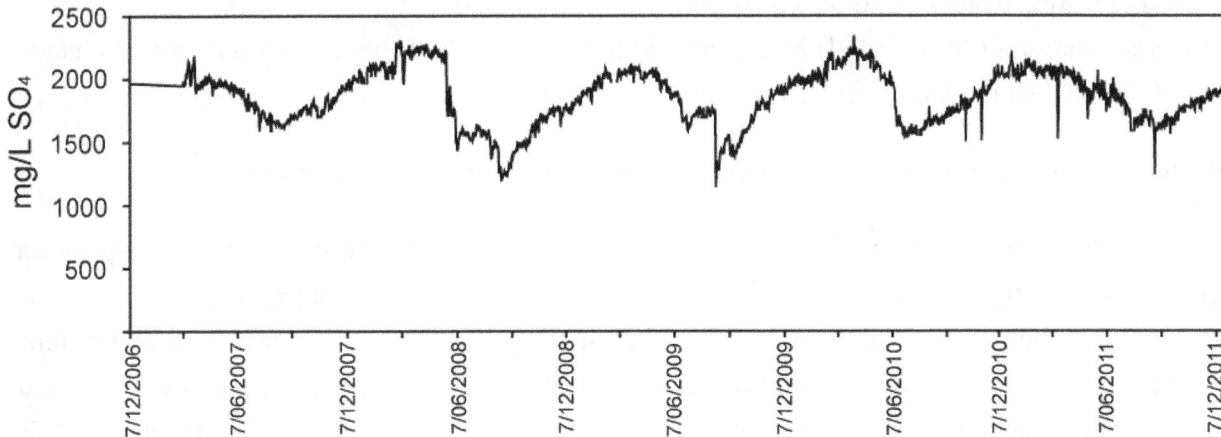

In some tailings impoundments the formation of AMD can be visualized during the operational phase in the dam area [42,43]. This is mainly the case when the dam is made of the coarser fraction of the tailings (e.g., hydro-cyclone separation). This results in a higher content of sulphide minerals in the dam material, which has also a coarser grain size (sandy material). Additionally, the dam must be maintained in an unsaturated condition for stability reasons, so that this area is an excellent environment for sulphide oxidation, which is visible by the precipitation of schwertmannite from the effluents at the foot of the dam [42,43]. The presence of schwertmannite directly at the outcrop of the tailings dam, suggests that acidic (pH 2–4) and ferric iron rich solutions are leaching from the tailings. If a ferrous iron rich neutral plume flows out from the dam, then iron oxidation will occur followed by hydrolysis and subsequent ferrihydrite precipitation [43]. If the ferrous iron rich plume is acidic, then temperature, pH, and microbiological activity will determine how fast the ferrous iron will be oxidized in the drainage stream [44,45] in order to be able to subsequently hydrolyze and precipitate as lepidocrocite, schwertmannite, jarosite or ferrihydrite, depending on the final geochemical conditions.

In general, it can be pointed out, that if an active tailings impoundment shows signs of acidification in the decantation pond during operation or even of AMD formation, then severe management problems can be assumed.

Summarizing, active tailings impoundments might have the following environmental problems:

1. Increased sulphate concentrations (between 1500 and 2000 mg/L), if gypsum and/or anhydrite are present in the ore mineralogy (e.g., porphyry coppers). The sulphate concentrations are controlled by the gypsum equilibrium. The sulphate concentrations can additionally increase with time in the tailings impoundment, depending on increasing input of major cations from weathering processes.

2. If oxyanions (e.g., arsenate, molybdate) are associated with Fe(III) hydroxides from the primary ore mineralogy, they will potentially be released in the alkaline flotation process.

3. During the flotation process and tailings transport, sulphide oxidation can begin, but will not be able to strongly influence the geochemical regime (*i.e.*, the pH will not drop dramatically). In the active tailings impoundment, when a thin, unsaturated zone develops in the dry season, then sulphide oxidation can lower pH conditions and increase the metal release in the uppermost part of the tailings.

4. In situations where tailings dams are constructed by coarse tailings material, sulphide oxidation might lead to the release of AMD from the unsaturated dam area. This might be visible by the precipitation of schwertmannite and/or ferrihydrite [42,43].

5. The precipitation of these Fe(III) hydroxides in the pore space of the tailings dam might change the permeability and so produce stability problems for the tailings dam.

4. Evolution of Post-Deposition Geochemical Processes in Tailings Impoundments

In order to study the evolution of sulphide oxidation in a natural environment after the operation has ceased, the Talabre tailings impoundment of the Chuquicamata porphyry copper mine was investigated [1]. Although the Talabre tailings impoundment is an active impoundment, its dimensions (52 km^2 surface area) and deposition technique allowed a study of tailings exposure at a well defined time frame under the hyper-arid conditions of the Atacama Desert. As the deposition point is periodically changed on the tailings surface of the impoundment and the tailings are disposed of into different basins, there was an exact register available of how long the tailings were exposed to the atmosphere, *i.e.*, weathering. This gave the possibility to select the samples sites from fresh tailings (actual discharge point at time of sampling) up to five years of exposure and track the mineralogical and geochemical changes over time. The mineralogy of the tailings is typical of porphyry copper systems, with pyrite as the major sulphide (1.75 wt %), followed by chalcopyrite and bornite. Minor sulphide fractions found in polished sections were enargite, covellite, chalcocite and sphalerite. There were no carbonates present in the mineral assemblage and the gangue mineralogy was dominated by quartz, K-feldspar, plagioclase, biotite, chlorite, muscovite and gypsum. Primary anhydrite was not found due to hydration to gypsum during flotation. Apatite, rutile, magnetite, hematite, and goethite occurred in trace amounts [1].

The key parameters, pH and Eh, evolved from alkaline (fresh tailings pH 9.1) towards acidic and from reducing to oxidizing conditions. After three years of oxidation the pH was still in the circumneutral range (pH 6.4–7.5), while after four years a drop to acidic conditions was observed (pH 4.7) at the surface (0–4 cm), leading to a pH of 3.9 after five years with the development of a well defined 29 cm thick oxidation zone (Figure 3A).

Associated with this geochemical change, the main element groups in this system showed their characteristic behaviour and distribution. The major cations and anions showed an increasing trend of enrichment towards the tailings surface, due to capillary transport in the hyper arid climate [1,46–48], with the fast precipitation of halite, gypsum, and Na-K-Mg-Ca sulphates and chlorides at the surface (mainly white efflorescent salts). Heavy metal cations like Cu, Zn, and Ni were not mobile in the neutral to alkaline pH conditions in the first years due to their sorption behaviour to iron oxides. However, after five years of oxidation, the drop of the pH in the oxidation zone resulted in increasingly high concentrations of Cu (up to 170 mg/L) and Zn (150 mg/L) in the pore water near the surface of the tailings. This was visible by the precipitation of greenish eriochalcite ($CuCl_2 \cdot 2H_2O$) on the tailings surface, as observed in other chloride-rich environments [47,48]. In contrast, arsenic and molybdenate, which are stable as oxyanions in solution, occurred in high concentrations in the pore water due to the alkaline conditions at the beginning of weathering. The origin of these elements is mainly due to high natural background concentrations of As in the area [49], desorption of oxyanions associated with Fe(III) hydroxides in the ore mineralogy, and increasing concentrations in the recycled industrial water due to evaporation. With decreasing pH by sulphide oxidation and hydrolysis of Fe(III) hydroxides in the oxidation zone, arsenate and molybdenate decrease their concentrations in the pore water of the oxidation zone to below detection limits due to the well know adsorption to the neo-formed sorbents (Fe(III) hydroxides). This is confirmed by sequential extraction data, showing a strong increase of As (175 mg/kg) and Mo (155 mg/kg) associated with the Fe(III) hydroxide fraction in the upper oxidation zone after five years of oxidation. Stable isotope data also clearly demonstrated that sulphate had its origin at the beginning from gypsum dissolution, while in the acid oxidation zone a clear change towards the supply of sulphate by sulphide oxidation is observed [1].

Figure 3. (**A**) Oxidation zone in the Talabre tailings impoundment after five years of oxidation (pH 3.9). Clearly visible the precipitation of Fe(III) hydroxides and the oxidation front [1]. (**B**) Precipitation of ferrihydrite in an active tailings impoundment due to the exposure of Fe(II)-rich waters to the atmosphere (Ocroyoc, Cerro de Pasco, Peru) [14]. And (**C**) outcrop of AMD (pH 3.15) at the foot of an active tailings dam with the precipitation of schwertmannite (Ojancos, Hochschild, Chile) [42].

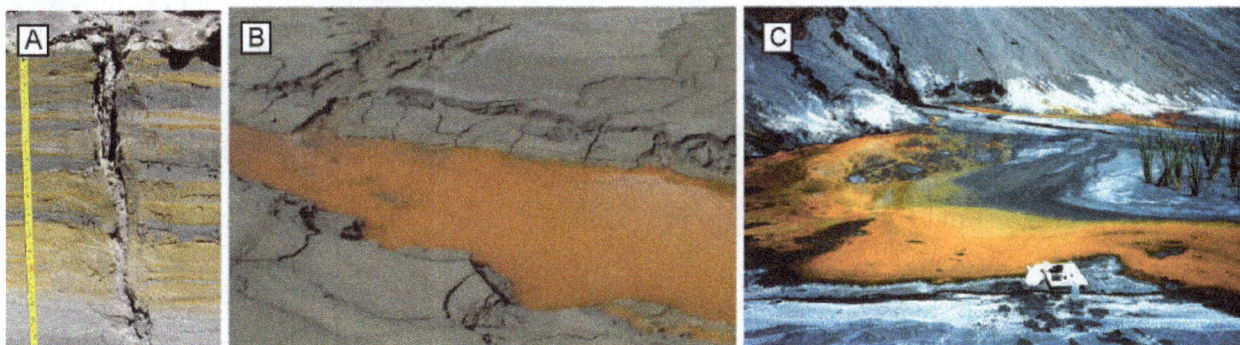

These findings explain why standard kinetic cell tests for AMD prediction (ASTM D5744-96) [50] do not correctly predict the behaviour of porphyry copper material [51,52]. As seen in the case of Talabre, the material needs at least 3–4 years in order to reach acidic pH conditions, and this without any buffering from carbonates. Therefore, the time frame proposed in the standard method of 25 cycles (half year or up to one year depending on the length of each cycle), is far too short in order to reach, *i.e.*, predict, acidic conditions in the porphyry copper system. While there is some improvement, *i.e.*, increased oxidation kinetics with new modified cell tests [53,54], they still have to be run for at least 2–3 years, until acid conditions are reached (in case the acid base accounting indicates an excess of acid potential; the usual case for porphyry copper deposits [46]). This increases the costs and time scale for mine waste characterization, which is not very attractive for the mining industry.

In the study of the Talabre tailings impoundment, another important process for tailings management could be observed. As the tailings deposition point returns periodically to the same place of deposition, where the tailings were exposed to oxidation over several years with the subsequent formation of the above described oxidation zone and formation of efflorescent salts on the surface, this re-deposition will have the following geochemical impact: As explained before, after 4–5 years a well defined acid oxidation zone has developed with the formation of secondary Fe(III) hydroxides (Figure 3A), which have the role of the sorbent for arsenic and molybdenum in these geochemical condition. With the new deposition of fresh alkaline tailings in the same place were the acid oxidation zone formed in an unsaturated zone of the tailings stratigraphy, the system is changed to saturated, alkaline reducing conditions. This will first dissolve all efflorescent salts and liberate the associated elements into the aqueous phase, but also it will initiate the reductive dissolution of Fe(III) hydroxides from the oxidation zone, which will liberate the associated As (up to 23 mg/L) and Mo (up to 16 mg/L) to the groundwater of the tailings impoundment [1].

5. Biogeochemical Iron Cycling at the Oxidation Front: The First Step in the Formation of Acid Mine Drainage (AMD)

Until now we have observed how the system evolves over time at the surface and its element-release sequence. In this section we will enter in more detail into the biogeochemical interactions occurring at the oxidation front and in the vertical stratigraphy, in oxidation zones that are well developed.

This is the case (for example), after 16 years of oxidation in the high mountain climate Piuquenes tailings impoundment, Chile [46,55–58]. Its oxidation zone reached pH 2.3–3 and nearly all sulphide minerals were oxidized (Eh = 750 mV), only some relics of pyrite and chalcopyrite remained (Figure 4). The secondary mineral assemblage was controlled by schwertmannite, jarosite, gypsum, and a vermiculite-type mixed layer mineral resulting from the alteration of biotite in the oxidation zone [46]. Below the oxidation front, a change from acidic-oxidizing conditions towards more reducing (500 mV, which is controlled by the Fe^{3+}/Fe^{2+} redox pair) and an increase to pH 4.5 (Gibbsite buffer) can be observed [55] (Figure 4B). Iron speciation in the pore water was dominated by ferric iron in the oxidation zone (up to 2000 mg/L), while directly below the oxidation front a ferrous iron plume of up to 4000 mg/L could be detected [55].

The above-mentioned increase of pH at the oxidation front should initiate the hydrolysis of the Fe^{3+} ions and the precipitation of Fe(III) hydroxides in this area of the profile. However, data from sequential extractions show the contrary, that at the oxidation front and below there were less secondary Fe(III) hydroxides precipitated than in the oxidation zone itself and the underlying primary zone [55]. This can be explained as follows (Figure 4A): At the oxidation front, main microbial activity was detected by Diaby et al. [57], due to the fact that sulphides are still available as energy source (in the oxidation zone they are mainly consumed and only ferric iron is available). In this study, the authors also found that Leptospirillum spp. are dominating the system and that the bacterial population was about 100 times greater at the oxidation front than above or below this horizon. However, Acidithiobacillus spp. and Acidiphillum spp. were also detected and seemed to be mainly responsible for iron reduction in this system, as Leptospirillum spp. is only able to oxidize ferrous to ferric iron. The $\delta^{18}O$ values of dissolved sulphate suggest that from the top of the oxidation zone downwards to the oxidation front, a change from initially atmospheric oxygen towards oxygen from water can be observed. This indicates that at the oxidation front sulphide oxidation takes place by ferric iron, while towards the tailings surface more atmospheric oxygen is involved [56].

Figure 4. (**A**) Schematic model of biogeochemical iron cycling at the sulphide oxidation front (modified after Dold *et al.* [55]). (**B**) Schematic iron speciation as a function of the tailings depth (modified after Dold *et al.* [55]). And (**C**) volume fraction of the different primary and secondary sulphide and ferric iron oxide minerals as a function of tailings depth obtained by reactive transport modelling by Peter Lichtner with the code FLOTRAN [59] for 50 years of oxidation based on pore water composition in the Piuquenes tailings impoundment (with permission). The mineral distribution modelled is confirmed by the detected mineralogy in this tailings profile [46].

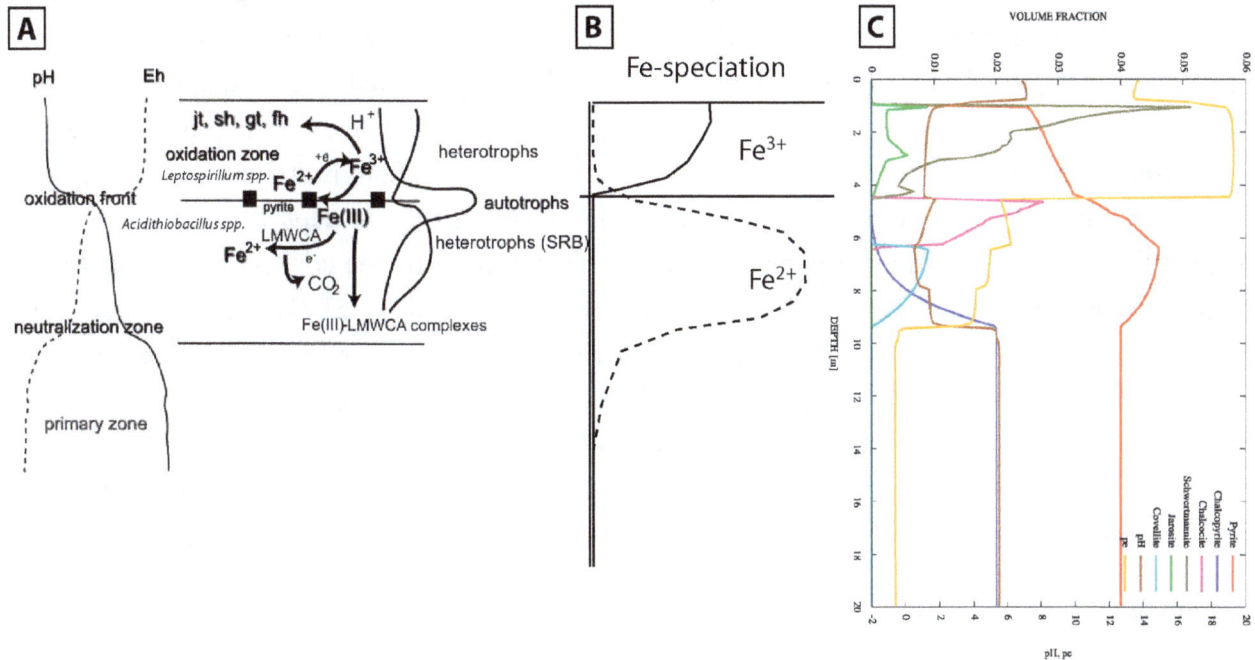

Sulphate reducing bacteria were also detected, and found to have their highest number below the oxidation front, so that some sulphate reduction can be expected. However, stable isotopic data suggest that due to the lack of increase of $\delta^{34}S$ shift towards heavier signature in this area of the profile, sulphate reduction is not occurring in a significant amount in this system, possibly due to the limited availability of organic matter. In the oxidation zone, no organic molecules like low molecular weight carboxylic acids (LMWCA) could be detected, so that the only organic matter is possibly dead bacteria cells available for organic carbon cycling. In contrast, below the oxidation front a peak of LMWCA, like acetate, formate, and pyruvate could be detected. Associated with this LMWCA peak, which is interpreted to be a result of the microbial activity around the oxidation front, the ferrous iron plume and an increase in CO_2 in the pore gas correlates directly [55,60]. These data suggest that the microbial community, in this case mainly *Acidthiobacillus ferrooxidans* and/or *Acidiphilium* spp. [61] use the monodentate LMWCA like acetate and formate as electron donors and ferric iron as electron sink, resulting in the reduction to ferrous iron and the formation of CO_2 [55]. This reduction increases the mobility of iron, as now the ferrous iron can migrate in the circumneutral pH conditions of the underlying tailings stratigraphy until it outcrops at the foot of the dam, where it will auto-oxidize and hydrolyze to form ferrihydrite (outcrop pH still neutral). Another process, which might enable the ferric iron to pass the geochemical barrier of the oxidation front, is via complexation by bidentate LMWCA like oxalate or pyruvate, which changes the solubility and therefore, these complexes might reach a lower tailings horizon, where then again the microbial community will reduce it to ferrous iron and CO_2. Thus, these processes explain why at and below the oxidation front less secondary Fe(III) hydroxides precipitate and instead a ferrous iron plume is formed due to iron reduction processes. This plume can now migrate in the system until it encounters more oxidizing condition or higher pH conditions (e.g., in contact with carbonate rich strata, which then promote the hydrolysis of ferrihydrite). This will be the first visible indication of AMD formation, although the main flow path in the tailings is still neutral (Figure 3B).

The change in redox at the oxidation front also triggers the replacement of chalcopyrite with covellite by copper, leached out from the overlying oxidation zone downwards (Figure 4B), due to general downwards-dominated movement of the released elements in the rainfall dominated alpine climate of Piuquenes [46]. The thickness of this copper enrichment is limited by a second pH increase towards pH around 5.5–6 (siderite buffer) at 3 m depth, as Cu is only mobile until pH 5 in freshwater and is therefore adsorbed at higher pH conditions [46]. As the oxidation front is defined by the drop of oxygen concentrations to zero in the pore gas of the tailings profile (70 cm depth), which correlates with a pH and redox switch, and the groundwater level was at 4 m depth, the copper enrichment zone is defined between oxidation front and siderite buffer (0.7–3 m depth) [55]. This means, that the general belief that supergene enrichment is associated with the groundwater level is not necessarily correct. It is defined by the oxidation front and the pH gradient induced by the neutralization reactions of the gangue mineralogy, which controls the thickness of the mobility window of copper (pH <5 and Eh <500 mV), necessary for the enrichment process. This is the case in fresh water systems, but in high-chlorine system Cu can be mobile at neutral pH as $Cu(II)Cl_2$ or $Cu(I)Cl_2^-$ complexes [47].

6. Consumption of the Neutralization Potential and Final Acid Flow

As discussed above, the resulting ferrous iron plume is the first sign of AMD that might outcrop. However the production of protons still goes on at the oxidation front and in the oxidation zone. These protons interact with the gangue mineralogy and will be partly neutralized, liberating other elements into solution from the dissolution processes of carbonates and silicates. Therefore, depending on the composition of the mineral assemblage of the gangue mineralogy a specific neutralization sequence can be observed across the tailings stratigraphy, which is controlled by the different buffering minerals. For example, in the Piuquenes tailings impoundment the carbonates present are dominated by siderite with traces of calcite. Thus, when the protons produced by sulphide oxidation migrate with the acid solution downwards, first calcite will buffer to around neutral pH until it is completely consumed or passivated by iron oxides. Then siderite will buffer the system to around pH 5.5, until it is consumed. Then the pH can drop further down to around pH 4.5, were the gibbsite buffer will maintain the pH until also this buffer is consumed. Finally, in the oxidation zone itself, the Fe(III) hydroxide assemblage will buffer the pH around the typical pH between 2 and 3 in this area. If it is close to pH 2 a dominance of jarosite can be expected, while if it is closer to pH 3 schwertmannite will control the system [4,46,55]. If there is still an excess of protons added to the system, in some cases even the jarosite buffer might be consumed and even negative pH can be reached as reported from Iron Mountain [62].

This sequence of pH values increases from 2–3, to 4.5, 5.5 and neutral correlates with a successive decrease in redox potential occurring in oxidised tailings, clearly defining the geochemical systems active in each zone, and controlling which elements can be mobilized downwards through the tailings stratigraphy.

Oxyanions like arsenate and molybdate are retained effectively by the Fe(III) hydroxides due to sorption at low pH conditions in the oxidation zone. Below, due to reduction of arsenate to arsenite or at very low pH condition arsenate will be completely protonized and therefore the mobility might be increased for arsenic under these specific conditions [63].

Heavy metals occur mainly as divalent cations, stable in solution and mobile at low-pH conditions. With increasing pH, they become adsorbed and therefore immobile [4]. Additionally, as observed above in the case of copper, replacement processes and reduction processes can precipitate the metals as secondary sulphides or hydroxides in a deeper part of the stratigraphy [46]. As the system will increasingly acidify, these secondary sulphides will be re-dissolved and so the acid oxidation zone migrates further down, increasing the mobility of the heavy metals. When the protons produced in the oxidation zone exceed the neutralization capacity of the gangue mineralogy below in the tailings stratigraphy, the situation can be reached where the whole flow path is under acid conditions, so that the acid, heavy metal rich solution can outcrop at the foot of the tailings dam, or infiltrate into the groundwater. This will be visible with a broad range of bright colours of the precipitates forming at the outcrop, as secondary heavy metal sulphate minerals can have blue, yellow, green, or red colours, depending on their composition (Figure 5). Therefore, when you observe bright colours at the foot of your tailings dam, you can expect an advanced system with acid flow path, or you have an active tailings dam built using the coarse tailings fraction and you are observing the effect of the sulphide oxidation in the unsaturated dam.

Figure 5. Acid flow precipitates: (**A**) Efflorescent salts surface of acid oxidation zone pH 2.5, Ite Bay, Peru. (**B**) Efflorescent salts at the Excelsior Waste rock dump, Cerro de Pasco, Peru. And (**C**) acid Effluent with chalcoalumite precipitation (light blue) at pH 4.9 and schwertmannite at pH 3.15 (orange-brown) at Ojancos, Copiapo, Chile.

7. Some Common Errors in AMD and Mine Waste Management

7.1. AMD Management → Fe³⁺-Rich Solutions

7.1. AMD Management → Fe^{3+}-Rich Solutions

In mines, where AMD occurs, the Fe^{3+}-rich solutions are sometimes pumped into the active mine tailings. This has to be avoided, as the input of ferric iron to sulphide rich material will efficiently oxidize the sulphides and produce 16 moles of protons per mole of pyrite oxidized (Equation (3)), with the result that the pH might drop quickly in the active tailings impoundment [14]. Therefore, mine management strategies need to prevent the contact of the Fe^{3+}-rich solution with any sulphide containing material.

7.2. Fe^{3+}-Rich Sludge or Mud from AMD Neutralization or Treatment Plants

Lately, due to increased efforts in the mining industry not to dispose AMD to the environment, many mines have implemented AMD neutralization or treatment plants. This process produces a certain volume of sludge or mud, which is mainly ferrihydrite, lepidocrocite, goethite [64], schwertmannite [65], depending on the process, with co-precipitated and/or adsorbed elements like arsenic, molybdenum or heavy metals. Thus, this sludge is now a hazardous waste material, which has to be managed properly. An often-used solution for its disposal and unfortunately performed in many mining operations is the deposition of iron oxide sludge in the active tailings impoundment.

The problem with this practice is highlighted here: The sludge of the treatment plant contains mainly Fe(III) hydroxides like ferrihydrite or schwertmannite, the two unstable Fe(III) hydroxides. If we dispose of this sludge together with the tailings from sulphide flotation, we add ferric iron to the sulphides and subsequently cover this sludge with more fresh tailings, so that they eventually end up in a reducing environment. Thus, the ferric iron from the sludge can oxidize the sulphides or undergo reductive dissolution; both processes will produce a ferrous iron plume (some times even acidic) in the tailings stratigraphy, which again will migrate down through the tailings stratigraphy of the active tailings impoundment. Therefore, even if we take (unrealistic) precautions to prevent sulphidic tailings from coming in contact with the atmosphere, e.g., by maintaining a water saturated tailings impoundment, so that only minimum sulphide oxidation can occur, and we cover the tailings directly after the operation has ceased to prevent any further oxidation, the tailings impoundment will one day produce AMD, when the ferrous plume formed due to the addition of the ferric sludge flows out from the foot of the tailings dam.

With these examples we have learned, that we should not mix mine waste from different geochemical systems. Do not mix sulphides with Fe(III) hydroxide sludge in a tailings impoundment, or you will increase the volume of the waste and create adverse geochemical reactions, increasing your long-term environmental management costs. The same is the case for hazardous materials, containing problem elements in the form of oxyanions (e.g., As, Mo, Cr, SO_4), as they should not be mixed with material that contains heavy metals (e.g., Cu, Zn, Ni, Cd, Pb) due to the reverse sorption behaviour. Confine your reactive waste separately in well-designed disposal facilities, so that the geochemical reactions can be controlled long-term and no hazardous elements can escape these systems to the hydrological system surrounding your operations. This will also help future generations to re-exploit these resources with better techniques than are available today.

8. Conclusions

Sulphide oxidation and the subsequent formation of acid mine drainage (AMD) in mine tailings impoundments is associated with a sequence of biogeochemical and mineral dissolution processes and can be classified in three main phases from the operational phase towards the final outcrop of AMD.

8.1. Operational Phase of a Tailings Impoundment: Neutral-Alkaline Oxyanions-Rich Effluents (Figure 6A)

During the operational phase of a sulphidic mine tailings impoundment, no sulphide oxidation should occur, when it is properly managed. This means it should be completely water saturated without exposure of the tailings to the atmosphere, and the system should maintain neutral to alkaline pH conditions (Figure 6A). If this is not the case, sulphide oxidation might start in the unsaturated parts of the tailings, as well as in the tailings dam, if it is built with the coarser fraction of the tailings themselves. This might lead to sulphide oxidation and AMD formation during the operational phase. In ore deposit types, which contain soluble sulphate minerals like gypsum-anhydrite (e.g., porphyry copper deposits), high SO_4 concentrations can be expected in the tailings controlled by the solubility of these minerals, with a typical range between 1500 and 2000 mg/L SO_4.

Additionally, if the ore deposit shows some pre-oxidation naturally or due to the exploitation process, such as block-caving, elements like arsenic or molybdenum, which are adsorbed onto Fe(III)hydroxides, might be desorbed in the alkaline flotation circuit and maintain elevated concentrations in the active tailings impoundment and its effluents. These processes might lead to the need for implementation of sulphate treatment plants or a facility for Mo treatment as in the case of the Carén tailings impoundment from the El Teniente mine, Chile. Mine tailings should not be used as a general waste dump for other industrial waste material, as this might produce severe environmental risks for the whole system and importantly might increase the environmental waste management costs. The visible signs of this stage are usually white precipitates on the surface or around leachates (at this stage the patient is starting to feel bad and have some problems, he becomes pale, but there is still time for prevention).

8.2. After Operation Ceases; Neutral Ferrous Plume Outcrop (Figure 6B)

When the active operational phase ceases, no water and tailings are deposited, which will lead to a drop in the groundwater level in the tailings impoundment and produce an unsaturated zone, where atmospheric oxygen can start the process of sulphide oxidation. This will lead over several years to the formation of an acid oxidation zone, where heavy metals leach out and oxyanions like As and Mo are adsorbed onto the secondary Fe(III) hydroxides formed due to sulphide oxidation. Additionally, due to reduction processes at the oxidation front a ferrous iron plume is formed in the stratigraphy of the tailings impoundment. This ferrous plume can then migrate through the still neutral stratigraphy downwards through the tailings. Once this ferrous plume (which might contain high sulphate concentrations and other oxyanions like As and Mo in solution) outcrops at the foot of the dam for example, the ferrous iron will auto-oxidize due to the neutral pH and precipitate as ferrihydrite (Bordeaux red). This outcrop of the neutral ferrous plume is the first visible sign of the AMD formation process. With subsequent hydrolysis producing ferrihydrite, the effluent will be acidified and the final pH will depend on the buffering capacity of the effluent.

Now the patient has still increased blood pressure (visible red head) and needs help, for prevention it might be too late, most likely long-term treatment is needed. Only by drillings and piezometers can this stage be detected in the tailings stratigraphy in time (it is like taking the blood pressure, if there is no visible sign).

Figure 6. The here shown tailings impoundment is a modern design, with integrated basement impermeabilization and internal drainage system. Most of the tailings impoundments around the world do not have impermeabilization and the contaminated solution will directly infiltrate into the groundwater. (**A**) During the operational phase the system is saturated and alkaline. There might be increased concentrations of oxyanions depending on the mineralogy of the ore. If the dam is built with the coarse fraction of the tailings, oxidation and acidification might start during the operational phase with first signs of AMD (Sh). Surface precipitates are white at this stage. (**B**) After operation has ceased, an acid oxidation zone will develop and a ferrous iron plume below the oxidation front can migrate at neutral pH conditions in the tailings stratigraphy. This neutral, ferrous iron-rich plume will produce ferrihydrite (Fh) at its outflow. And (**C**) acid production due to sulphide oxidation continues and the neutralization potential will be completely consumed, resulting in an acid flow in the tailings mobilizing heavy metal cations and resulting in the formation of AMD with multi-colour precipitates (mainly metal sulphates and/or chlorides).

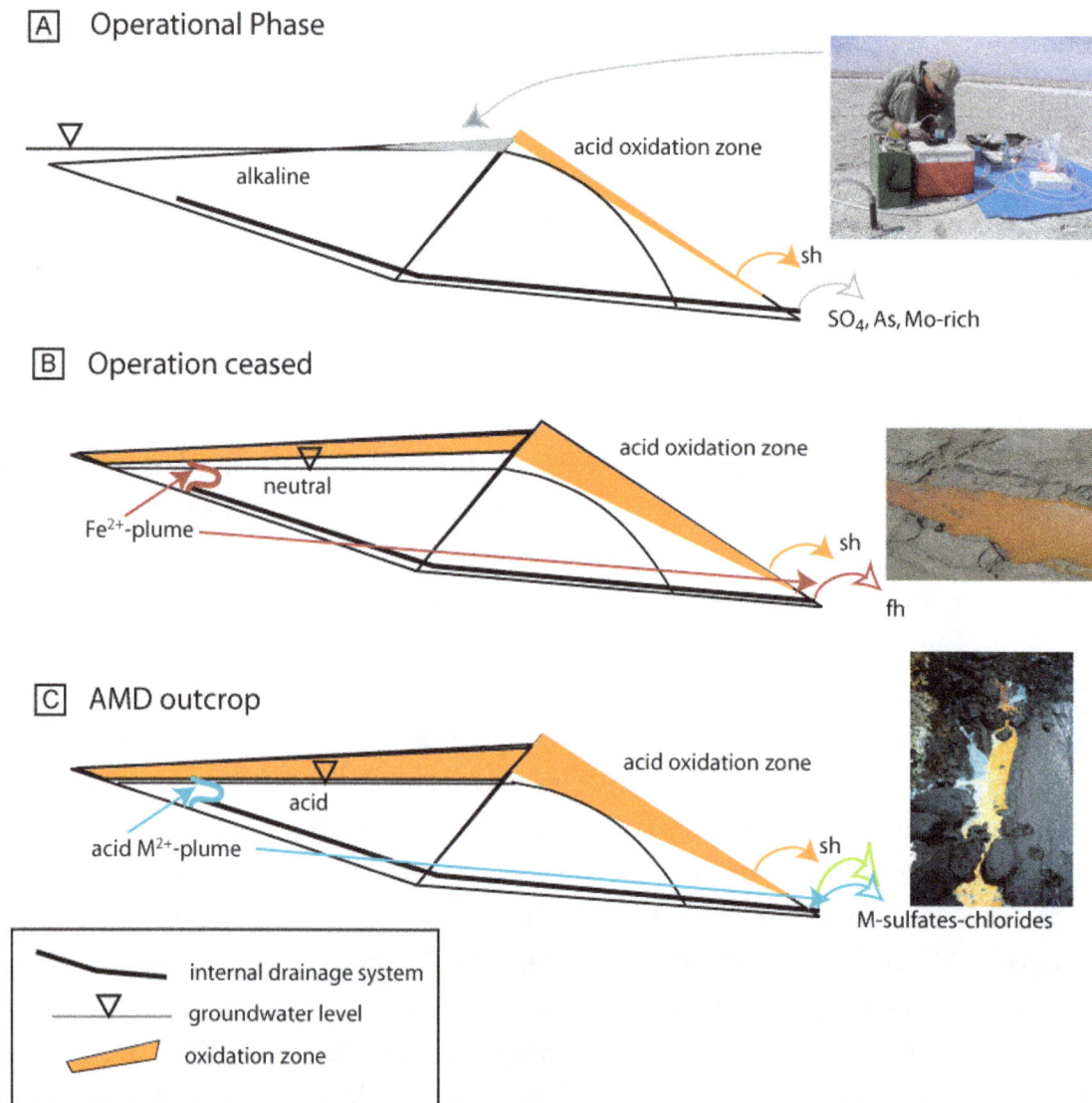

8.3. Final AMD Appearance (Acid Flow, Heavy Metal-Rich Effluent; Figure 6C)

If sulphide oxidation continues and the neutralization potential of the underlying gangue mineralogy is consumed, an acid flow will become established in the tailings. This enables heavy metals like Cu, Zn, Ni, Pb, and Cd to be mobilized through the tailings and outcrop at the foot of the dam or infiltrate into the groundwater, if no impermeable liners have been installed. The efflorescent salts resulting from this acid flow are brightly coloured, blue, green, yellow, white, or red depending on their elemental composition.

This is the final stage of AMD formation and the patient is now extremely ill (you can see it clearly in his green, yellow, blue face), where only final long-term treatment might mitigate the environmental damage. Prevention is here not possible any more, in some cases some drastic remediation with complete saturation of the system might help to alive the symptoms, if there is enough suitable water available [41,66] and the dam stability is not an issue.

Only proper studies can detect in time, at which stage an impoundment is present and predict how the evolution will continue. This is the key knowledge required in order to control and manage these systems properly long-term.

Acknowledgments

I would like to thank all the people, who supported our research throughout these years, mainly in the laboratories of the University of Bremen, Germany; University of Geneva and Lausanne in Switzerland; University of Waterloo, Canada; University of Concepcion, and University of Chile, Chile; and the Institute of Environmental Assessment and Water Research IDAEA-CSIC (Consejo Superior de Investigaciones Cientificas), Barcelona, Spain. Many thanks are due to all the students and colleagues, working in these projects. Special thanks is due to the management CODELCO (Division Norte, El Salvador, Andina, El Teniente), Chile; Anglo American Sur, Chile, Southern Peru Copper Corporation, Cerro de Pasco, Peru, for access to their properties, their support, and interest in our research. I would like to thank the reviewers and Gordon Southam for thorough corrections and helpful comments.

Conflicts of Interest

The author declares no conflict of interest.

References

1. Smuda, J.; Dold, B.; Spangenberg, J.E.; Friese, K.; Kobek, M.R.; Bustos, C.A.; Pfeifer, H.-R. Element cycling during the transition from alkaline to acidic environment in an active porphyry copper tailings impoundment, Chuquicamata, Chile. *J. Geochem. Explor.* **2014**, *140*, 23–40.

2. Azam, S.; Li, Q. Tailings dam failures: A review of the last one hundred years. *Geotech. N.* **2010**, *28*, 50–53.

3. Dold, B. Submarine tailings disposal (STD)—A review. *Minerals* **2014**, *4*, 642–666.

4. Dold, B. Basic Concepts in Environmental Geochemistry of Sulfide Mine-Waste Management. In *Waste Management*; Kumar, S., Ed.; InTech: Rijeka, Croatia, 2010; pp. 173–198.

5. Sato, M. Oxidation of sulfide ore bodies, II. Oxidation mechanisms of sulfide minerals at 25 °C. *Econ. Geol.* **1960**, *55*, 1202–1231.

6. Nordstrom, D.K. Aqueous Pyrite Oxidation and the Consequent Formation of Secondary Iron Minerals. In *Acid Sulfate Weathering*; Kittrick, J.A., Fanning, D.S., Eds.; Soil Science Society of America: Madison, WI, USA, 1982; pp. 37–56.

7. Moses, C.O.; Nordstrom, D.K.; Herman, J.S.; Mills, A.L. Aqueous pyrite oxidation by dissolved oxygen and by ferric iron. *Geochim. Cosmochim. Acta* **1987**, *51*, 1561–1571.

8. Blowes, D.W.; Reardon, E.J.; Jambor, J.L.; Cherry, J.A. The formation and potential importance of cemented layers in inactive sulfide mine tailings. *Geochim. Cosmochim. Acta* **1991**, *55*, 965–978.

9. Jambor, J.L. Mineralogy of Sulfide-Rich Tailings and Their Oxidation Products. In *Short Course Handbook on Environmental Geochemistry of Sulfide Mine-Waste*; Jambor, J.L., Blowes, D.W., Eds.; Mineralogical Association of Canada: Quebec, QC, Canada, 1994; Volume 22, pp. 59–102.

10. Rimstidt, J.D.; Chermak, J.A.; Gagen, P.M. Rates of Reaction of Galena, Spalerite, Chalcopyrite, and Asenopyrite with Fe(III) in Acidic Solutions. In *Environmental Geochimistry of Sulfide Oxidation*; Alpers, C.N., Blowes, D.W., Eds.; American Chemical Society: Washington, DC, USA, 1994; Volume 550, pp. 2–13.

11. Rimstidt, J.D.; Vaughan, D.J. Pyrite oxidation: A state-of-the-art assessment of the reaction mechanism. *Geochim. Cosmochim. Acta* **2003**, *67*, 873–880.

12. Huminicki, D.M.C.; Rimstidt, J.D. Iron oxyhydroxide coating of pyrite for acid mine drainage control. *Appl. Geochem.* **2009**, *24*, 1626–1634.

13. Kwong, Y.T.J. *Prediction and Prevention of Acid Rock Drainage from a Geological and Mineralogical Perspective*; MEND Project 1.32.1; Mine Environment Neutral Drainage (MEND): Ottawa, Canada, 1993.

14. Dold, B.; Wade, C.; Fontbote, L. Water management for acid mine drainage control at the polymetallic Zn-Pb-(Ag-Bi-Cu) deposit of Cerro de Pasco, Peru. *J. Geochem. Explor.* **2009**, *100*, 133–141.

15. Dutrizac, J.E.; Jambor, J.L. Jarosites and their application in hydrometallurgy. *Rev. Mineral. Geochem.* **2000**, *40*, 405–452.

16. Cornell, R.M.; Schwertmann, U. *The Iron Oxides*, 2nd ed.; Wiley-VCH: Weinheim, Germany, 2003.

17. Bigham, J.M.; Schwertmann, U.; Traina, S.J.; Winland, R.L.; Wolf, M. Schwertmannite and the chemical modeling of iron in acid sulfate waters. *Geochim. Cosmochim. Acta* **1996**, *60*, 2111–2121.

18. Jerz, J.K.; Rimstidt, J.D. Pyrite oxidation in moist air. *Geochim. Cosmochim. Acta* **2004**, *68*, 701–714.

19. Ehrlich, H.L. *Geomicrobiology*; Marcel Dekker: New York, NY, USA, 1996.

20. Nordstrom, D.K.; Jenne, E.A.; Ball, J.W. Redox Equilibria of Iron in Acid Mine Waters. In *Chemical Modeling in Aqueous Systems*; American Chemical Society: Washington, DC, USA, 1979; Volume 93, pp. 51–79.

21. Ritchie, A.I.M. Sulfide Oxidation Mechanisms: Controls and Rates of Oxygen Transport. In *Short Course Handbook on Environmental Geochemistry of Sulfide Mine-Waste*; Jambor, J.L., Blowes, D.W., Eds.; Mineralogical Association of Canada: Quebec, QC, Canada, 1994; Volume 22, pp. 201–244.

22. Bryner, L.C.; Walker, R.B.; Palmer, R. Some factors influencing the biological and non-biological oxidation of sulfide minerals. *Trans. Soc. Min. Eng. AIME* **1967**, *238*, 56–65.

23. Singer, P.C.; Stumm, W. Acid mine drainage: The rate-determining step. *Science* **1970**, *167*, 1121–1123.

24. Johnson, D.B. Biodiversity and ecology of acidophilic microorganisms. *FEMS Microbiol. Ecol.* **1998**, *27*, 307–317.

25. Johnson, D.B. Importance of Microbial Ecology in the Development of New Mineral Technologies. In *Biohydrometallurgy and the Environment Toward the Mining of the 21st Century*; Amils, R., Ballester, A., Eds.; Elsevier: Amsterdam, The Netherlands, 1999; Volume 9A, pp. 645–656.

26. Norris, P.R.; Johnson, D.B. Acidophilic Microorganisms. In *Extremophiles: Microbial Life in Extreme Environments*; Horikoshi, K., Grant, W.D., Eds.; John Wiley & Sons: Hoboken, NJ, USA, 1998; pp. 133–154.

27. Johnson, D.B.; Hallberg, K.B. The microbiology of acidic mine waters. *Res. Microbiol.* **2003**, *154*, 466–473.

28. Nordstrom, D.K.; Southam, G. Geomicrobiology of Sulfide Mineral Oxidation. In *Geomicrobiology*; Banfield, J.F., Nealson, K.H., Eds.; Mineralogical Society of America: Chantilly, VA, USA, 1997; Volume 35, pp. 361–390.

29. Mielke, R.; Pace, D.; Porter, T.; Southam, G. A critical stage in the formation of acid mine drainage: Colonization of pyrite by Acidithiobacillus ferrooxidans under pH-neutral conditions. *Geobiology* **2003**, *1*, 81–90.

30. Dockrey, J.W.; Lindsay, M.B.; Mayer, U.; Beckie, R.D.; Norlund, K.L.I.; Warren, L.A.; Southam, G. Acidic microenvironments in waste rock characterized by neutral drainage: Bacteria–mineral interactions at sulfide surfaces. *Minerals* **2014**, *4*, 170–190.

31. Carson, C.D.; Fanning, D.S.; Dixon, J.B. Alfisols and Ultisols with Acid Sulfate Weathering Features in Texas. In *Acid Sulfide Weathering*; Kittrick, J.A., Fanning, D.S., Hossner, L.R., Eds.; Soil Science Society of America: Madison, WI, USA, 1982; Volume 10, pp. 127–146.

32. Barker, W.W.; Welch, S.A.; Chu, S.; Banfield, J.F. Experimental observations of the effects of bacteria on aluminosilicate weathering. *Am. Mineral.* **1998**, *83*, 1551–1563.

33. Dold, B. Sustainability in metal mining: From exploration, over processing to mine waste management. *Rev. Environ. Sci. Biotechnol.* **2008**, *7*, 275–285.

34. Bruckard, W.J.; Sparrow, G.J.; Woodcock, J.T. A review of the effects of the grinding environment on the flotation of copper sulphides. *Int. J. Miner. Process.* **2011**, *100*, 1–13.

35. Weiss, N.L. *SME Mineral Processing Handbook*; Society of Mining Engineers of the American Institute of Mining, Metallurgical, and Petroleum Engineers: Englewood, CO, USA, 1985.

36. Moreno, P.A.; Aral, H.; Cuevas, J.; Monardes, A.; Adaro, M.; Norgate, T.; Bruckard, W. The use of seawater as process water at Las Luces copper-molybdenum beneficiation plant in Taltal (Chile). *Miner. Eng.* **2011**, *24*, 852–858.

37. Ramos, O.; Castro, S.; Laskowski, J.S. Copper-molybdenum ores flotation in sea water: Floatability and frothability. *Miner. Eng.* **2013**, *53*, 108–112.

38. Fourie, A. Paste and Thickened Tailings: Has the Promise Been Fulfilled? In Proceedings of the GeoCongress 2012: State of the Art and Practice in Geotechnical Engineering, Oakland, CA, USA, 25–29 March 2012; pp. 4126–4135.

39. Tariq, A.; Yanful, E.K. A review of binders used in cemented paste tailings for underground and surface disposal practices. *J. Environ. Manag.* **2013**, *131*, 138–149.

40. Smuda, J.; Dold, B.; Spangenberg, J.E.; Pfeifer, H.R. Geochemistry and stable isotope composition of fresh alkaline porphyry copper tailings: Implications on sources and mobility of elements during transport and early stages of deposition. *Chem. Geol.* **2008**, *256*, 62–76.

41. Dold, B.; Diaby, N.; Spangenberg, J.E. Remediation of a marine shore tailings deposit and the importance of water-rock interaction on element cycling in the coastal aquifer. *Environ. Sci. Technol.* **2011**, *45*, 4876–4883.

42. Dold, B.; Fontboté, L. A mineralogical and geochemical study of element mobility in sulfide mine tailings of Fe oxide Cu-Au deposits from the Punta del Cobre belt, northern Chile. *Chem. Geol.* **2002**, *189*, 135–163.

43. Sima, M.; Dold, B.; Frei, L.; Senila, M.; Balteanu, D.; Zobrist, J. Sulfide oxidation and acid mine drainage formation within two active tailings impoundments in the Golden Quadrangle of the Apuseni Mountains, Romania. *J. Hazard. Mater.* **2011**, *189*, 624–639.

44. Dold, B.; Gonzalez-Toril, E.; Aguilera, A.; Lopez-Pamo, E.; Bucchi, F.; Cisternas, M.E.; Amils, R. Acid rock drainage and rock weathering in Antarctica: Important sources for iron cycling in the Southern Ocean. *Environ. Sci. Technol.* **2013**, *47*, 6129–6136.

45. Sanchez Espana, J.; Lopez Pamo, E.; Santofimia Pastor, E. The oxidation of ferrous iron in acidic mine effluents from the Iberian Pyrite Belt (Odiel Basin, Huelva, Spain): Field and laboratory rates. *J. Geochem. Explor.* **2007**, *92*, 120–132.

46. Dold, B.; Fontboté, L. Element cycling and secondary mineralogy in porphyry copper tailings as a function of climate, primary mineralogy, and mineral processing. *J. Geochem. Explor.* **2001**, *74*, 3–55.

47. Dold, B. Element flows associated with marine shore mine tailings deposits. *Environ. Sci. Technol.* **2006**, *40*, 752–758.

48. Bea, S.A.; Ayora, C.; Carrera, J.; Saaltink, M.W.; Dold, B. Geochemical and environmental controls on the genesis of efflorescent salts on coastal mine tailings deposits: A discussion based on reactive transport modeling. *J. Contam. Hydrol.* **2010**, *111*, 65–82.

49. Romero, L.; Alonso, H.; Campano, P.; Fanfani, L.; Cidu, R.; Dadea, C.; Keegan, T.; Thornton, I.; Farago, M. Arsenic enrichment in waters and sediments of the Rio Loa (Second Region, Chile). *Appl. Geochem.* **2003**, *18*, 1399–1416.

50. ASTM International. *ASTM D5744-96: Standard Test Method for Accelerated Weathering of Solid Materials Using a Modified Humidity Cell*; ASTM International: West Conshohocken, PA, USA, 1996.

51. Filipek, L.H.; van Wyngarden, T.J.; Papp, C.S.E.; Curry, J. A Multi-Phase Appraoch to Predict Acid Production from Porphyry Copper-Gold Waste Rock in an Arid Montane Environment. In *The Environmental Geochemistry of Ore Deposits, Part B: Case Studies and Research Topics. Reviews in Economic Geology*; Filipek, L.H., Plumlee, G.S., Eds.; Society of Economic Geologists: Littleton, CO, USA, 1999; Volume 6, pp. 433–444.

52. Weibel, L.; Dold, B.; Cruz, J. Application and Limitation of Standard Humidity Cell Tests at the Andina Porphyry Copper Mine, CODELCO, Chile. In Proceedings of the 11th Biennial SGA Meeting, Antofagasta, Chile, 26–29 September 2011.

53. Dold, B.; Weibel, L.; Cruz, J. New Modified Humidity Cells Test for Acid Rock Drainage Prediction in Porphyry Copper Deposits. In Proceedings of the 2nd International Seminar on Environmental Issues in the Mining Industry, Santiago, Chile, 23–25 November 2011.

54. Dold, B.; Weibel, L. Biogeometallurgical pre-mining characterization of ore deposits: An approach to increase sustainability in the mining process. *Environ. Sci. Pollut. Res.* **2013**, *20*, 7777–7786.

55. Dold, B.; Blowes, D.W.; Dickhout, R.; Spangenberg, J.E.; Pfeifer, H.-R. Low molecular weight carboxylic acids in oxidizing porphyry copper tailings. *Environ. Sci. Technol.* **2005**, *39*, 2515–2521.

56. Dold, B.; Spangenberg, J.E. Sulfur speciation and stable isotope trends of water-soluble sulfates in mine tailings profiles. *Environ. Sci. Technol.* **2005**, *39*, 5650–5656.

57. Diaby, N.; Dold, B.; Holliger, C.; Pfeifer, H.R.; Johnson, D.B.; Hallberg, K.B. Microbial communities in a porphyry copper tailings impoundment and their impact on the geochemical dynamics of the mine waste. *Environ. Microbiol.* **2007**, *9*, 298–307.

58. Spangenberg, J.; Dold, B.; Vogt, M.-L.; Pfeifer, H.R. The stable hydrogen and oxygen isotope composition of waters from mine tailings in different climatic environments. *Environ. Sci. Technol.* **2007**, *41*, 1870–1876.

59. Lichtner, P.C. *FLOTRAN User's Manual*; Report LA-UR-02-2349; Los Alamos National Laboratory: Los Alamos, NM, USA, 2001.

60. Johnson, D.B.; McGinness, S. Ferric iron reduction by acidophilic heterotrophic bacteria. *Appl. Environ. Microbiol.* **1991**, *57*, 207–211.

61. Johnson, D.B.; Bridge, T.A.M. Reduction of ferric iron by acidophilic heterotrophic bacteria: Evidence for constitutive and inducible enzyme systems in *Acidiphilium* spp. *J. Appl. Microbiol.* **2002**, *92*, 315–321.

62. Nordstrom, D.K.; Alpers, C.N.; Ptacek, C.J.; Blowes, D.W. Negative pH and extremely acidic mine waters from Iron Mountain, California. *Environ. Sci. Technol.* **2000**, *34*, 254–258.

63. Alarcon, R.; Gaviria, J.; Dold, B. Liberation of adsorbed and co-precipitated arsenic from jarosite, schwertmannite, ferrihydrite, and goethite in seawater. *Minerals* **2014**, *4*, 603–620.

64. Genty, T.; Bussiere, B.; Potvin, R.; Benzaazoua, M.; Zagury, G.J. Dissolution of calcitic marble and dolomitic rock in high iron concentrated acid mine drainage: Application to anoxic limestone drains. *Environ. Earth Sci.* **2012**, *66*, 2387–2401.

65. Hedrich, S.; Lönsdorf, H.; Kleeberg, R.; Heide, G.; Seifert, J.; Schlömann, M. Schwertmannite formation adjacent to bacterial cells in a mine water treatment plant and in pure cultures of *Ferrovum myxofaciens*. *Environ. Sci. Technol.* **2011**, *45*, 7685–7692.

66. Diaby, N.; Dold, B. Evolution of geochemical and mineralogical parameters during *in situ* remediation of a marine shore tailings deposit by the implementation of a wetland cover. *Minerals* **2014**, *4*, 578–602.

7

Chemoorganotrophic Bioleaching of Olivine for Nickel Recovery [†]

Yi Wai Chiang [1,2,]*, **Rafael M. Santos** [3], **Aldo Van Audenaerde** [3], **Annick Monballiu** [4],
Tom Van Gerven [3] **and Boudewijn Meesschaert** [2,4]

[1] School of Engineering, University of Guelph, Guelph, ON N1G 2W1, Canada
[2] Department of Microbial and Molecular Systems, KU Leuven, Leuven 3001, Belgium;
E-Mail: boudewijn.meesschaert@biw.kuleuven.be
[3] Department of Chemical Engineering, KU Leuven, Leuven 3001, Belgium;
E-Mails: rafael.santos@alumni.utoronto.ca (R.M.S.); aldovanaudenaerde@gmail.com (A.V.A.);
tom.vangerven@cit.kuleuven.be (T.V.G.)
[4] Laboratory for Microbial and Biochemical Technology (Lab μBCT), KU Leuven @ Brugge-Oostende
(Kulab), Oostende 8400, Belgium; E-Mail: annick.monballiu@vives.be

* Author to whom correspondence should be addressed; E-Mail: chiange@uoguelph.ca

[†] Note: Contents of this paper also appear in the conference proceedings of the MetSoc of CIM's 7th
International Hydrometallurgy Symposium, Victoria, BC, Canada, 22–25 June 2014.

Abstract: Bioleaching of olivine, a natural nickel-containing magnesium-iron-silicate, was conducted by applying chemoorganotrophic bacteria and fungi. The tested fungus, *Aspergillus niger*, leached substantially more nickel from olivine than the tested bacterium, *Paenibacillus mucilaginosus*. *Aspergillus niger* also outperformed two other fungal species: *Humicola grisae* and *Penicillium chrysogenum*. Contrary to traditional acid leaching, the microorganisms leached nickel preferentially over magnesium and iron. An average selectivity factor of 2.2 was achieved for nickel compared to iron. The impact of ultrasonic conditioning on bioleaching was also tested, and it was found to substantially increase nickel extraction by *A. niger*. This is credited to an enhancement in the fungal growth rate, to the promotion of particle degradation, and to the detachment of the stagnant biofilm around the particles. Furthermore, ultrasonic conditioning enhanced the selectivity of *A. niger* for nickel over iron to a value of 3.5. Pre-carbonating the olivine mineral, to enhance mineral liberation and change metal speciation, was also attempted, but did not result in improvement as a consequence of the mild pH of chemoorganotrophic bioleaching.

Keywords: bioleaching; chemoorganotrophic; fungi; bacteria; olivine; nickel; ultrasonic treatment; mineral carbonation

---

1. Introduction

The increasing demand and diminishing availability of raw materials requires us to look beyond conventional resources; particularly, the importance of low-grade ores is expected to increase in the near future [1]. An example is the depletion of accessible high-grade sulfidic ores, which makes it necessary to seek more abundant but lower grade ores, often rich in siliceous mafic minerals (olivine, pyroxene, amphibole and biotite). The processing of such ores, however, is linked to high processing costs when using traditional extraction routes (e.g., high pressure acid leaching (HPAL) and ferro-nickel smelting).

An alternative, potentially more sustainable approach is the application of biohydrometallugy, wherein microorganisms act as renewable chemical producers [1] of substances that deteriorate and dissolve minerals, thereby liberating the immobilized metals into solution (leachate). Biogenic substances produced include not only organic acids [2], but also chelates, mineral acids, and for certain bacteria, ammonia or amines [3]. The three main mechanisms that can act in the solubilization of metals are: acidolysis, complexolysis and redoxolysis [4].

Several types of microorganisms have been tested for leaching of mineral ores. These can be classified into two groups: autotrophic bacteria and archaea, and heterotrophic bacteria, archaea and fungi [5]. Autotrophic microorganisms use carbon dioxide as their carbon source, whereas heterotrophic microorganisms use organic compounds as a carbon source. Additionally, a distinction can be made between chemoorganotrophic and chemolithotrophic microorganisms. The former obtain energy through oxidation of organic compounds, while the latter use reduced inorganic compounds as their energy resource [6]. Here a distinction can be made between the microorganisms utilized in the present work, which do not depend on the presence of sulfur, and those typically used industrially for bioleaching of sulfidic ores, which utilize the intrinsic sulfur.

Bioleaching is influenced by a wide range of parameters including physicochemical parameters, microbiological factors of the leaching environment, and the properties of the solids to be leached. These will influence both the growth of the microorganisms as well as their leaching behavior [1,7–9]. Most importantly, the microorganism must be able to leach the material, and also be resistant to the metals that are leached out. In terms of types of minerals, silicates and saprolites have been found to leach more readily than limonites [6]. In terms of the toxic effect of heavy metals, these can have an adverse impact on the microorganism's survival due to four reasons: (i) the blocking of essential functional groups of enzymes; (ii) the displacement of essential metals; (iii) the induction of conformational changes of polymers; and (iv) the influence on the membrane integrity and transport processes [10]. For the leaching of nickel laterite ores it has been found that *Aspergillus niger* and *Penicillium funiculosum* are tolerant to practical levels of nickel (0.1 g/L) concentration [2].

In an effort to accelerate bioleaching, recent research has provided evidence for the beneficial effects of controlled sonication on the leachability of metals catalyzed by living cells. Anjum *et al.* [11] tested sonobioleaching on black shale and concluded that it enhanced metal recovery and reduced the time needed for the maximum recovery. They also found that the production of acids was higher for ultrasonically treated *Aspergillus niger*. The acid yields with ultrasound treatment amounted to 6.17% citric acid, 4.68% malic acid, 2.36% oxalic acid and 0.052% tartaric acid. In comparison, not sonicated *A. niger* yielded 5.25% citric acid, 3.45% malic acid, 0.94% oxalic acid and 0.09% tartaric acid.

Swamy *et al.* [12] found that for the leaching of nickel from laterites, ultrasound increased and accelerated the recovery from 92% in 20 days to 95% in 14 days. The recovery of iron, on the other hand, reduced from 12.5% to 0.16%, indicating that ultrasound also greatly increases the selectivity of nickel over iron. Swamy *et al.* [12] also investigated the influence of certain parameters of the ultrasonic treatment such as frequency, intensity and sonication time on the leaching results. The metal recovery was highest when a daily sonication time of 30 min was used. Maximum extractions were achieved at the lower frequency of 20 kHz because more citric and oxalic acids were produced at this frequency. The optimal intensity was found to be 1.5 W/cm^2. Further increasing this intensity was postulated to lead to cell disruption, causing a decrease in acid production and consequently lowering the nickel extraction.

Bioleaching may also benefit from mineral processing steps that increase specific surface area, alter the mineralogy or otherwise liberate metal-rich mineral from the gangue matrix. Santos *et al.* [13] explored an approach wherein carbon dioxide was used to promote mineral alterations that led to improved extractability of nickel from olivine ($(Mg,Fe)_2SiO_4$). Olivine is an abundant silicate mineral within the Earth's crust that contains minor amount of nickel and chromium, and is a precursor to lateritic nickel ores that are commercially explored. Carbonation pre-treatment was found to promote mineral liberation and concentration of metals in physically separable phases. In that study, olivine was fully carbonated at high CO_2 partial pressures (35 bar) and optimal temperature (200 °C) with the addition of pH buffering agents. The main products of the carbonation reaction included amorphous colloidal silica, chromium-rich metallic particles, and iron-substituted magnesite ($Mg_{1-x}Fe_xCO_3$). The percentage of nickel extracted from carbonated olivine significantly increased compared to leaching from untreated olivine. Using HCl, 100% of nickel could be leached from carbonated olivine, while only 66% nickel was recovered from untreated olivine using the same acid concentration (2.6 N).

The present work investigates the possibility of bioleaching nickel from olivine. The microorganisms utilized were the bacterium *Paenibacillus mucilaginosus*, and the fungi *Aspergillus niger*, *Penicillium chrysogenum* and *Humicola grisea*. These microorganisms were selected based on the prior work of Chiang *et al.* [1], who screened several bacteria and fungi for bioleaching of various alkaline materials. In that study, the bioleaching performance of fungi was more comparable, hence three fungi and one bacteria were selected for the present study. Buford *et al.* [3] report that the three selected fungal species are commonly associated in the natural environment with sandstone, marble and granite rock substrata, indicating their adaptability to silicate materials. The soil-inhabiting bacterium has been commonly applied in bioleaching of silica-rich materials, e.g., Yao *et al.* [9]. The extent of leaching with these different microorganisms is compared. The effect of carbonation as a mineral alteration/liberation pre-treatment step in the recovery of nickel is investigated. The influence of ultrasonic treatment on both the growth of the microorganisms as well as on the leaching are studied.

2. Experimental Section

2.1. Materials

The microorganisms were acquired from culture collections: *P. mucilaginosus* bacterium was obtained from the China Center of Industrial Culture Collection (CICC, Beijing, China); *A. niger*, *P. chrysogenum* and *H. grisea* fungi were acquired from DSMZ (Braunschweig, Germany). The fungi strains were maintained on potato dextrose agar (PDA, containing infusion from potatoes, 2% glucose and 1.5% agar). The *P. mucilaginosus* strain was maintained on nutrient agar (NA) with 0.001% $MnSO_4 \cdot H_2O$ for sporulation enhancement. All growth media components were obtained from Sigma Aldrich (Bornem, Belgium). Each strain was incubated at 30 °C for growth and preserved at 4 °C for storage.

Two types of olivine mineral were tested to investigate the effect of pre-treatment on the leaching behavior: fresh olivine (GL30, Eurogrit B.V., Papendrecht, The Netherlands) and fully carbonated olivine. The fresh olivine was milled using a centrifugal mill (Retsch ZM100, 1400 rpm, 80 μm sieve mesh), resulting in a material with 86 vol % below 80 μm, and the average mean diameter (D{4,3}, determined by laser diffraction) of 34.8 μm. The chemical composition of the olivine, determined following chemical digestion by inductively coupled plasma mass spectrometry (ICP-MS, X Series, Thermo Electron Corporation, Waltham, MA, USA) and by wavelength dispersive X-ray fluorescence (WDXRF, PW2400, Panalytical, Almelo, The Netherlands), was: 27.5 wt % Mg, 20.7 wt % Si, 3.7 wt % Fe, 0.27 wt % Ni, 0.24 wt % Cr, and 0.17 wt % Al. The carbonated olivine consisted of a mixture of iron-substituted magnesite, amorphous colloidal silica, and chromium-rich metallic particles (more details in Santos *et al.* [13] and Van Audenaerde [14]). It was obtained by reacting 200 grams of fresh milled olivine in 800 mL 1 M NaCl aqueous solution for 72 h, at 200 °C, with 35 bar CO_2 partial pressure. The carbonation conversion achieved, determined by thermogravimetric analysis, was 99.0%.

2.2. Bioleaching Methodology

Bioleaching experiments consisted of two phases: the growth phase and the leaching phase. At the beginning of the growth phase, the microorganisms were inoculated in 400 mL nutrient broth in 1 L cotton wad-sealed Erlenmeyer flasks and placed on magnetic stirring plates inside an incubator set to 30 °C to grow. Once the growth phase was completed (after seven days), olivine solids were added (50 g/L) to the microbial broth and to a reference sterile nutrient broth, to start the leaching phase. Two methods of bioleaching were used: microbially-assisted leaching, and biogenic substance leaching. In microbially-assisted leaching, the living microorganisms were present during the leaching phase, and assisted with the extraction of metals. In biogenic substance leaching, the microbial broth was autoclaved (15 min at 120 °C) at the end of the growth phase to sterilize the broth, and filtered to remove the dead biomass; this way, only the soluble biogenic substances produced by the microorganisms were responsible for the subsequent leaching. It should be noted that autoclaving may denature some biogenic substances, such as enzymes, though it is expected that the main substances responsible for bioleaching, such as organic acids, remained unaltered. During the leaching phase the flasks were agitated on a shaking table at 160 rpm under a temperature controlled hood set to 25 °C. Liquid samples were collected and analyzed during the leaching phase to monitor the metal concentrations at various times. The pH was measured at various intervals during the growth and the leaching phases to track biogenic acid production and consumption. Viable microbial concentrations in solution were determined by spread plating serial dilutions onto nutrient agar plates using a sterile Drigalski spatula and counting colony forming units (CFU) after incubation for 24 h. Upon completion of the leaching phase (7 to 21 days duration), the liquid and solids were separated by centrifugation. The liquids were analyzed for soluble metals concentrations using ICP-MS, whereas the solids were analyzed by wet laser diffraction (LD, Mastersizer S, Malvern Instruments, Malvern, UK). Leaching experiments were carried out in duplicate and mean values are reported. Owing to occasional variability caused by uneven inoculation or biomass pelletization (particularly with fungi), and to the low number of replicates, reported values and trends should be treated qualitatively.

The experiments that tested the influence of sonication on bioleaching used an ultrasonic bath (Elmasonic S300 (H), Elma Hans Schmidbauer, Singen, Germany) operated at a non-adjustable frequency of 37 kHz, 300 W power, and 0.2 W/cm^2 sonication intensity. Flasks containing microbial broth were ultrasonically treated during both the growth and leaching phases daily by placing them inside the bath for 15 min.

3. Results and Discussion

3.1. Bioleaching of Fresh Olivine by Bacterium versus Fungus

Bioleaching results for fresh olivine by *A. niger* and *P. mucilaginosus* are presented in Figure 1. The leaching extent with *A. niger* was substantially better than that with *P. mucilaginosus*. *A. niger* leached 12.8% nickel in the microbially-assisted subset after 21 days, compared to 11.4% with only the biogenic substances. For *P. mucilaginosus*, these same extractions were only 3.3% and 2.7%, respectively. In fact, these extraction extents were comparable to the nutrient reference for *P. mucilaginosus*, indicating that the bacterium or its biogenic products did not improve olivine leaching. Figure 2 suggests that biogenic acids produced by *A. niger* were largely responsible for its better bioleaching performance. This is consistent with the finding of Castro *et al.* [5], who compared various bacteria and fungi for the leaching of nickel and zinc from calamine and garnierite silicates, and concluded that fungi were more effective due to the production of citric and oxalic acids together with other organic metabolites (e.g., amino acids, peptides and proteins [10]). Therefore, the subsequent set of experiments focused on the leaching behavior of various fungi towards olivine.

From Figure 1a, it also appears that live *A. niger* provides enhanced leaching compared to the experiment performed with sterile biogenic substances. Burgstaller and Schinner [10] have noted that cell-particle contact may contribute to the enhancement of fungal bioleaching, for example by stimulating biogenic acid production. Finally, it is notable from Figure 1 that elemental leaching was incongruent, with higher yields of nickel leaching compared to magnesium and iron, the two main constituents of olivine. Leaching of silicon was further reduced compared to these elements, as can be expected by the low solubility of silicic acid at these conditions, while chromium and aluminum remained largely unaffected, likely due to their presence primarily within metal-rich crystals [13] and their differing geochemical characteristics.

Figure 1. (**a**) Elemental leaching extent by *Aspergillus niger* from fresh olivine over 21 days; and (**b**) elemental leaching extent by *Paenibacillus mucilaginosus* from fresh olivine over 21 days.

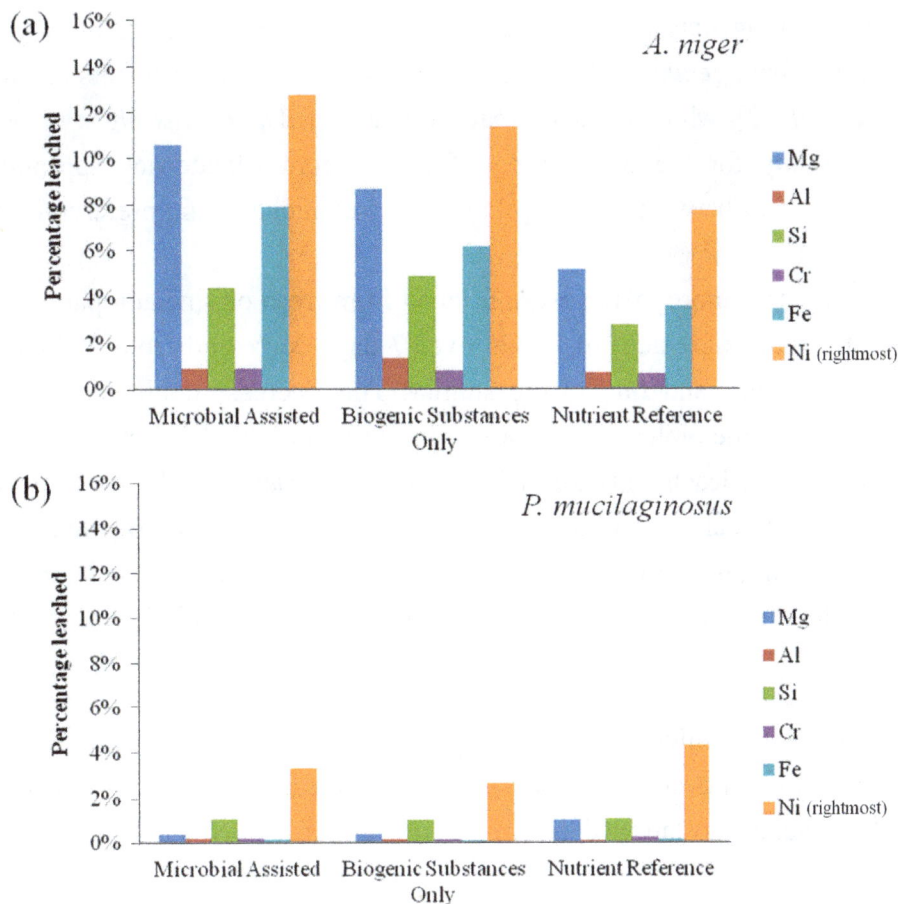

Figure 2. Variation of pH, measured weekly, during bioleaching of fresh olivine over 21 days by *Aspergillus niger* and *Paenibacillus mucilaginosus* and their respective nutrient broth references.

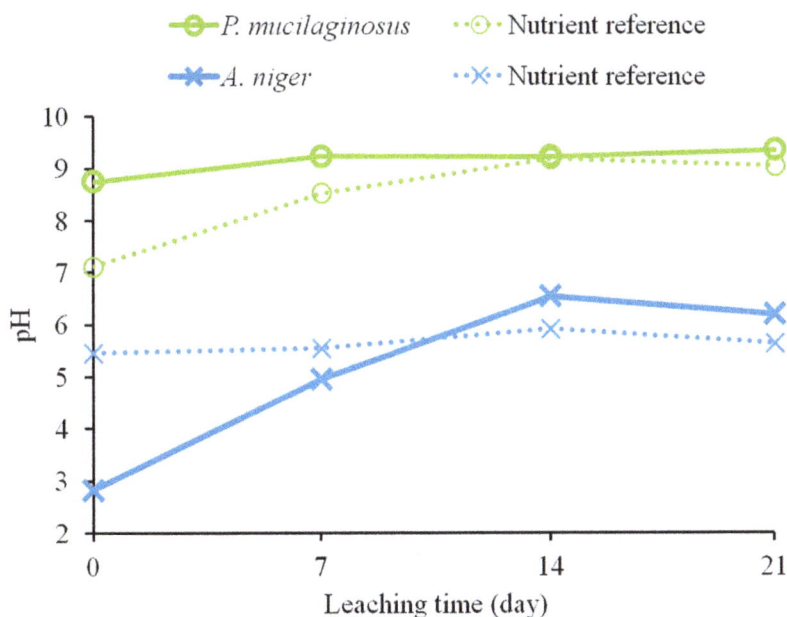

3.2. Bioleaching of Fresh and Carbonated Olivine by Various Fungi

Bioleaching results from fresh and carbonated olivine using various fungi are presented in Figure 3. *Aspergillus niger* was found to be the preferential type of fungi for leaching of both types of olivine, followed by *Penicillium chrysogenum*. *A. niger* leached 7.8% of nickel from fresh olivine after 7 days, compared to 5.3% for *P. chrysogenum* and only 2.8% for *H. grisae*. These results are in agreement with findings of Valix *et al.* [2], who concluded that the strain of *Aspergillus niger* is more effective than *Penicillium funiculosum* for the bioleaching of nickel from silicate-rich saprolite ores. The leaching selectivity did not vary between the fungal species, with nickel being preferentially leached in all cases.

From Figure 3 it is clear that more metals were leached from fresh olivine compared to carbonated olivine. For *A. niger*, the leaching extent of nickel over 7 days decreased from 7.8% to 4.2%. The decreases for the other elements and fungi were similar. This decrease could, however, be due to the mild process conditions of the bioleaching process. In our related work (Santos *et al.* [13]), it was also noticed that at low absolute leaching (namely low acid concentrations), olivine carbonation brings little benefit, while at higher acidity carbonation acts as a mineral liberation, and thus leaching enhancing treatment. Improving the absolute percentages of the bioleaching, by optimizing the process conditions further than done in the present study, might help carbonated olivine to leach better than fresh olivine.

Figure 3. (**a**) Elemental leaching extent from fresh olivine by sterile biogenic substances derived from different fungi over 7 days; and (**b**) elemental leaching extent from carbonated olivine by sterile biogenic substances derived from different fungi over 7 days.

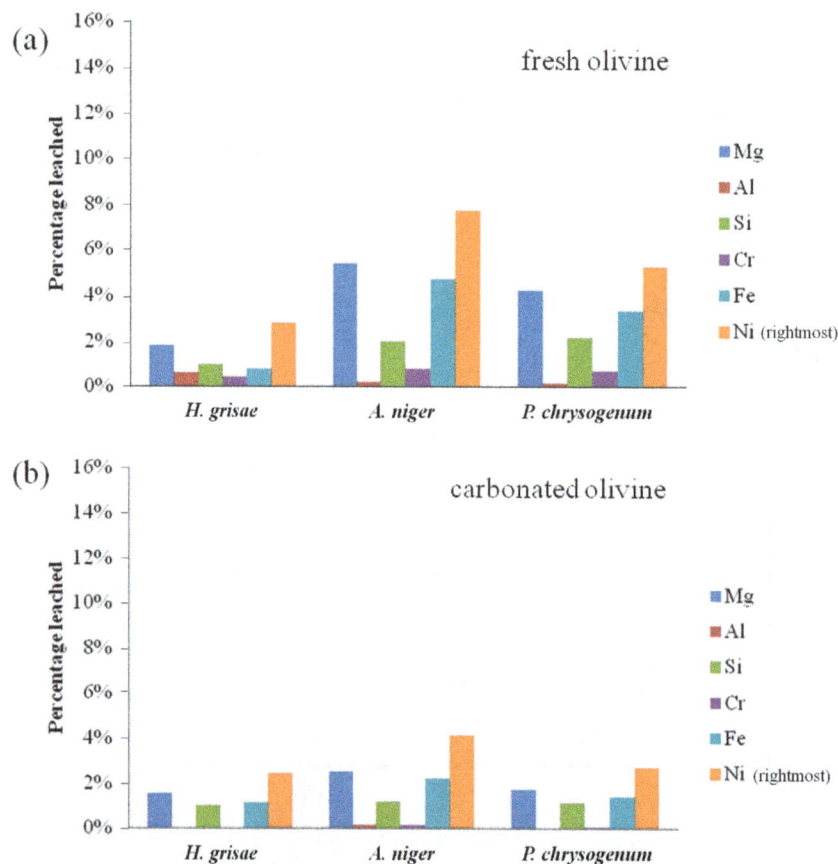

3.3. Ultrasound-Assisted Fungal Bioleaching

The influence of sonication on the leaching of fresh olivine by *Aspergillus niger* is shown in Figure 4. Treating the fungus both during the growth phase as well as during the leaching phase clearly enhanced the leaching. Nickel leaching after 17 days increased from 9.9% without ultrasonic conditioning to 15.7% with ultrasonic conditioning. This enhancement can be due to a wide combination of effects. One of the most important being that sonication increases the microbial growth, which will in turn lead to an increased production of biogenic substances [11]. Ultrasound can disrupt fungal pellets and cause the biomass to grow mainly as dispersed hyphae [15]; this affects the broth rheology, since solution viscosity depends on the morphology of the suspended biomass, and can promote the production of fungal metabolites that require freely dispersed fungal morphology. The latter effect was confirmed by pH measurements of the fungal broth at the end of the growth phase. The pH of the broth under standard conditions was 2.8, whereas the pH of ultrasonically treated broth was 2.3. Plating indicated a viable fungal concentration above 10^5 CFU/mL in both cases. The reduction in pH may be explained by an increase in the production of biogenic organic acids, as measured by Anjum *et al.* [11]. Yao *et al.* [9] pointed out, however, that pH effect is not the only driving force for the accelerated dissolution of minerals. Organic compounds other than acids can act as ligands to form surface or aqueous complexes; the formation of surface complexes can weaken the chemical bonds in the bulk material, leading to an accelerated elemental release.

During the leaching phase, when solids are present, an additional effect of sonication that can promote leaching is the abrasion of the surface of olivine particles by the collapsing cavitations and micro-jets formed by ultrasound, which can lead to particle/agglomerate breakage or the removal of passivating depleted layers that cover the unreacted particle core [16]. The size reduction of olivine particles was confirmed by laser diffraction analysis after bioleaching, as presented in Figure 5. The volume-moment mean diameter (D{4,3}) of the bioleached olivine under standard conditions was 158.2 μm, whereas using ultrasonic conditioning this value essentially halved to 79.2 μm. Based on the shape of the particle size distribution, it appears that the dispersion of large (~200–1000 μm) agglomerates, likely bonded by biofilm, was the dominant sonication effect on morphology.

Figure 4. Elemental leaching extent from fresh olivine in untreated and ultrasonically conditioned *Aspergillus niger* broth over 17 days; error bars indicate range of duplicates.

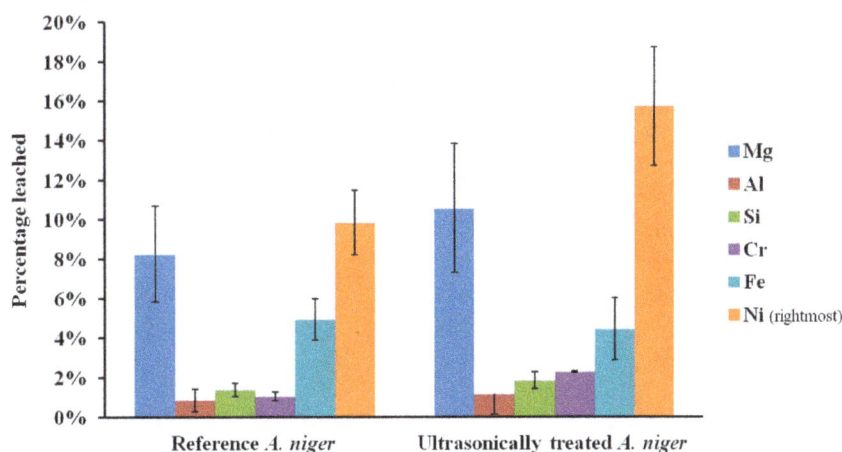

Figure 5. (**a**) Particle size distribution of fresh olivine after bioleaching with *Aspergillus niger* without ultrasonic conditioning; and (**b**) with ultrasonic conditioning.

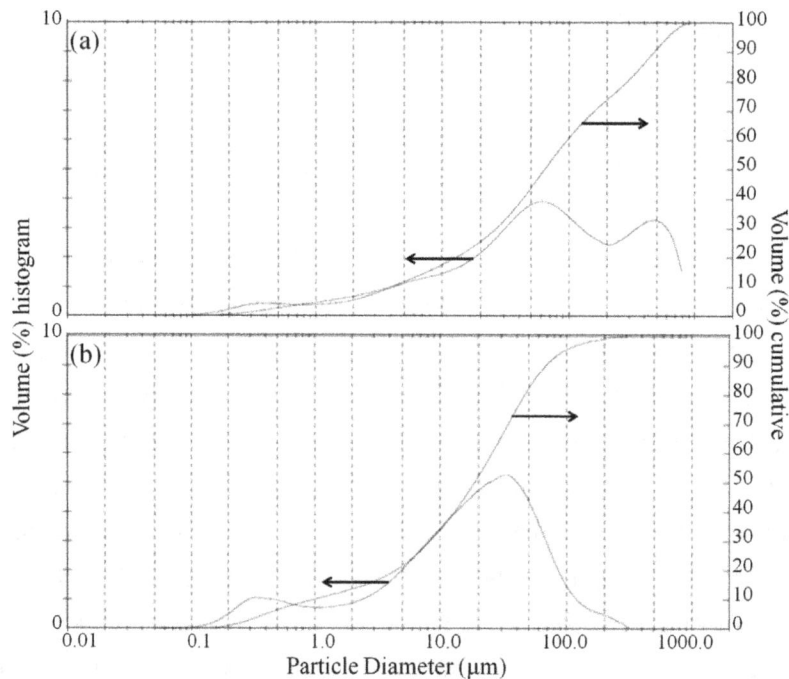

3.4. Leaching Selectivity

In all bioleaching experiments conducted, nickel leached preferentially over iron and magnesium. In conventional acid leaching, however, there is no preferential leaching of nickel, as noted by Sanemasa *et al.* [17], and also confirmed in our related study on the leaching of fresh and carbonated olivine [13]. Preferential leaching of nickel over iron and other metals, observed here in batch bioleaching and reported by McDonald and Whittington [18] to occur during heap leaching of certain laterite ores, makes further processing of the leach liquor easier and could potentially reduce the recovery cost. The preferential leaching is also clear in the case when only sterile biogenic substances are used for the leaching. It thus shows that both the microorganisms as well as the biogenic substances that they produce contribute to the selectivity. Likewise, sonication was seen to improve the Ni/Fe leaching ratio, wherein the leaching of fresh olivine by *A. niger* increased from 2.2 without ultrasonic conditioning to 3.5 under intermittent sonication. Sukla *et al.* [19] have reported similar effect for bioleaching of iron-rich (and nickel richer) lateritic ore, although the geochemistry of that ore differs significantly from that of silicate-based olivine, which explains the negligible iron dissolution observed in that study. The selectivity mechanism may follow two routes: (i) nickel may be preferentially chelated by the biogenic substances, and or (ii) some elements, for example magnesium, may re-precipitate in the form of inorganic or organic compounds such as (hydr)oxides, carbonates and oxalates [3].

4. Conclusions

In conclusion, it was found that the tested fungus *Aspergillus niger* leached substantially more nickel from olivine than the tested bacterium, *Paenibacillus mucilaginosus*, and also outperformed the fungi *Penicillium chrysogenum* and *Humicola grisea*. The production of greater quantities of organic acids by the fungi, and the resulting lower pH, can be attributed to this distinction. Contrary to conventional acid leaching, the microorganisms and their biogenic substances were found to leach nickel preferentially over magnesium and iron, main components of olivine. On average, a selectivity factor of 2.2 was achieved for nickel compared to iron. This suggests that nickel was chelated preferentially by soluble biogenic substances. Carbonating the olivine, however, did not improve the bioleaching performance. It appears that at low acidity levels found during bioleaching, the mobility of nickel became reduced in carbonated olivine, but this effect reverses once the pH is lowered (as reported by Santos et al. [13] and Van Audenaerde [14]), likely as a result of carbonate decomposition.

The impact of ultrasonic conditioning on bioleaching was tested and showed to substantially increase the bioleaching extent by fungi by over 50%, and to further contribute to nickel selectivity over iron, reaching a factor of 3.5. Ultrasound appeared to control fungal flocculation and reduce olivine particle agglomeration, thus improving fungal growth during the growth stage and nickel leaching during the leaching stage. The very low application rate of ultrasound (15 min every 24 h) required to achieve these benefits contributes to maintaining low processing costs of bioleaching. The assistance of ultrasonic conditioning can be further optimized by adjusting the energy intensity (e.g., 100–2000 W/L), the frequency (e.g., 16–100 kHz) of the ultrasonic waves, and the duration and the frequency of the treatment. Leaching efficiency can also likely be improved by reducing the particle size of the olivine ore, performing bioleaching for longer durations, and in continuous rather than batch mode. Further study is required to more rigorously assert the qualitative trends herein reported. In particular, care should be exercised to avoid conditions that stimulate fungal pelletization, which contributes to experimental variability. Pelletization can occur due to the presence of filamentous mycelium in the inoculum and shearing forces during mixing [20].

The findings of this study represent a first step in verifying that olivine can potentially become a commercially exploitable source of nickel in the future. It has also been demonstrated that chemoorganotrophic bioleaching can represent an alternative approach to more conventional chemolithotrophic bioleaching, especially when considering bioleaching of non-sulfidic ores or when looking to avoid the formation of environmentally hazardous sulfuric acid.

Acknowledgments

The KU Leuven Industrial Research Fund (IOF) is gratefully acknowledged for funding the Knowledge Platform on Sustainable Materialization of Residues from Thermal Processes into Products (SMaRT-Pro^2) in which this work was performed.

Author Contributions

Yi Wai Chiang conceptualized and managed the research, co-supervised the Master student and co-wrote the paper. Rafael M. Santos characterized materials and samples, co-supervised the Master student and co-wrote the paper. Aldo Van Audenaerde performed experiments, analyzed the data and co-wrote the paper. Annick Monballiu assisted with experiments and managed the microbial collection. Tom Van Gerven and Boudewijn Meesschaert supervised the personnel and managed the laboratory facilities.

Conflicts of Interest

The authors declare no conflict of interest.

References

1. Chiang, Y.W.; Santos, R.M.; Monballiu, A.; Ghyselbrecht, K.; Martens, J.A.; Mattos, M.L.T.; Van Gerven, T.; Meesschaert, B. Effects of bioleaching on the chemical, mineralogical and morphological properties of natural and waste-derived alkaline materials. *Miner. Eng.* **2013**, *48*, 116–125.

2. Valix, M.; Usai, F.; Malik, R. Fungal bioleaching of low grade laterite ores. *Miner. Eng.* **2001**, *14*, 197–203.

3. Burford, E.P.; Fomina, M.; Gadd, G.M. Fungal involvement in bioweathering and biotransformation of rocks and minerals. *Mineral. Mag.* **2003**, *67*, 1127–1155.

4. Brandl, H. Microbial Leaching of Metals. In *Biotechnology Set*, 2nd ed.; Rehm, H.J., Reed, G., Eds.; Wiley-VCH Verlag GmbH: Weinheim, Germany, 2001; pp. 191–224.

5. Castro, I.M.; Fietto, J.L.R.; Vieira, R.X.; Tropia, M.J.M.; Campos, L.M.M.; Paniago, E.B.; Brandão, R.L. Bioleaching of zinc and nickel from silicates using *Aspergillus niger* cultures. *Hydrometallurgy* **2000**, *57*, 39–49.

6. Simate, G.S.; Ndlovu, S.; Walubita, L.F. The fungal and chemolithotrophic leaching of nickel laterites—Challenges and opportunities. *Hydrometallurgy* **2010**, *103*, 150–157.

7. Chandraprabha, M.N.; Natarajan, K.A. Microbially induced mineral beneficiation. *Miner. Process. Extr. Metall. Rev.* **2010**, *31*, 1–29.

8. Willscher, S.; Bosecker, K. Studies on the leaching behaviour of heterotrophic microorganisms isolated from an alkaline slag dump. *Hydrometallurgy* **2003**, *71*, 257–264.

9. Yao, M.; Lian, B.; Teng, H.H.; Tian, Y.; Yang, X. Serpentine dissolution in the presence of bacteria *Bacillus mucilaginosus*. *Geomicrobiol. J.* **2013**, *30*, 72–80.

10. Burgstaller, W.; Schinner, F. Leaching of metals with fungi. *J. Biotechnol.* **1993**, *27*, 91–116.

11. Anjum, F.; Bhatti, H.N.; Ghauri, M.A. Enhanced bioleaching of metals from black shale using ultrasonics. *Hydrometallurgy* **2010**, *100*, 122–128.

12. Swamy, K.M.; Sukla, L.B.; Narayana, K.L.; Kar, R.N.; Panchanadikar, V.V. Use of ultrasound in microbial leaching of nickel from laterites. *Ultrason. Sonochem.* **1995**, *2*, S5–S9.

13. Santos, R.M.; Van Audenaerde, A.; Chiang, Y.W.; Iacobescu, R.I.; Knops, P.; Van Gerven, T. Enhanced Nickel Extraction from Ultrabasic Silicate Ores Using Mineral Carbonation pre-Treatment. In Proceedings of the 7th International Hydrometallurgy Symposium, Victoria, BC, Canada, 22–25 June 2014.

14. Van Audenaerde, A. Sustainable Nickel Extraction from Olivine via Carbonation and Bioleaching Treatments. Master's Thesis, KU Leuven, Leuven, Belgium, 2013.

15. Herrán, N.S.; López, J.L.C.; Pérez, J.A.S.; Chisti, Y. Effects of ultrasound on culture of Aspergillus terreus. *J. Chem. Technol. Biotechnol.* **2008**, *83*, 593–600.

16. Santos, R.M.; François, D.; Mertens, G.; Elsen, J.; Van Gerven, T. Ultrasound-intensified mineral carbonation. *Appl. Therm. Eng.* **2013**, *57*, 154–163.

17. Sanemasa, I.; Yoshida, M.; Ozawa, T. The dissolution of olivine in aqueous solution of inorganic acids. *Bull. Chem. Soc. Jpn.* **1972**, *45*, 1741–1746.

18. McDonald, R.G.; Whittington, B.I. Atmospheric acid leaching of nickel laterites review: Part I. Sulphuric acid technologies. *Hydrometallurgy* **2008**, *91*, 35–55.

19. Sukla, L.B.; Swamy, K.M.; Narayana, K.L.; Kar, R.N.; Panchanadikar, V.V. Bioleaching of Sukinda laterite using ultrasonics. *Hydrometallurgy* **1995**, *37*, 387–391.

20. Van Suijdam, J.C.; Kossen, N.W.F.; Paul, P.G. An inoculum technique for the production of fungal pellets. *Eur. J. Appl. Microbiol. Biotechnol.* **1980**, *10*, 211–221.

8

Interaction of Natural Organic Matter with Layered Minerals: Recent Developments in Computational Methods at the Nanoscale

Jeffery A. Greathouse [1,*], Karen L. Johnson [2] and H. Christopher Greenwell [3]

[1] Geochemistry Department, Sandia National Laboratories, P.O. Box 0754, Albuquerque, NM 87185-0754, USA

[2] School of Engineering and Computing Sciences, Durham University, South Road, Durham DH1 3LE, UK; E-Mail: karen.johnson@durham.ac.uk

[3] Department of Earth Sciences, Durham University, South Road, Durham DH1 3LE, UK; E-Mail: chris.greenwell@durham.ac.uk

* Author to whom correspondence should be addressed; E-Mail: jagreat@sandia.gov;

Abstract: The role of mineral surfaces in the adsorption, transport, formation, and degradation of natural organic matter (NOM) in the biosphere remains an active research area owing to the difficulties in identifying proper working models of both NOM and mineral phases present in the environment. The variety of aqueous chemistries encountered in the subsurface (e.g., oxic *vs.* anoxic, variable pH) further complicate this field of study. Recently, the advent of nanoscale probes such as X-ray adsorption spectroscopy and surface vibrational spectroscopy applied to study such complicated interfacial systems have enabled new insight into NOM-mineral interfaces. Additionally, due to increasing capabilities in computational chemistry, it is now possible to simulate molecular processes of NOM at multiple scales, from quantum methods for electron transfer to classical methods for folding and adsorption of macroparticles. In this review, we present recent developments in interfacial properties of NOM adsorbed on mineral surfaces from a computational point of view that is informed by recent experiments.

Keywords: mineral; surface; layered mineral; manganese oxides; manganese; molecular modeling; density functional theory; molecular dynamics; simulation; natural organic matter; humic acid; fulvic acid

1. Introduction to Natural Organic Matter-Mineral Systems

Natural, non-living, organic matter, often referred to as natural organic matter (NOM), plays a critical role in many biogeochemical processes, many of which have been broadly reviewed by Senesi *et al.* [1]. NOM is presently a relevant topic in terms of its role in global carbon budgets and cycling [2], organic matter preservation and conversion in petroleum systems [3], the stabilization and degradation of other contaminants [4] and within industrial contexts of water treatment [5], and particularly fouling in desalination plants [6].

Structurally, NOM is extremely complex with a three-dimensional macromolecular structure and consisting of a diverse group of organic molecules [7]. At the moment, there is no consensus on the primary binding mechanism holding the aggregated molecules that comprise NOM in place. However, the nature of the major functional groups present in NOM has been well characterized, with groups such as carboxyl, hydroxyl, phenolic, alcohol, carbonyl and methoxy, all present. Within these, carboxyl and phenolic groups are considered the most important functional groups responsible for the adsorption of NOM onto metal oxides [8]. Most NOM in sediments is present as molecularly-uncharacterized carbon (MUC) [9] and occurs in soil humus (globally estimated to be ~1.6×10^{18} g of carbon (gC)) and recently deposited marine deposits (1.0×10^{18} gC). MUC is so-called because it is not easily characterized as recognizable distinct organic molecules such as amino acids, sugars, lipids or lignin. Organic matter stabilized at mineral surfaces within sediments is, generally, less available to microorganisms and therefore plays a role in the overall removal of fixed CO_2 from the atmosphere [10].

The role of minerals in the stabilization of NOM in soils has been reviewed by von Lützow *et al.* [11], and two key pathways for the role of minerals in NOM preservation are proposed: (i) spatial inaccessibility against decomposer organisms due to, for example, occlusion or intercalation within the structure of layered minerals; and (ii) stabilization through interaction with mineral surfaces. More generally, Weber *et al.* [12] proposed a model of two types of natural organic matter within soils and sediments, "soft carbon" (amorphous or hydrolysable carbon) which is extractable at low temperatures with acids and/or bases and "hard carbon" (condensed or non-hydrolysable carbon) which is not extractable with acids and/or bases but is released at high temperatures. This "hard" stable carbon pool is thought to be composed of "humin" or "protokerogen" bound to mineral surfaces [13]. There is no current consensus on how organic matter becomes partitioned into hard and soft pools. Theories include geopolymerization, selective preservation of refractory molecules and physical protection of organic matter by mineral occlusion [2,10]. Geopolymerization (humification) is a general term to describe the oxidative polymerization of low molecular weight monomers and is thought to be catalyzed by mineral surfaces [14].

Though the interactions of NOM with mineral surfaces is fundamentally important within many natural processes [15], hitherto research in the area has seen a disconnect between the macro-scale and nano-scale. In other areas of research where analogous materials are present, for example within crude oil formation [16,17], clay-polymer nanocomposite materials [18] and heterogeneous catalysis [19,20], there have been concerted efforts to understand macroscopic phenomena in terms of nano-scale surface interactions between the organic molecules and the mineral surface. These have been undertaken using nano-scale surface chemistry methods and, increasingly, computer simulation. The latter has become an essential adjunct to understanding complex chemical structures and reactivity at surfaces, often beyond the resolution of analytical methods, or simply not possible to analyze without drastically perturbing the system of interest. NOM presents a severe challenge to analysis using computer simulation and closely coupled experiments owing to its complexity, high molecular weight and diversity, which leads to difficulty in both experimental characterization of NOM and building molecular models of NOM, though present day computational chemistry resources have now matured sufficiently to begin to make such problems tractable [21].

In this review, we seek to give an introductory overview of the present state of the art in modeling of NOM, or representative fragments, at mineral surfaces as well as identifying areas that will present interesting research challenges into the future. In the next section an introduction to computational chemistry is given, which is followed by a summary of simulation studies of NOM and NOM-mineral systems. This is followed by a brief overview of experimental studies on natural and model systems to understand NOM-mineral interactions. The article concludes with an analysis of a future challenge in this area—studying and simulating the surface structure and reactions of NOM at redox active mineral surfaces, which requires an understanding of electron transfer coupled to other physico-chemical processes.

2. Introduction to Molecular Simulation Methods

The application of molecular simulation methods to minerals and their interfaces has been reviewed elsewhere [21–25]. A key factor in considering which approach is employed in computational chemistry is the size of the system of interest, the type of information desired, and the time scale of the process of interest (see Figure 1). Inevitably, there is a trade off in accuracy against accessible time and length scales, though multi-scale simulation methods are being developed, which employ either hierarchical [26] or on-the-fly parameter exchange between length and time-scales [27] to try and address these challenges.

Figure 1. Modeling and simulation methods and corresponding system sizes and times. Reprinted with permission from [21].

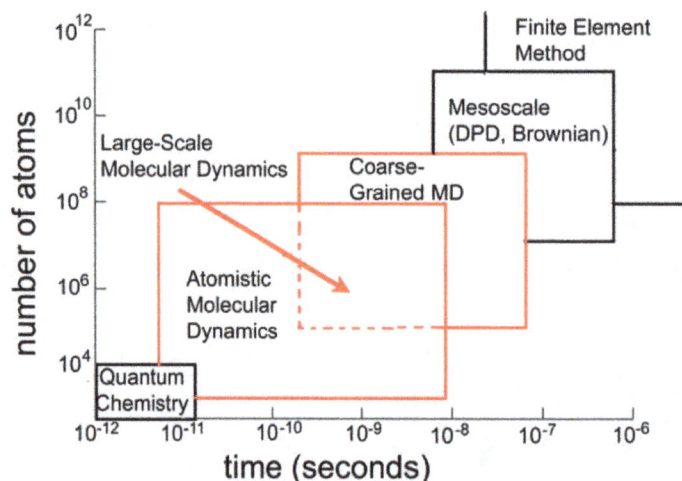

Electronic structure methods such as Hartree-Fock and density functional theory (DFT) include electronic effects (*i.e.*, electron transfer, bond breaking/forming, UV spectroscopy) and represent the highest level of theory, firmly based in quantum mechanics (QM), which can be applied to chemical systems. However, owing to the high computational cost of such methods, their use on NOM or

NOM-surface applications are restricted to small organic molecules or surface cluster (a cluster is a small, distinct group of representative atoms) models containing representative functional groups, or relevant model compounds where key functionality is included. Such calculations reveal structural properties such as interatomic distances and angles or binding energies that can be used to investigate the effect of organic functional groups on surface interactions. However, one limitation of cluster-based approaches is that the charge on the atoms present can be artificially high, and it is challenging to represent the right degree of substitution in extended mineral structures [28]. Periodic models eliminate the need for small clusters to represent mineral surfaces, as well as allow the use of simpler plane waves over localized basis sets, and DFT results for bulk minerals such as lattice parameters and mechanical properties can be directly compared with experiment for validation [29–32].

Classical simulation methods (so called as they apply classical, non-quantum mechanics based, Newtonian mechanics) require the use of interatomic potential energy expressions (commonly called a force field (FF)) that have been parameterized to reproduce structural and energetic properties from experiment, or quantum calculations. This allows for the simulation of larger and more realistic systems and remains the method of choice for mineral interfaces with bulk aqueous solutions or organic liquids. However, one disadvantage is that to employ these parameters for systems outside the original systems studied, or under different conditions of temperature and/or pressure can lead to inconsistent or inaccurate results. In molecular dynamics (MD) simulation, the time evolution of a molecular system is determined by solving the relevant equations of motion for the specific thermodynamic ensemble of interacting particles/atoms/molecules. The temperature is usually held constant through a thermostat that controls atom velocities, although isoenthalpic methods are available. As such MD is a method that is usually readily comparable to experimental systems, usually carried out under constant pressure conditions.

3. Simulation Studies of Organic-Mineral Structures

Though QM simulation methods are hampered by the small atom numbers that can be studied, cluster models of clay mineral surfaces have been used in DFT calculations of organic adsorption. A larger mineral surface cluster was achieved using a molecular mechanics/quantum mechanics (QM/MM) hybrid calculation of acetate binding on a cluster model of kaolinite-type mineral surfaces [33]. More recently, periodic DFT methods have been used to study the binding of organic acids and larger hydrophobic molecules to mineral surfaces [16,34–40], as well as the modeling of reactions of organic molecules at mineral surfaces, including decarboxylation of natural organic matter [16,41].

For larger, more complex, organic molecules, and in particular biomolecular compounds, classical, FF-based simulation methods have enabled the computational chemist to probe interfacial structures and dynamics. Several general FFs have been developed that are transferrable to a wide range of organic and bioorganic molecules [42–44] and aqueous solutions of these molecules. Some of these FFs were parameterized to reproduce the bulk properties of organic liquids [45], which is critical when simulating organic phases near a mineral surface. Several FFs have been developed specifically for use with bulk and layered minerals and their aqueous interfaces, ranging from a computationally simple nonbonded approach (ClayFF [46]), bonded layers (Teppen et al. [47], INTERFACE [48]), and reactive methods (ReaxFF [49]). While mineral-water interfaces have been carefully considered when developing these FFs, the simulation of mineral-organic interfaces presents a challenge because no general FF has been developed specifically for mineral-organic interfaces. Instead, MD simulations of such interfaces usually combine a general organic FF with a mineral FF. This untested approach can yield successful comparisons with experiment. In simulating the adsorption of small alcohol and thiol compounds onto $Al(OH)_3$, Greathouse *et al.* [50] showed that trends in simulated adsorption enthalpies were in excellent agreement with experiment (Figure 2).

Figure 2. Adsorption enthalpy of C1–C3 alcohols and thiols on $Al(OH)_3$ (**top**); and graphite (**middle**); surfaces, compared with chemisorption experiments of alcohol adsorption (**bottom**). Reprinted with permission from [50]. Copyright 2012 American Chemical Society.

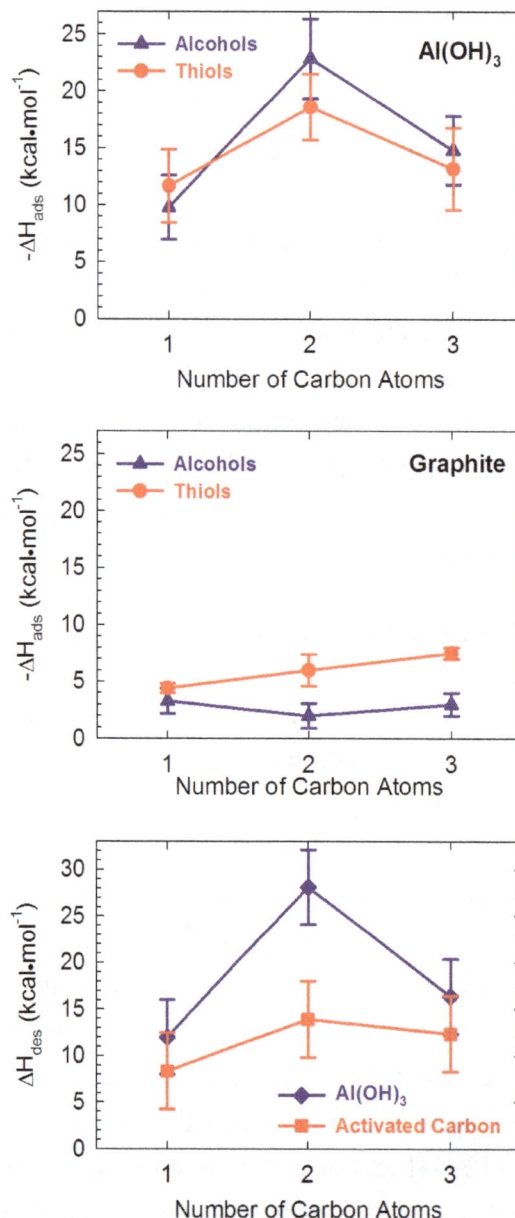

MD simulations have also been used to study the adsorption of small organic molecules [51–54], polymers [55–57], (bio)organic acids [58–60], or entire nucleic acid strands [61–65] onto mineral surfaces. The latter simulations have been particularly noteworthy owing to the size and complexity of the biomolecules studied, with Thyveetil *et al.* [61] using a high performance computing grid to simulate hydrated whole deoxyribonucleic acid (DNA) plasmids of 480 base pairs within a periodic mineral cell of 50 nm lateral dimensions, and containing 1,157,038 atoms in total. This methodology was further used to probe the structures of different nucleic acids (ribonucleic acid—RNA, peptide nucleic acid (PNA) and DNA) on layered hydroxide and aluminosilicates surfaces [64]. An interesting insight gained from bringing MD techniques to bear on these systems was the novel discovery that a clay mineral surface can rapidly accelerate folding in charged biomolecules, here RNA but feasibly NOM, relative to the same molecule in solution, as illustrated in Figure 3 by Swadling *et al.* [63]. Minerals can also encourage the unfolding of macromolecules due to the binding of functional groups on mineral surfaces [14].

Figure 3. Simulation snapshot taken from simulations of a folded RNA (nitrogen and phosphorous atoms shown) strand tethered to a montmorillonite surface (silicon atoms shown). Atoms are colored as follows: Si (magenta), Ca (gray), N (blue), P (yellow). Water molecules are hidden to aid in viewing. Reprinted with permission from [63]. Copyright 2010 American Chemical Society.

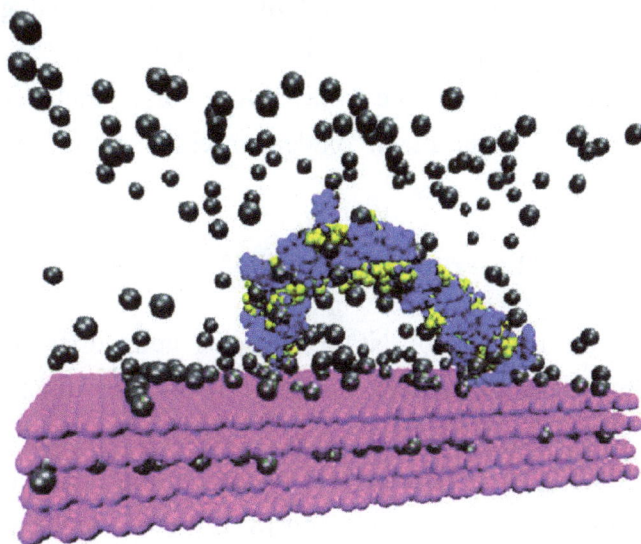

4. Experimental Studies of Natural Organic Matter-Mineral Structures

While spectroscopic methods have been used to investigate the structure of NOM adsorbed on mineral surfaces [66–74], using simple organic acids as proxies for NOM can provide valuable insight into possible adsorption mechanisms [75,76]. Carboxylic functional groups are particularly relevant, not only to NOM but also as ligands responsible for binding and transporting potentially toxic elements, owing to many mineral surfaces carrying areas of net positive charge and poly-carboxylate species bridging between these surfaces and dissolved metals during flocculation processes. Furthermore, once formed, metal carboxylates are extremely stable and can rapidly remove carbon into the geosphere. One well-studied example of geoploymerization is the Maillard reaction [77], a sugar-amino-acid condensation reaction, which forms compounds known as melanoidins, catalyzed by redox active minerals, for example manganese oxides such as birnessite [78,79].

The association between stabilized organic matter and clay content in sediments is well documented (e.g., [80]) but far from understood. Kennedy and Wagner [3] have recently argued that smectite is particularly important in governing carbon preservation in marine sediments. Zhang *et al.* [68] noted that pH increases in clay suspensions after fulvic acid (FA)/humic acid (HA) adsorption suggesting that ligand exchange occurred and FA/HA-clay complexes formed. This was particularly noted for the 2:1 type clay minerals such as smectite and vermiculite, which have surface complexed cations.

More fundamental studies have also been undertaken for NOM-mineral interfaces. Janot *et al.* [81] have recently studied the adsorption behavior and reactivity of purified HA on α-alumina, varying the pH, ionic strength and surface coverage. The authors characterized the complexes formed at a variety of scales, including UV-Vis spectroscopy, spectrophotometric titration and size exclusion chromatography. From the HA concentration in the supernatant and mass balance calculations, "titration curves" were experimentally proposed for the adsorbed fractions, and other measurements allowed insight into protonation state [81]. Such detailed experiments are essential to inform initial configurations for computer simulation.

The role of iron and aluminum oxides, which are adsorbed onto clay particles, is known to play a part in MUC stabilization [68,82–85], although the mechanisms for stabilization have not been determined. Zhang *et al.* [68] also noted that the higher concentration of iron cations in vermiculite facilitated ligand exchange and cation bridging of carboxyl and hydroxyl functional groups, thus improving adsorption capacity. Finally, Chorover [78] and Jokic [79] have looked at the role of synthetic manganese oxides in stabilizing dissolved organic matter (DOM) and concluded that although they are not as good at adsorbing DOM as other metal oxides or clays, they have a large capacity for fractionating DOM at their surfaces.

5. Simulations of Natural Organic Matter (NOM) and NOM-Mineral Interfaces

The broad range of structural features of NOM particles adds difficulty in creating realistic molecular models of these compounds. Fortunately, much of the ion coordination activity in NOM has been shown to occur through several key functional groups, including primarily carboxylate groups but also alcohol/phenol groups [86,87]. A recent review paper outlines current understanding of the important effect of cation bridging in NOM aggregation [87]. Indeed, molecular modeling studies of NOM-mineral interfacial systems indicates that cation bridging between NOM and surface sites predominates over direct interactions between NOM functional groups and the surface [88,89], although humic substances can adsorb directly to clay surfaces in confined interlayer environments [90]. In a recent review of soil humic substances, Sutton and Sposito concluded that NOM consists of a "supramolecular association" of small organic molecules linked by short-range interactions dominated by amide-amide interactions [91]. Regardless of its classification as a single large macromolecule or a cluster of small molecules, the interaction of NOM with hydrophilic sites on mineral surfaces occurs through key functional groups. In fact, various quantum and classical (FF) modeling methods have been used to show that simpler organic compounds containing these key functional groups successfully reproduce the structural, cation binding, and thermodynamic properties of NOM [92–96].

While a number of NOM models based on experimental measurements have been proposed, two recent reviews summarize the few NOM models that have been used in molecular simulations [88,97]. The Temple-Northeastern-Birmingham (TNB) model [7,86,98] of a NOM fragment is an average structure based on spectroscopic and analytical data. It is a good computational analog of the Suwannee

River NOM often used in experimental studies [99–101]. Other NOM models have been proposed based on pyrolysis measurements [102,103], lignin-carbohydrate complexes [104,105], and fulvic acid models based on oxysuccinic acid structures [106]. Representative structures of these models are shown in Figure 4.

Kalinichev *et al.* have used MD simulation to test ion bridging models of NOM aggregation in aqueous solution [107,108] and at polyethersulfone membrane surfaces [101]. In aqueous solution, the residence time of calcium ion coordination to NOM carboxylic groups was found to be 0.5 ns, more than an order of magnitude longer than for sodium ion coordination [107,108]. Aggregation of NOM particles occurred in calcium solutions due to ion bridging, but not in sodium solutions where ion bridging did not occur [107,108]. Ion bridging between NOM particles was also found to predominate near polyethersulfone membrane surfaces, while no evidence of NOM surface complexes was seen [101].

Figure 4. Repeating units of proposed natural organic matter (NOM) models: (**a**) Temple-Northeastern-Birmingham (TNB) [86]; (**b**) lignin-carboxylate complex [104]; and (**c**) cutin-lignin-tannin complex [106]. Atoms are colored as followed: C (gray), H (white), O (red), N (blue).

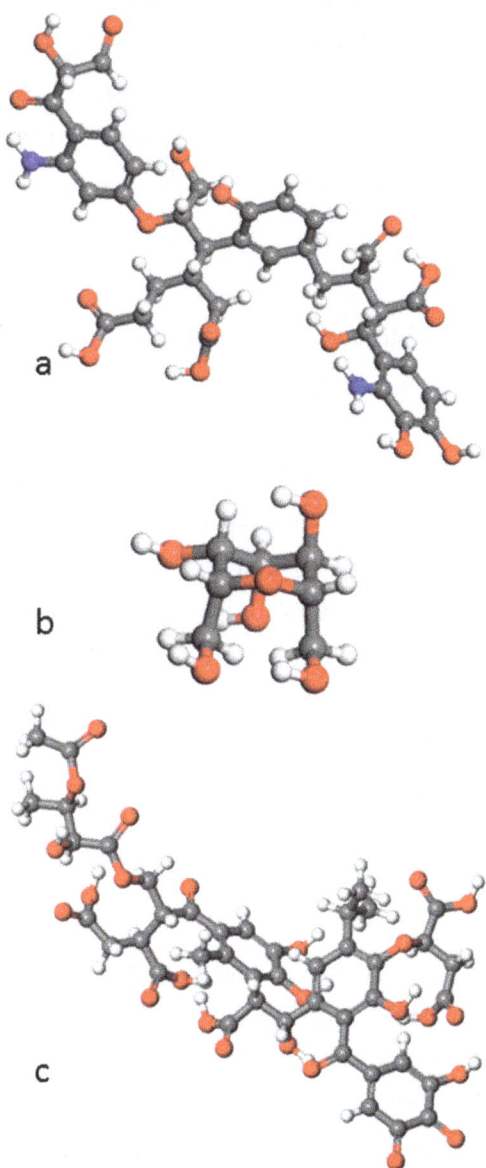

Kalinichev [97] also compared structural features of Ca-NOM complexes using three different organic FFs, and found that all three FFs gave approximately the same structure and binding energy for calcium ion binding at carboxylate groups. This is an important finding in the development of specific organic-mineral FF parameters, and it is consistent with experimental results showing that metal complexation behavior in NOM is similar to well-characterized polymers with similar functionalization [109–113]. The fact that different organic FF parameters yield similar structural and energetic properties for surface complexes indicates a very broad energy minimum for such complexes. Therefore, DFT calculations of such surface complexes might have convergence difficulties if a wide range of conformations have equivalent energy.

6. Redox Active Mineral-Natural Organic Matter—A Future Challenge

Whereas simulations have started to add insight into the interactions of natural organic matter, and related model compounds with minerals, and in particular layered minerals, an area that is of key importance is the role of redox active minerals in, firstly, adsorbing organic material, and, secondly, acting as an oxidant to enable the degradation of NOM both at the metal oxide surface and within the mineral matrix when NOM becomes occluded. In this context, iron- and manganese-containing minerals have an important role to play. The environmental significance of dissolved organic matter (DOM) oxidation coupled to Fe(III) and Mn(IV) reduction has been well documented [114], but there remain large uncertainties on how and where organic matter is transformed into different fractions and how these fractions take part in the slow and fast carbon cycles.

6.1. Role of Iron Based Minerals

Lalonde *et al.* [84] have recently highlighted the importance of Fe oxides in stabilizing NOM in marine sediments, but no mechanism was presented for how stabilization occurs. Additionally, it is not clear whether the stabilized organic matter is associated with an iron phase or instead consists of precipitated iron-organic matter complexes. However, ligand exchange between carboxyl groups of NOM and hydroxyl groups on hydrous aluminum and iron oxide surfaces is thought to be the dominant mechanism for NOM adsorption in fresh water sediments [115]. Kaiser *et al.* [116] found that sorption of dissolved NOM derived from the oxidative decomposition of lignocellulose to Al and Fe oxyhydroxides involved relatively strong interactions between surface metals and acidic, particularly aromatic, organic ligands. They observed that the sorption of a large fraction of the NOM was hardly reversible and, as such, cycles of adsorptive and desorptive processes strongly favored the accumulation of the more recalcitrant lignin-derived NOM.

From a computational perspective, a number of recent studies have examined electron transfer processes in iron bearing minerals, which necessitate the use of quantum methods. For example Geatches *et al.* have looked at how electron transfer occurs in different iron rich nontronite clay forms, using density functional theory [117] and further considering the effect of spin [41], while, recently, Alexandrov *et al.* [118] studied exchange and conduction processes, also in nontronite. Alexandrov and Rosso [119] further investigated the adsorption and oxidation of iron (II) at the surface of iron rich clay minerals. Wander *et al.* [120] have studied the dynamics of charge hopping in mixed-valent iron layered hydroxides (commonly called green rusts). Though these studies do not include NOM, the potential for studying decomposition of NOM using these techniques is apparent. In a move to start investigating organic-mineral reactions, Geatches *et al.* [41] have recently studied the decomposition of fatty acid model compounds at the basal surfaces of ferruginous clay.

Owing to the importance of electron transfer processes in mixed valent iron systems, there have been few classical molecular dynamics studies as, inherently, such phenomena could not be captured. However, there have been studies of the adsorption of biomolecules at iron oxide surfaces, for example Qiang *et al.* [121] have recently studied the interactions between chitosan and Fe_3O_4 crystal surfaces. To try and bridge between the time and size scale limitations when studying iron bearing minerals, attempts have been made to develop a reactive force field, where parameters may change according to sets of rules being met [122].

6.2. Role of Manganese Oxides

Manganese oxide is one of the most powerful oxidizing agents found on Earth [123]. The most common form of manganese oxide in both soil and marine sediments is birnessite [124]. Birnessite ($[Na,Ca,K]_x Mn_2O_4 \cdot 1.5H_2O$), is often simplified and denoted as δ-MnO_2 in the geochemical literature. Birnessite precipitates in soils mainly as a result of Mn^{2+} oxidation promoted by bacteria and fungi, and similarly to bacteriogenic ferrihydrite, it forms highly reactive biofilms [125–128] as well as coatings [129,130], nodules and concretions [131].

Solving the crystal structure of birnessite is complicated not only by vacancies in the octahedral sheet, but also by multiple oxidation states of manganese present, namely Mn(III) and Mn(IV). As a result, very few papers on computational chemistry of birnessite have been published. Only one paper on classical MD simulations of birnessite has appeared, but the reported interlayer structure agreed with diffraction experiments [132]. Quantum methods have been used to investigate the effect of vacancy disordering on structural and energetic properties [133,134] as well as cation adsorption [133,135–137].

Since Mn is present at 1/100th of the concentration of Fe [138], δ-MnO_2 tends to be overlooked. And yet this is a questionable approximation, because firstly, δ-MnO_2 tends to be present as fine-grained aggregates [129,130], it exerts an influence far out of proportion to its total concentration. Secondly, since δ-MnO_2 is a much stronger oxidant than iron oxides, it is more rapidly reduced in anoxic environments taking part in more oxidative reactions than iron. Interestingly, Mn(III) oxides have higher redox potentials of 1.51 V (*cf.* Mn(IV) oxides 1.23 V). The manganese oxidation state in soil is controlled by redox potential and pH. Of the three possible oxidation states that manganese can have when present in soils (II, III and IV), it was thought that the occurrence of Mn(II) in the environment is thermodynamically favored in the absence of oxygen at low pH, whereas Mn(III) and Mn(IV) are favored under aerobic conditions at high pH [138]. However, Madison *et al.* [139] has just revised the redox sequence for sediments and soils to include Mn(III) in solution which can act as both an oxidant and a reductant. This new redox sequence has significant implications for the role of manganese in natural organic matter (NOM) interactions. Since Mn(III) is present in suboxic zones, this extends the reach of manganese's ability to transform NOM. Unlike iron, which can be removed in anoxic environments as a sulfide, manganese oxides are effectively recycled at oxic/anoxic interfaces [114], and soluble manganese is released which usually ends up re-precipitating at the boundary, or if present as Mn(III), may go on to oxidize NOM in its soluble form. For this reason, δ-MnO_2 may play an important role in NOM transformations in soil and marine sediments even though it is present at small concentrations.

Birnessite has a low point of zero charge (~pH 2) [78], and the oxidation of many polar organic compounds by birnessite increases with decreasing pH [140–144]. The increased oxidation of organic compounds at low pH has been attributed to:

- Charging of any amide groups present in the organic contaminant resulting in electrostatic attraction to the oxide surface [140];
- Increased sorption of neutral phenolic groups;
- Decrease of the negative charge on MnO_2;
- Increased redox potential of the MnO_2/Mn^{2+} couple [145]; and
- Enhanced removal of Mn^{2+} from the oxide surface, exposing new reactive sites [146].

Banerjee and Nesbitt [147,148] have described the reduction of birnessite by both oxalate and humate complexes. X-ray photoelectron spectroscopy (XPS) data confirmed that it is the carboxylic group which reacts with the mineral surface and the redox reaction proceeds via a one electron transfer from Mn(IV) to Mn(III). There is no evidence for reduction of Mn(III) to Mn(II) suggesting that a strong Mn(III) carboxyl surface complex has formed. Durand *et al.* [149] studied the binding mechanism of formic acid using XPS, fourier transform infrared spectroscopy (FTIR) and thermogravimetric analysis on amorphous manganese oxides and concluded that the carboxyl group binds to the mineral in a bidentate configuration on an electron neutral surface. However where surface defects such as oxygen vacancies are present, a bridging configuration predominates (Figure 5).

Figure 5. Possible surface complexes for carboxylate directly bound to surface metal atoms (blue) in monodentate, bridging, and bidentate configurations.

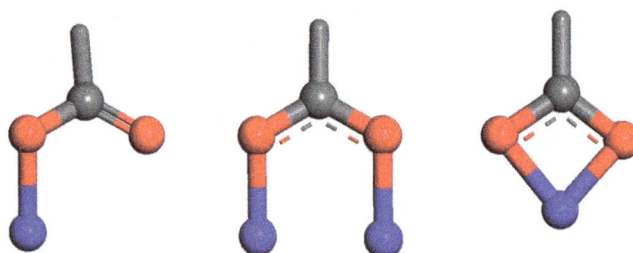

7. Concluding Remarks

Through advances in computer simulation hardware and software, as well as analytical technology, the study of molecules as complex as natural organic matter have become accessible, and computational chemists working alongside geoscientists are beginning to add insight into the nano-scale structures of NOM in solution and at mineral surfaces. Through studying simplified NOM macromolecular structures, as well as fragments of NOM and synthetic polymers of relevant functionality, increasing insight is being gained of how NOM interacts at mineral surfaces. Electronic structure simulations can reveal reaction pathways and key interfacial interactions, while classical simulations give structure and dynamics at a molecular-macromolecular scale.

Despite these advances, NOM-mineral interfaces present several areas where much work is still needed, and which are present research challenges to ongoing experimental and computational efforts.

Additionally, while here we are interested in mineral-NOM interactions, there is also an increasing awareness of the importance of ternary mineral-NOM-metal interactions, which present further challenges [150]. While high-resolution X-ray adsorption spectroscopy quantifies nearest neighbor structure at the atomistic level, there is a need for more such experimental studies of organic interactions with mineral surfaces. Additionally, NOM can often be present in only small quantities (e.g., monolayer) at a mineral surface and in irregular patterns, complicating the study of NOM-mineral interfaces using such methods.

The emergent interest in organo-manganese oxide systems in technological applications also bodes well for increased experimental research on how model organic molecules interact with manganese oxide surfaces. Since manganese oxides act as cation exchange minerals, the exchange of inorganic cations such as sodium and potassium for organo-ammonium species can be attained to afford new materials with enhanced adsorption [151] or electrical [152] properties, as well as potentially interesting and novel crystal morphologies [153]. In the latter case, Ahmed and Huang [153] were able to use polyethylene glycol-melamine-formaldehyde to grow birnessite nanotubes and nanorods. Further materials applications have involved carbon nanotube manganese oxide composites for electrode applications [154].

A significant computational challenge is the coupling of differing phenomena, allowing access to structure, dynamics and reactivity at interfaces. A second computational challenge is extending the timescale of simulations to enable the complex macromolecules in NOM to reach equilibrated, low energy structures. For the first of these challenges, the elements are there; research has been carried out on electron transfer processes in minerals, and more is understood regarding the nature and composition of NOM. The second challenge is presently limited to a few enhanced sampling MD methods [155,156]. Biased kinetic Monte Carlo methods can also be used to quickly search through a constrained set of possible reaction coordinates, though this technique has, hitherto, been mainly used to probe small molecules at catalyst surfaces [157].

Just a decade ago, a simulation of NOM systems would have required access to considerable high performance computing resources, whereas now, owing to rapid developments in computational performance as well as efficient codes, NOM systems are able to be simulated, and such approaches will both be informed by, and inform, data from the latest generation of experiments. For example, synchrotron-based mass spectrometry allows the simultaneous collection of molecular level data on organic species, and information on the mineral structure on which they are adsorbed [158]. Neutron diffraction of layered minerals intercalated with polymers is another promising experimental approach to obtain interfacial structure [159]. Through collaborative research using methods such as these, we presently stand at a point where research into NOM interactions is likely to grow rapidly.

Acknowledgments

This work is supported by the U.S. Department of Energy, Office of Basic Energy Sciences, Geosciences Research Program. Sandia National Laboratories is a multi-program laboratory managed and operated by Sandia Corporation, a wholly owned subsidiary of Lockheed Martin Corporation, for the U.S. Department of Energy's National Nuclear Security Administration under contract DE-AC04-94AL85000. Support from Durham University (Department of Earth Sciences and Institute of Advanced Study) as well as the EU are gratefully acknowledged.

Author Contributions

All authors were involved in writing and revising all parts of the manuscript.

Conflicts of Interest

The authors declare no conflict of interest.

References

1. Senesi, N.; Xing, B.; Huang, P.M. *Biophysico-Chemical Processes Involving Natural Nonliving Organic Matter in Environmental Systems*; John Wiley and Sons: Hoboken, NJ, USA, 2009.

2. Keil, R.G.; Mayer, L.M. Mineral matrices and organic matter. In *Treatise on Geochemistry*, 2nd ed.; Holland, H.D., Turekian, K.K., Eds.; Elsevier: Oxford, UK, 2014; pp. 337–359.

3. Kennedy, M.J.; Wagner, T. Clay mineral continental amplifier for marine carbon sequestration in a greenhouse ocean. *Proc. Natl. Acad. Sci. USA* **2011**, *108*, 9776–9781.

4. Polubesova, T.; Chefetz, B. DOM-affected transformation of contaminants on mineral surfaces: A review. *Crit. Rev. Env. Sci. Technol.* **2014**, *44*, 223–254.

5. Matilainen, A.; Vepsalainen, M.; Sillanpaa, M. Natural organic matter removal by coagulation during drinking water treatment: A review. *Adv. Colloid Interface Sci.* **2010**, *159*, 189–197.

6. Al-Amoudi, A.S. Factors affecting natural organic matter (NOM) and scaling fouling in NF membranes: A review. *Desalination* **2010**, *259*, 1–10.

7. Jansen, S.A.; Malaty, M.; Nwabara, S.; Johnson, E.; Ghabbour, E.; Davies, G.; Varnum, J.M. Structural modeling in humic acids. *Mater. Sci. Eng. C Bio. S.* **1996**, *4*, 175–179.

8. Snoeyink, V.L.; Jenkins, D. *Water Chemistry*; John Wiley and Sons: New York, NY, USA, 1980.

9. Hedges, J.I.; Eglinton, G.; Hatcher, P.G.; Kirchman, D.L.; Arnosti, C.; Derenne, S.; Evershed, R.P.; Kogel-Knabner, I.; de Leeuw, J.W.; Littke, R.; *et al.* The molecularly-uncharacterized component of nonliving organic matter in natural environments. *Org. Geochem.* **2000**, *31*, 945–958.

10. Burdige, D.J. Preservation of organic matter in marine sediments: Controls, mechanisms, and an imbalance in sediment organic carbon budgets? *Chem. Rev.* **2007**, *107*, 467–485.

11. Von Lutzow, M.; Kogel-Knabner, I.; Ekschmitt, K.; Matzner, E.; Guggenberger, G.; Marschner, B.; Flessa, H. Stabilization of organic matter in temperate soils: Mechanisms and their relevance under different soil conditions—A review. *Eur. J. Soil Sci.* **2006**, *57*, 426–445.

12. Weber, W.J., Jr.; McGinley, P.M.; Katz, L.E. A distributed reactivity model for sorption by soils and sediments. 1. Conceptual basis and equilibrium assessments. *Environ. Sci. Technol.* **1992**, *26*, 1955–1962.

13. Vandenbroucke, M.; Largeau, C. Kerogen origin, evolution and structure. *Org. Geochem.* **2007**, *38*, 719–833.

14. Collins, M.J.; Bishop, A.N.; Farrimond, P. Sorption by mineral surfaces—Rebirth of the classical condensation pathway for kerogen formation. *Geochim. Cosmochim. Acta* **1995**, *59*, 2387–2391.

15. Brown, G.E.; Calas, G. Mineral-aqueous solution interfaces and their impact on the environment. *Geochem. Perspect.* **2012**, *1*, 483–742.

16. Geatches, D.L.; Clark, S.J.; Greenwell, H.C. Role of clay minerals in oil-forming reactions. *J. Phys. Chem. A* **2010**, *114*, 3569–3575.

17. Wu, L.M.; Zhou, C.H.; Keeling, J.; Tong, D.S.; Yu, W.H. Towards an understanding of the role of clay minerals in crude oil formation, migration and accumulation. *Earth Sci. Rev.* **2012**, *115*, 373–386.

18. Chen, B.; Evans, J.R.G.; Greenwell, H.C.; Boulet, P.; Coveney, P.V.; Bowden, A.A.; Whiting, A. A critical appraisal of polymer-clay nanocomposites. *Chem. Soc. Rev.* **2008**, *37*, 568–594.

19. Sels, B.F.; de vos, D.E.; Jacobs, P.A. Hydrotalcite-like anionic clays in catalytic organic reactions. *Catal. Rev. Sci. Eng.* **2001**, *43*, 443–488.

20. Debecker, D.P.; Gaigneaux, E.M.; Busca, G. Exploring, tuning, and exploiting the basicity of hydrotalcites for applications in heterogeneous catalysis. *Chem. Eur. J.* **2009**, *15*, 3920–3935.

21. Suter, J.L.; Anderson, R.L.; Greenwell, H.C.; Coveney, P.V. Recent advances in large-scale atomistic and coarse-grained molecular dynamics simulation of clay minerals. *J. Mater. Chem.* **2009**, *19*, 2482–2493.

22. Geysermans, P.; Noguera, C. Advances in atomistic simulations of mineral surfaces. *J. Mater. Chem.* **2009**, *19*, 7807–7821.

23. Greenwell, H.C.; Jones, W.; Coveney, P.V.; Stackhouse, S. On the application of computer simulation techniques to anionic and cationic clays: A materials chemistry perspective. *J. Mater. Chem.* **2006**, *16*, 708–723.

24. Anderson, R.L.; Ratcliffe, I.; Greenwell, H.C.; Williams, P.A.; Cliffe, S.; Coveney, P.V. Clay swelling—A challenge in the oilfield. *Earth Sci. Rev.* **2010**, *98*, 201–216.

25. Cygan, R.T.; Greathouse, J.A.; Heinz, H.; Kalinichev, A.G. Molecular models and simulations of layered minerals. *J. Mater. Chem.* **2009**, *19*, 2470–2481.

26. Lyubartsev, A.; Tu, Y.Q.; Laaksonen, A. Hierarchical multiscale modelling scheme from first principles to mesoscale. *J. Comput. Theor. Nanosci.* **2009**, *6*, 951–959.

27. Moras, G.; Choudhury, R.; Kermode, J.R.; CsÁnyi, G.; Payne, M.C.; de Vita, A. Hybrid quantum/classical modeling of material systems: The "learn on the fly" molecular dynamics scheme. In *Trends in Computational Nanomechanics: Transcending Length and Time Scales*; Dumitrica, T., Ed.; Springer Netherlands: Dordrecth, The Netherlands, 2010; Volume 9, pp. 1–23.

28. Boulet, P.; Greenwell, H.C.; Stackhouse, S.; Coveney, P.V. Recent advances in understanding the structure and reactivity of clays using electronic structure calculations. *J. Mol. Struct.* **2006**, *762*, 33–48.

29. Larentzos, J.P.; Greathouse, J.A.; Cygan, R.T. An ab initio and classical molecular dynamics investigation of the structural and vibrational properties of talc and pyrophyllite. *J. Phys. Chem. C* **2007**, *111*, 12752–12759.

30. Ortega-Castro, J.; Hernandez-Haro, N.; Hernandez-Laguna, A.; Sainz-Diaz, C.I. DFT calculation of cystallographic properties of dioctahedral 2:1 phyllosilicates. *Clay Miner.* **2008**, *43*, 351–361.

31. Berghout, A.; Tunega, D.; Zaoui, A. Density functional theory (DFT) study of the hydration steps of $Na^+/Mg^{2+}/Ca^{2+}/Sr^{2+}/Ba^{2+}$-exchanged montmorillonites. *Clays Clay Miner.* **2010**, *58*, 174–187.

32. Voora, V.K.; Al-Saidi, W.A.; Jordan, K.D. Density functional theory study of pyrophyllite and M-montmorillonites (M = Li, Na, K, Mg, and Ca): Role of dispersion interactions. *J. Phys. Chem. A* **2011**, *115*, 9695–9703.

33. Gerzabek, M.H.; Aquino, A.J.A.; Haberhauer, G.; Tunega, D.; Lischka, H. Molecular modelling-opportunities for soil research. *Bodenkultur* **2001**, *52*, 133–146.

34. Aquino, A.J.A.; Tunega, D.; Gerzabek, M.H.; Lischka, H. Modeling catalytic effects of clay mineral surfaces on peptide bond formation. *J. Phys. Chem. B* **2004**, *108*, 10120–10130.

35. Tunega, D.; Gerzabek, M.H.; Haberhauer, G.; Totsche, K.U.; Lischka, H. Model study on sorption of polycyclic aromatic hydrocarbons to goethite. *J. Colloid Interface Sci.* **2009**, *330*, 244–249.

36. Liu, X.D.; Lu, X.C.; Wang, R.C.; Zhou, H.Q.; Xu, S.J. Surface complexes of acetate on edge surfaces of 2:1 type phyllosilicate: Insights from density functional theory calculation. *Geochim. Cosmochim. Acta* **2008**, *72*, 5896–5907.

37. Geatches, D.L.; Jacquet, A.; Clark, S.J.; Greenwell, H.C. Monomer adsorption on kaolinite: Modeling the essential ingredients. *J. Phys. Chem. C* **2012**, *116*, 22365–22374.

38. Aquino, A.J.A.; Tunega, D.; Haberhauer, G.; Gerzabek, M.H.; Lischka, H. Adsorption of organic substances on broken clay surfaces: A quantum chemical study. *J. Comput. Chem.* **2003**, *24*, 1853–1863.

39. Kwon, K.D.; Vadillo-Rodriguez, V.; Logan, B.E.; Kubicki, J.D. Interactions of biopolymers with silica surfaces: Force measurements and electronic structure calculation studies. *Geochim. Cosmochim. Acta* **2006**, *70*, 3803–3819.

40. Lee, B.H.; Lee, S.K. Effect of lattice topology on the adsorption of benzyl alcohol on kaolinite surfaces: Quantum chemical calculations of geometry optimization, binding energy, and NMR chemical shielding. *Am. Mineral.* **2009**, *94*, 1392–1404.

41. Geatches, D.L.; Clark, S.J.; Greenwell, H.C. DFT plus U investigation of the catalytic properties of ferruginous clay. *Am. Mineral.* **2013**, *98*, 132–140.

42. Dauber-Osguthorpe, P.; Roberts, V.A.; Osguthorpe, D.J.; Wolff, J.; Genest, M.; Hagler, A.T. Structure and energetics of ligand binding to proteins: Escherichia coli dihydrofolate reductase-trimethoprim, a drug-receptor system. *Proteins Struct. Funct. Genet.* **1988**, *4*, 31–47.

43. Jorgensen, W.L.; Tiradorives, J. The OPLS potential functions for proteins. Energy minimizations for crystals of cyclic peptides and crambin. *J. Am. Chem. Soc.* **1988**, *110*, 1657–1666.

44. Brooks, B.R.; Bruccoleri, R.E.; Olafson, B.D.; States, D.J.; Swaminathan, S.; Karplus, M. CHARMM—A program for macromolecular energy, minimization, and dynamics calculations. *J. Comput. Chem.* **1983**, *4*, 187–217.

45. Jorgensen, W.L.; Maxwell, D.S.; TiradoRives, J. Development and testing of the OPLS all-atom force field on conformational energetics and properties of organic liquids. *J. Am. Chem. Soc.* **1996**, *118*, 11225–11236.

46. Cygan, R.T.; Liang, J.-J.; Kalinichev, A.G. Molecular models of hydroxide, oxyhydroxide, and clay phases and the development of a general force field. *J. Phys. Chem. B* **2004**, *108*, 1255–1266.

47. Teppen, B.J.; Rasmussen, K.R.; Bertsch, P.M.; Miller, D.M.; Schafer, L. Molecular dynamics modeling of clay minerals. 1. Gibbsite, kaolinite, pyrophillite, and beidellite. *J. Phys. Chem. B* **1997**, *101*, 1579–1587.

48. Heinz, H.; Lin, T.J.; Mishra, R.K.; Emami, F.S. Thermodynamically consistent force fields for the assembly of inorganic, organic, and biological nanostructures: The INTERFACE force field. *Langmuir* **2013**, *29*, 1754–1765.

49. Pitman, M.C.; van Duin, A.C.T. Dynamics of confined reactive water in smectite clay-zeolite composites. *J. Am. Chem. Soc.* **2012**, *134*, 3042–3053.

50. Greathouse, J.A.; Hart, D.B.; Ochs, M.E. Alcohol and thiol adsorption on (oxy)hydroxide and carbon surfaces: Molecular dynamics simulation and desorption experiments. *J. Phys. Chem. C* **2012**, *116*, 26756–26764.

51. Veteska, M.; Pospisil, M.; Melanova, K.; Benes, L.; Zima, V. Structure analysis of hydrotalcite intercalated with pyrenetetrasulfonate; experiments and molecular modelling. *J. Mol. Model.* **2008**, *14*, 1119–1129.

52. Kovar, P.; Pospisil, M.; Kafunkova, E.; Lang, K.; Kovanda, F. Mg-Al layered double hydroxide intercalated with porphyrin anions: Molecular simulations and experiments. *J. Mol. Model.* **2010**, *16*, 223–233.

53. Praus, P.; Veteska, M.; Pospisil, M. Adsorption of phenol and aniline on natural and organically modified montmorillonite: Experiment and molecular modelling. *Mol. Simul.* **2011**, *37*, 964–974.

54. Liu, C.; Li, H.; Johnston, C.T.; Boyd, S.A.; Teppen, B.J. Relating clay structural factors to dioxin adsorption by smectites: Molecular dynamics simulations. *Soil Sci. Soc. Am. J.* **2012**, *76*, 110–120.

55. Greenwell, H.C.; Bowden, A.A.; Chen, B.Q.; Boulet, P.; Evans, J.R.G.; Coveney, P.V.; Whiting, A. Intercalation and *in situ* polymerization of poly (alkylene oxide) derivatives within M$^+$-montmorillonite (M = Li, Na, K). *J. Mater. Chem.* **2006**, *16*, 1082–1094.

56. Anderson, R.L.; Greenwell, H.C.; Suter, J.L.; Coveney, P.V.; Thyveetil, M.A. Determining materials properties of natural composites using molecular simulation. *J. Mater. Chem.* **2009**, *19*, 7251–7262.

57. Kulhankova, L.; Tokarsky, J.; Peikertova, P.; Kutlakova, K.M.; Ivanek, L.; Capkova, P. Montmorillonite intercalated by conducting polyanilines. *J. Phys. Chem. Solids* **2012**, *73*, 1530–1533.

58. Kumar, P.P.; Kalinichev, A.G.; Kirkpatrick, R.J. Hydration, swelling, interlayer structure, and hydrogen bonding in organolayered double hydroxides: Insights from molecular dynamics simulation of citrate-intercalated hydrotalcite. *J. Phys. Chem. B* **2006**, *110*, 3841–3844.

59. Kumar, P.P.; Kalinichev, A.G.; Kirkpatrick, R.J. Molecular dynamics simulation of the energetics and structure of layered double hydroxides intercalated with carboxylic acids. *J. Phys. Chem. C* **2007**, *111*, 13517–13523.

60. Kalinichev, A.G.; Kumar, P.P.; Kirkpatrick, R.J. Molecular dynamics computer simulations of the effects of hydrogen bonding on the properties of layered double hydroxides intercalated with organic acids. *Philos. Mag.* **2010**, *90*, 2475–2488.

61. Thyveetil, M.A.; Coveney, P.V.; Greenwell, H.C.; Suter, J.L. Computer simulation study of the structural stability and materials properties of DNA-intercalated layered double hydroxides. *J. Am. Chem. Soc.* **2008**, *130*, 4742–4756.

62. Thyveetil, M.A.; Coveney, P.V.; Greenwell, H.C.; Suter, J.L. Role of host layer flexibility in DNA guest intercalation revealed by computer simulation of layered nanomaterials. *J. Am. Chem. Soc.* **2008**, *130*, 12485–12495.

63. Swadling, J.B.; Coveney, P.V.; Greenwell, H.C. Clay minerals mediate folding and regioselective interactions of RNA: A large-scale atomistic simulation study. *J. Am. Chem. Soc.* **2010**, *132*, 13750–13764.

64. Swadling, J.B.; Coveney, P.V.; Greenwell, H.C. Stability of free and mineral-protected nucleic acids: Implications for the RNA world. *Geochim. Cosmochim. Acta* **2012**, *83*, 360–378.

65. Swadling, J.B.; Suter, J.L.; Greenwell, H.C.; Coveney, P.V. Influence of surface chemistry and charge on mineral-RNA interactions. *Langmuir* **2013**, *29*, 1573–1583.

66. Lee, S.S.; Fenter, P.; Park, C.; Nagy, K.L. Fulvic acid sorption on muscovite mica as a function of pH and time using *in situ* X-ray reflectivity. *Langmuir* **2008**, *24*, 7817–7829.

67. Lee, S.S.; Park, C.; Fenter, P.; Sturchio, N.C.; Nagy, K.L. Competitive adsorption of strontium and fulvic acid at the muscovite-solution interface observed with resonant anomalous X-ray reflectivity. *Geochim. Cosmochim. Acta* **2010**, *74*, 1762–1776.

68. Zhang, L.C.; Luo, L.; Zhang, S.Z. Integrated investigations on the adsorption mechanisms of fulvic and humic acids on three clay minerals. *Colloids Surf. A* **2012**, *406*, 84–90.

69. Ha, J.Y.; Yoon, T.H.; Wang, Y.G.; Musgrave, C.B.; Brown, G.E. Adsorption of organic matter at mineral/water interfaces: 7. ATR-FTIR and quantum chemical study of lactate interactions with hematite nanoparticles. *Langmuir* **2008**, *24*, 6683–6692.

70. Yoon, T.H.; Johnson, S.B.; Brown, G.E. Adsorption of organic matter at mineral/water interfaces. IV. Adsorption of humic substances at boehmite/water interfaces and impact on boehmite dissolution. *Langmuir* **2005**, *21*, 5002–5012.

71. Conte, P.; Abbate, C.; Baglieri, A.; Negre, M.; De Pasquale, C.; Alonzo, G.; Gennari, M. Adsorption of dissolved organic matter on clay minerals as assessed by infra-red, CPMAS C-13 NMR spectroscopy and low field T-1 NMR relaxometry. *Org. Geochem.* **2011**, *42*, 972–977.

72. Kang, S.; Xing, B. Humic acid fractionation upon sequential adsorption onto goethite. *Langmuir* **2008**, *24*, 2525–2531.

73. Spagnuolo, M.; Jacobson, A.R.; Baveye, P. Electron paramagnetic resonance analysis of the distribution of a hydrophobic spin probe in suspensions of humic acids, hectorite, and aluminum hydroxide-humate-hectorite complexes. *Environ. Toxicol. Chem.* **2005**, *24*, 2435–2444.

74. Kaiser, K.; Guggenberger, G.; Haumaier, L.; Zech, W. Dissolved organic matter sorption on subsoils and minerals studied by C-13-NMR and drift spectroscopy. *Eur. J. Soil Sci.* **1997**, *48*, 301–310.

75. Evanko, C.R.; Dzombak, D.A. Influence of structural features on sorption of NOM-analogue organic acids to goethite. *Environ. Sci. Technol.* **1998**, *32*, 2846–2855.

76. Ali, M.A.; Dzombak, D.A. Competitive sorption of simple organic acids and sulfate on goethite. *Environ. Sci. Technol.* **1996**, *30*, 1061–1071.

77. Maillard, L.C. Formation of humic matters by the effect of polypeptides on sugars. *C. R. Hebd. Seances Acad. Sci.* **1913**, *156*, 1159–1160.

78. Chorover, J.; Amistadi, M.K. Reaction of forest floor organic matter at goethite, birnessite and smectite surfaces. *Geochim. Cosmochim. Acta* **2001**, *65*, 95–109.

79. Jokic, A.; Frenkel, A.I.; Vairavamurthy, M.A.; Huang, P.M. Birnessite catalysis of the Maillard reaction: Its significance in natural humification. *Geophys. Res. Lett.* **2001**, *28*, 3899–3902.

80. Keil, R.G.; Montlucon, D.B.; Prahl, F.G.; Hedges, J.I. Sorptive preservation of labile organic-matter in marine-sediments. *Nature* **1994**, *370*, 549–552.

81. Janot, N.; Reiller, P.E.; Zheng, X.; Croue, J.-P.; Benedetti, M.F. Characterization of humic acid reactivity modifications due to adsorption onto alpha-Al$_2$O$_3$. *Water Res.* **2012**, *46*, 731–740.

82. Schnitzer, M.; Kodama, H. Interactions between organic and inorganic components in particle-size fractions separated from 4 soils. *Soil Sci. Soc. Am. J.* **1992**, *56*, 1099–1105.

83. Saidy, A.R.; Smernik, R.J.; Baldock, J.A.; Kaiser, K.; Sanderman, J. The sorption of organic carbon onto differing clay minerals in the presence and absence of hydrous iron oxide. *Geoderma* **2013**, *209*, 15–21.

84. Lalonde, K.; Mucci, A.; Ouellet, A.; Gelinas, Y. Preservation of organic matter in sediments promoted by iron. *Nature* **2012**, *483*, 198–200.

85. Wiseman, C.L.S.; Puttmann, W. Soil organic carbon and its sorptive preservation in central germany. *Eur. J. Soil Sci.* **2005**, *56*, 65–76.

86. Davies, G.; Fataftah, A.; Cherkasskiy, A.; Ghabbour, E.A.; Radwan, A.; Jansen, S.A.; Kolla, S.; Paciolla, M.D.; Sein, L.T.; Buermann, W.; *et al.* Tight metal binding by humic acids and its role in biomineralization. *J. Chem. Soc. Dalton Trans.* **1997**, 4047–4060.

87. Kunhi Mouvenchery, Y.; Kucerik, J.; Diehl, D.; Schaumann, G.E. Cation-mediated cross-linking in natural organic matter: A review. *Rev. Env. Sci. Biotechnol.* **2012**, *11*, 41–54.

88. Schaumann, G.E.; Thiele-Bruhn, S. Molecular modeling of soil organic matter: Squaring the circle? *Geoderma* **2011**, *166*, 1–14.

89. Perry, T.D.; Cygan, R.T.; Mitchell, R. Molecular models of alginic acid: Interactions with calcium ions and calcite surfaces. *Geochim. Cosmochim. Acta* **2006**, *70*, 3508–3532.

90. Sutton, R.; Sposito, G. Molecular simulation of humic substance-Ca-montmorillonite complexes. *Geochim. Cosmochim. Acta* **2006**, *70*, 3566–3581.

91. Sutton, R.; Sposito, G. Molecular structure in soil humic substances: The new view. *Environ. Sci. Technol.* **2005**, *39*, 9009–9015.

92. Aquino, A.J.A.; Tunega, D.; Pasalic, H.; Haberhauer, G.; Gerzabek, M.H.; Lischka, H. The thermodynamic stability of hydrogen bonded and cation bridged complexes of humic acid models—A theoretical study. *Chem. Phys.* **2008**, *349*, 69–76.

93. Aquino, A.J.A.; Tunega, D.; Schaumann, G.E.; Haberhauer, G.; Gerzabek, M.H.; Lischka, H. Stabilizing capacity of water bridges in nanopore segments of humic substances: A theoretical investigation. *J. Phys. Chem. C* **2009**, *113*, 16468–16475.

94. Aquino, A.J.A.; Tunega, D.; Schaumann, G.E.; Haberhauer, G.; Gerzabek, M.H.; Lischka, H. The functionality of cation bridges for binding polar groups in soil aggregates. *Int. J. Quantum Chem.* **2011**, *111*, 1531–1542.

95. Aquino, A.J.A.; Tunega, D.; Pasalic, H.; Schaumann, G.E.; Haberhauer, G.; Gerzabek, M.H.; Lischka, H. Molecular dynamics simulations of water molecule-bridges in polar domains of humic acids. *Environ. Sci. Technol.* **2011**, *45*, 8411–8419.

96. Aquino, A.J.A.; Tunega, D.; Pasalic, H.; Schaumann, G.E.; Haberhauer, G.; Gerzabek, M.H.; Lischka, H. Study of solvent effect on the stability of water bridge-linked carboxyl groups in humic acid models. *Geoderma* **2011**, *169*, 20–26.

97. Kalinichev, A.G., Molecular models of natural organic matter and its colloidal aggregation in aqueous solutions: Challenges and opportunities for computer simulations. *Pure Appl. Chem.* **2013**, *85*, 149–158.

98. Sein, L.T.; Varnum, J.M.; Jansen, S.A. Conformational modeling of a new building block of humic acid: Approaches to the lowest energy conformer. *Environ. Sci. Technol.* **1999**, *33*, 546–552.

99. Xu, X.; Kalinichev, A.G.; Kirkpatrick, R.J. ^{133}Cs and ^{35}Cl NMR spectroscopy and molecular dynamics modeling of Cs^+ and Cl^- complexation with natural organic matter. *Geochim. Cosmochim. Acta* **2006**, *70*, 4319–4331.

100. Kalinichev, A.G.; Kirkpatrick, R.J. Molecular dynamics simulation of cationic complexation with natural organic matter. *Eur. J. Soil Sci.* **2007**, *58*, 909–917.

101. Ahn, W.Y.; Kalinichev, A.G.; Clark, M.M. Effects of background cations on the fouling of polyethersulfone membranes by natural organic matter: Experimental and molecular modeling study. *J. Membr. Sci.* **2008**, *309*, 128–140.

102. Schulten, H.R.; Schnitzer, M. Chemical model structures for soil organic matter and soils. *Soil Sci.* **1997**, *162*, 115–130.

103. Schulten, H.R.; Schnitzer, M. A state-of-the-art structural concept for humic substances. *Naturwissenschaften* **1993**, *80*, 29–30.

104. Shevchenko, S.M.; Bailey, G.W.; Akim, L.G. The conformational dynamics of humic polyanions in model organic and organo-mineral aggregates. *J. Mol. Struct.* **1999**, *460*, 179–190.

105. Shevchenko, S.M.; Bailey, G.W. Modeling sorption of soil organic matter on mineral surfaces: Wood-derived polymers on mica. *Supramol. Sci.* **1998**, *5*, 143–157.

106. Leenheer, J.A.; Brown, G.K.; MacCarthy, P.; Cabaniss, S.E. Models of metal binding structures in fulvic acid from the Suwannee River, Georgia. *Environ. Sci. Technol.* **1998**, *32*, 2410–2416.

107. Iskrenova-Tchoukova, E.; Kalinichev, A.G.; Kirkpatrick, R.J. Metal cation complexation with natural organic matter in aqueous solutions: Molecular dynamics simulations and potentials of mean force. *Langmuir* **2010**, *26*, 15909–15919.

108. Kalinichev, A.G.; Iskrenova-Tchoukova, E.; Ahn, W.-Y.; Clark, M.M.; Kirkpatrick, R.J. Effects of Ca^{2+} on supramolecular aggregation of natural organic matter in aqueous solutions: A comparison of molecular modeling approaches. *Geoderma* **2011**, *169*, 27–32.

109. Kirishima, A.; Tanaka, K.; Niibori, Y.; Tochiyama, O. Complex formation of calcium with humic acid and polyacrylic acid. *Radiochim. Acta* **2002**, *90*, 555–561.

110. Plaschke, M.; Rothe, J.; Armbruster, M.K.; Denecke, M.A.; Naber, A.; Geckeis, H. Humic acid metal cation interaction studied by spectromicroscopy techniques in combination with quantum chemical calculations. *J. Synchrotron Radiat.* **2010**, *17*, 158–165.

111. Roger, G.M.; Durand-Vidal, S.; Bernard, O.; Meriguet, G.; Altmann, S.; Turq, P. Characterization of humic substances and polyacrylic acid: A high precision conductimetry study. *Colloids Surf. A* **2010**, *356*, 51–57.

112. Crea, F.; Giacalone, A.; Gianguzza, A.; Piazzese, D.; Sammartano, S. Modelling of natural and synthetic polyelectrolyte interactions in natural waters by using SIT, Pitzer and ion pairing approaches. *Mar. Chem.* **2006**, *99*, 93–105.

113. Crea, F.; de Stefano, C.; Gianguzza, A.; Pettignano, A.; Piazzese, D.; Sammartano, S. Acid-base properties of synthetic and natural polyelectrolytes: Experimental results and models for the dependence on different aqueous media. *J. Chem. Eng. Data* **2009**, *54*, 589–605.

114. Lovley, D.R. Dissimilatory Fe(III) and Mn(IV) reduction. *Microbiol. Rev.* **1991**, *55*, 259–287.

115. Gu, B.H.; Schmitt, J.; Chen, Z.H.; Liang, L.Y.; McCarthy, J.F., Adsorption and desorption of natural organic-matter on iron-oxide - mechanisms and models. *Environ. Sci. Technol.* **1994**, *28*, 38–46.

116. Kaiser, K.; Guggenberger, G. The role of DOM sorption to mineral surfaces in the preservation of organic matter in soils. *Org. Geochem.* **2000**, *31*, 711–725.

117. Geatches, D.L.; Clark, S.J.; Greenwell, H.C. Iron reduction in nontronite-type clay minerals: Modelling a complex system. *Geochim. Cosmochim. Acta* **2012**, *81*, 13–27.

118. Alexandrov, V.; Neumann, A.; Scherer, M.M.; Rosso, K.M. Electron exchange and conduction in nontronite from first-principles. *J. Phys. Chem. C* **2013**, *117*, 2032–2040.

119. Alexandrov, V.; Rosso, K.M. Insights into the mechanism of Fe(II) adsorption and oxidation at Fe-clay mineral surfaces from first-principles calculations. *J. Phys. Chem. C* **2013**, *117*, 22880–22886.

120. Wander, M.C.F.; Rosso, K.M.; Schoonen, M.A.A. Structure and charge hopping dynamics in green rust. *J. Phys. Chem. C* **2007**, *111*, 11414–11423.

121. Qiang, L.H.; Li, Z.F.; Zhao, T.Q.; Zhong, S.L.; Wang, H.Y.; Cui, X.J. Atomic-scale interactions of the interface between chitosan and Fe_3O_4. *Colloids Surf. A* **2013**, *419*, 125–132.

122. Aryanpour, M.; van Duin, A.C.T.; Kubicki, J.D. Development of a reactive force field for iron-oxyhydroxide systems. *J. Phys. Chem. A* **2010**, *114*, 6298–6307.

123. Dismukes, G.C. The metal centers of the photosynthetic oxygen-evolving complex. *Photochem. Photobiol.* **1986**, *43*, 99–115.

124. Burns, R.G.; Burns, V.M. Mineralogy. In *Marine Manganese Deposits*; Glasby, G.P., Ed.; Elsevier: Amsterdam, The Netherlands, 1977; pp. 185–248.

125. O'Reilly, S.E.; Hochella, M.F. Lead sorption efficiencies of natural and synthetic Mn and Fe-oxides. *Geochim. Cosmochim. Acta* **2003**, *67*, 4471–4487.

126. Sposito, G. *The Chemistry of Soils*, 2nd ed.; Oxford University Press: Oxford, UK, 2008.

127. Sparks, D.L. *Environmental Soil Chemistry*, 2nd ed.; Academic Press: San Diego, CA, USA, 2003.

128. Villalobos, M.; Lanson, B.; Manceau, A.; Toner, B.; Sposito, G. Structural model for the biogenic Mn oxide produced by pseudomonas putida. *Am. Mineral.* **2006**, *91*, 489–502.

129. Essington, M.E. *Soil and Water Chemistry: An. Integrative Approach*; CRC Press: Boca Raton, FL, USA, 2003.

130. Post, J.E. Manganese oxide minerals: Crystal structures and economic and environmental significance. *Proc. Natl. Acad. Sci. USA* **1999**, *96*, 3447–3454.

131. Gasparatos, D. Sequestration of heavy metals from soil with Fe-Mn concretions and nodules. *Environ. Chem. Lett.* **2013**, *11*, 1–9.

132. Cygan, R.T.; Post, J.E.; Heaney, P.J.; Kubicki, J.D. Molecular models of birnessite and related hydrated layered minerals. *Am. Mineral.* **2012**, *97*, 1505–1514.

133. Kwon, K.D.; Refson, K.; Sposito, G. Surface complexation of Pb(II) by hexagonal birnessite nanoparticles. *Geochim. Cosmochim. Acta* **2010**, *74*, 6731–6740.

134. Kwon, K.D.; Refson, K.; Sposito, G. On the role of Mn(IV) vacancies in the photoreductive dissolution of hexagonal birnessite. *Geochim. Cosmochim. Acta* **2009**, *73*, 4142–4150.

135. Kwon, K.D.; Refson, K.; Sposito, G. Understanding the trends in transition metal sorption by vacancy sites in birnessite. *Geochim. Cosmochim. Acta* **2013**, *101*, 222–232.

136. Pena, J.; Kwon, K.D.; Refson, K.; Bargar, J.R.; Sposito, G. Mechanisms of nickel sorption by a bacteriogenic birnessite. *Geochim. Cosmochim. Acta* **2010**, *74*, 3076–3089.

137. Kwon, K.D.; Refson, K.; Sposito, G. Zinc surface complexes on birnessite: A density functional theory study. *Geochim. Cosmochim. Acta* **2009**, *73*, 1273–1284.

138. Tebo, B.M.; Bargar, J.R.; Clement, B.G.; Dick, G.J.; Murray, K.J.; Parker, D.; Verity, R.; Webb, S.M. Biogenic manganese oxides: Properties and mechanisms of formation. *Annu. Rev. Earth Planet. Sci.* **2004**, *32*, 287–328.

139. Madison, A.S.; Tebo, B.M.; Mucci, A.; Sundby, B.; Luther, G.W. Abundant porewater Mn(III) is a major component of the sedimentary redox system. *Science* **2013**, *341*, 875–878.

140. Laha, S.; Luthy, R.G. Oxidation of aniline and other primary aromatic-amines by manganese-dioxide. *Environ. Sci. Technol.* **1990**, *24*, 363–373.

141. Stone, A.T. Reductive dissolution of manganese(III/IV) oxides by substituted phenols. *Environ. Sci. Technol.* **1987**, *21*, 979–988.

142. Ulrich, H.J.; Stone, A.T. Oxidation of chlorophenols adsorbed to manganese oxide surfaces. *Environ. Sci. Technol.* **1989**, *23*, 421–428.

143. Zhang, H.C.; Huang, C.H. Oxidative transformation of triclosan and chlorophene by manganese oxides. *Environ. Sci. Technol.* **2003**, *37*, 2421–2430.

144. Zhang, H.C.; Huang, C.H. Reactivity and transformation of antibacterial N-oxides in the presence of manganese oxide. *Environ. Sci. Technol.* **2005**, *39*, 593–601.

145. Zhang, H.C.; Huang, C.H. Oxidative transformation of fluoroquinolone antibacterial agents and structurally related amines by manganese oxide. *Environ. Sci. Technol.* **2005**, *39*, 4474–4483.

146. Klausen, J.; Haderlein, S.B.; Schwarzenbach, R.P. Oxidation of substituted anilines by aqueous MnO_2: Effect of co-solutes on initial and quasi-steady-state kinetics. *Environ. Sci. Technol.* **1997**, *31*, 2642–2649.

147. Banerjee, D.; Nesbitt, H.W. XPS study of dissolution of birnessite by humate with constraints on reaction mechanism. *Geochim. Cosmochim. Acta* **2001**, *65*, 1703–1714.

148. Banerjee, D.; Nesbitt, H.W. XPS study of reductive dissolution of birnessite by oxalate: Rates and mechanistic aspects of dissolution and redox processes. *Geochim. Cosmochim. Acta* **1999**, *63*, 3025–3038.

149. Durand, J.P.; Senanayake, S.D.; Suib, S.L.; Mullins, D.R. Reaction of formic acid over amorphous manganese oxide catalytic systems: An *in situ* study. *J. Phys. Chem. C* **2010**, *114*, 20000–20006.

150. Reiller, P.E. Modelling metal-humic substances-surface systems: Reasons for success, failure and possible routes for peace of mind. *Miner. Mag.* **2012**, *76*, 2643–2658.

151. Wang, N.-H.; Lo, S.-L. Preparation, characterization and adsorption performance of cetyltrimethylammonium modified birnessite. *Appl. Surf. Sci.* **2014**, *299*, 123–130.

152. Myeongjin, K.; Myeongyeol, Y.; Youngjae, Y.; Jooheon, K. Capacitance behavior of composites for supercapacitor applications prepared with different durations of graphene/nanoneedle MnO_2 reduction. *Microelectron. Reliab.* **2014**, *54*, 587–594.

153. Ahmed, K.A.M.; Huang, K. Synthesis, characterization and catalytic activity of birnessite type potassium manganese oxide nanotubes and nanorods. *Mater. Chem. Phys.* **2012**, *133*, 605–610.

154. Xia, H.; Wang, Y.; Lin, J.; Lu, L. Hydrothermal synthesis of MnO_2/CNT nanocomposite with a CNT core/porous MnO_2 sheath hierarchy architecture for supercapacitors. *Nanoscale Res. Lett.* **2012**, *7*, 1–10.

155. Jaramillo-Botero, A.; Nielsen, R.; Abrol, R.; Su, J.; Pascal, T.; Mueller, J.; Goddard, W.A. First-principles-based multiscale, multiparadigm molecular mechanics and dynamics methods for describing complex chemical processes. In *Multiscale Molecular Methods in Applied Chemistry*; Kirchner, B., Vrabec, J., Eds.; Springer-Verlag: Berlin, Germany, 2012; Volume 307, pp. 1–42.

156. Abrams, C.; Bussi, G. Enhanced sampling in molecular dynamics using metadynamics, replica-exchange, and temperature-acceleration. *Entropy* **2014**, *16*, 163–199.

157. Stamatakis, M.; Vlachos, D.G. Unraveling the complexity of catalytic reactions via kinetic Monte Carlo simulation: Current status and frontiers. *ACS Catal.* **2012**, *2*, 2648–2663.

158. Liu, S.Y.; Kleber, M.; Takahashi, L.K.; Nico, P.; Keiluweit, M.; Ahmed, M. Synchrotron-based mass spectrometry to investigate the molecular properties of mineral-organic associations. *Anal. Chem.* **2013**, *85*, 6100–6106.

159. Smalley, M. *Clay Swelling and Colloid Stability*; Taylor and Francis Group: Boca Raton, FL, USA, 2006.

9

Flotation Bubble Delineation Based on Harris Corner Detection and Local Gray Value Minima

Weixing Wang [1,2] and Liangqin Chen [1,*]

[1] School of Physics & Information Engineering, Fuzhou University, 350108 Fuzhou, China; E-Mail: wxwwx@fzu.edu.cn
[2] Royal Institute of Technology, 100 44 Stockholm, Sweden

* Author to whom correspondence should be addressed; E-Mail: odiechen@fzu.edu.cn.

Academic Editor: Kota Hanumantha Rao

Abstract: Froth image segmentation is an important and basic part in an online froth monitoring system in mineral processing. The fast and accurate bubble delineation in a froth image is significant for the subsequent froth surface characterization. This paper proposes a froth image segmentation method combining image classification and image segmentation. In the method, an improved Harris corner detection algorithm is applied to classify froth images first. Then, for each class, the images are segmented by automatically choosing the corresponding parameters for identifying bubble edge points through extracting the local gray value minima. Finally, on the basis of the edge points, the bubbles are delineated by using a number of post-processing functions. Compared with the widely used Watershed algorithm and others for a number of lead zinc froth images in a flotation plant, the new method (algorithm) can alleviate the over-segmentation problem effectively. The experimental results show that the new method can produce good bubble delineation results automatically. In addition, its processing speed can also meet the online measurement requirements.

Keywords: froth image; bubble delineation; classification; segmentation; gray value minima; Harris corner

1. Introduction

Froth flotation is a selective separation process that is widely used in mineral processing to extract valuable minerals. Froth is a three-phase structure comprising air bubbles, solids and water [1]. The modeling and the control of flotation processes are challenging because of the inherently chaotic nature of the underlying microscopic phenomena. The lack of sufficiently accurate and reliable process measurements intensifies the difficulty. In fact, the control of flotation process depends heavily on various experiences of human operators by looking at the appearance of the froth [2]. The performance thus depends on the operator's experience and is limited by the absence of physical, quantitative methods for measurement and characterization of the froth [3]. Therefore, methods based on machine vision and image processing have been developed for observation and analysis of froth images, including the application for extraction of bubble size, shape and other physical features [4–6]. Moreover, the correlation between these features and the flotation performance also has been studied [7,8].

This paper aims to extract the bubble size and shape based on image segmentation techniques. Here it should be noted that the bubbles (in this paper) are the bubbles on a froth surface. Although there exists remarkable difference between the surface bubble size distribution and the underlying bubble size distribution [9], Wang and Neethling have found that an empirical formula for the relationship between the above two kinds of bubbles can be used to obtain the underlying bubble size distribution from the surface bubble size distribution effectively [10].

Hence, machine vision based on bubble film size measurement becomes a reasonable solution to interpret the surface bubble structure automatically [5]. The surface bubble size and shape are two of the dominant visual features closely relating to the reagent addition [6,11]. And the surface bubble size can also be used as an effective measurement of bubble stability for the reason that it reflects the extent of bubble coalescence [12]. Zanin *et al.* used the bubble size on the froth top as one of indicators for froth stability [13]. Furthermore, many researchers have reported the relationship between the bubble size, water recovery, and froth recovery, *etc.* [9,14]. Hence, by using image processing and analysis, the bubble size estimation in flotation has received a considerable research attention over the years. However, the performance of the existing froth image segmentation methods is not satisfactory due to the uneven illumination and different noise. A froth image usually contains thousands of color mineral liquid bubbles of different sizes and shapes, with weak intensity boundaries. A bubble is mainly spherical or polyhedral in shape; therefore, the light makes the top part of a bubble with high brightness and the boundary region with low brightness. All the characteristics of a froth image make bubble delineation hard. The traditional edge detection methods often only detect the edges of particles, light reflection areas and black holes. How to achieve an efficient froth image segmentation method has become a major task in this research field.

Wang *et al.* and Citir *et al.* proposed their image segmentation algorithms through the usage of bubble edges and sizes [15,16], and their algorithms have succeeded in extracting the bubbles edges. An improved particle swarm optimization algorithm was applied to achieve a more suitable threshold in a froth edge extraction method [17]. A set of Watershed algorithms combined with clustering pre-segmentation and high-low scale distance reconstruction were used to segment color froth images [18], which can conquer the over-segmentation and under-segmentation problems in some cases. However, each of the above algorithms is only suitable for a special case, or it may need human intervention that cannot be automatically applied in the industrial flotation scale online. Some methods based on texture spectrum were also used to estimate bubble sizes [3]. Two recent literature reviews have been carried out by Aldrich *et al.* and Shean and Cilliers, respectively [19,20]; the previous froth image segmentation methods can be classified into two categories. One is based on mathematical morphology, in which, the Watershed algorithm is the representative one [21]. The Watershed algorithm is a morphological method based on a simulation of the water rising from a set of markers, which is well known as an excellent image segmentation method for the images of the complex and densely packed objects. A Watershed method can obtain a good image segmentation result when the processed image is of more uniform bubble size and shape distributions, but for the images with the large variations of the object sizes and shapes, it is easy to encounter the over-segmentation or under-segmentation problems [22]. In addition, the Watershed algorithm has the disadvantage in computational complexity (a huge number of repeating operations). The other is based on edge detection. The advantages of the method based on edge detection are (1) of a high processing speed; and (2) not affected by the variations of bubble size and shape, but such an algorithm is sensitive to the bubble surface noise in an image of large bubbles. Wang *et al.* proposed an improved valley edge detection algorithm based on multi-scale analysis [23]. Although the improved algorithm can produce good results for most froth images, the under-segmentation and over-segmentation problems still exist in some cases.

Currently, there are several commercial froth image processing systems, such as FrothMaster (Outokumpu, Espoo, Finland), SmartFroth (UCT, Cape Town, South Africa) and VisioFroth (Metso, Helsinki, Finland). For the bubble size and shape measurement, each of the systems uses its own method according to its flotation image properties, for instance, average size estimation by using statistical analysis for unclear bubble images (e.g., Iron flotation), and the bubble size delineation and measurement by applying the Watershed algorithm combining color information for the images where the bubbles are showing up clearly (e.g., Copper flotation), *etc.* There is no standard image segmentation algorithm for all the cases. The implementation of a long-term fully automated flotation control system is difficult due to the image segmentation problem.

As requirements from Jin Dong lead zinc flotation plant in China, we designed and developed a new system for flotation monitoring. The key task in the system is image segmentation-bubble delineation. In the testing, the results produced by most existing froth image segmentation algorithms are not satisfactory and their processing speeds are also far from the requirement of an online flotation control system, therefore, a new bubble delineation method is proposed in this study, which is the combination of image classification and image segmentation. The studied method is tested for a large number of lead froth images. The test results show that the new method can produce good image segmentation results than others, and our algorithm has the advantages in low computational complexity and it can effectively reduce the over-segmentation and under-segmentation problems.

2. Froth Image Segmentation on Improved Local Gray Value Minima

2.1. Characteristics of Froth Images

As shown in Figures 1 and 2, due to bubble's 3D geometry and the light reflection, there exist one or more high gray value regions in each bubble, called "white spot areas". The boundary regions among the adjacent bubbles have low gray values, called "boundary areas", where the local gray value minima points can be considered as bubble boundary pixels. For images of large bubbles, the number of bubbles in an image is small, and the white spot areas are large regions and the boundary regions are long on average (see Figure 1); for images of small and fine bubbles, the number of bubbles in an image is great, and the white spot areas are small regions and the boundary regions are short narrow on average (see Figure 2). It is noted that all of the images used in this paper are the lead froth images obtained from Jin Dong flotation plant in China. In the plant, the raw ore mainly contains lead and zinc, and also includes sulfur, iron, silver, copper and other sulfide minerals, and the Gangue minerals include quartz, feldspar, pyroxene, garnet, chlorite and other silicate minerals. Hence, for the image segmentation, the flotation images are complicated; we have to study a new image segmentation algorithm.

For the above reasons, before image segmentation, a froth image can be classified into the class of large or non-large bubbles. For an image of large bubbles, a large template is used in the subsequent local gray value minima detection process; the small template is used for an image of non-large bubbles.

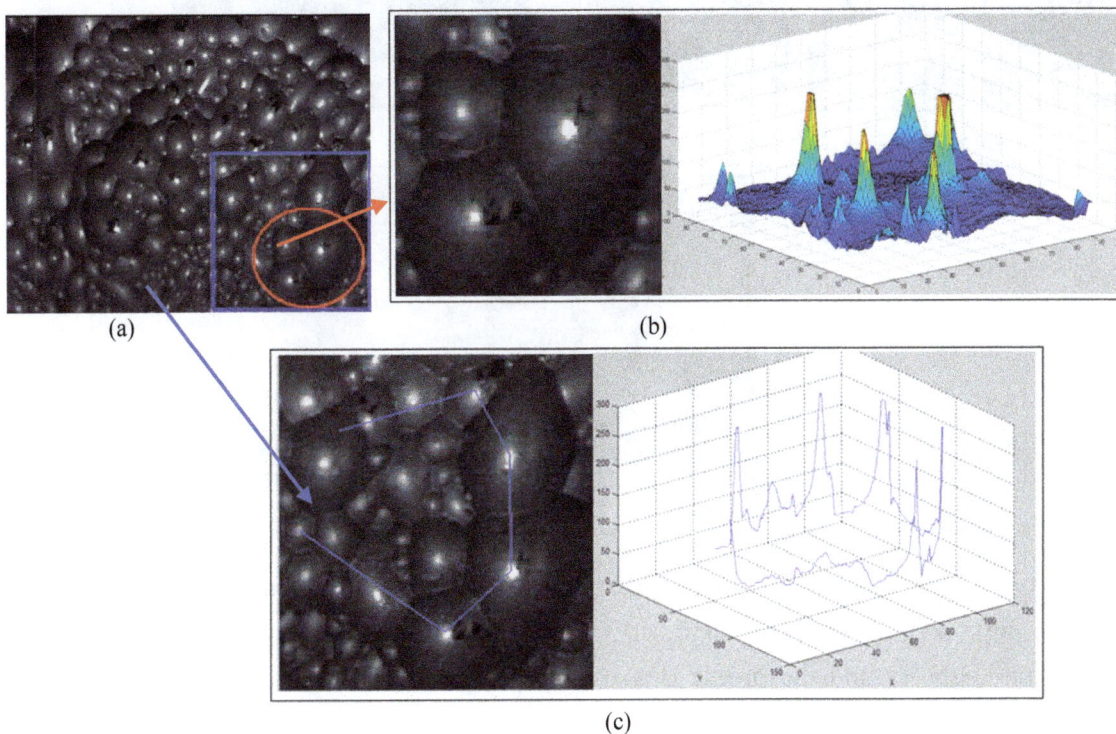

Figure 1. Gray value characteristics of the image of large bubbles. (**a**) a large bubble image; (**b**) **left**: a part of the image (a); **right**: the 3D surface of the left image; (**c**) **left**: large bubbles and the profile line; **right**: the gray level histogram of the profile corresponding to the left image.

2.2. Edge Detection on Improved Local Gray Value Minima

According to the above analyses of the gray value variations in a froth image, it can be concluded that if all the local gray value minima points are detected, the information of the minima points can be used for bubble delineation on edges.

For a large bubble, its boundary region is a long border area, and there are often multiple local valley points in its boundary area. Moreover, because of the rough surface of a large bubble, the gray value distribution on the bubble surface is quite uneven, as shown in Figure 3, in other words, a number of the local gray value minima points exist in the area, and these points belong to the non-bubble boundary points. For this kind of information, once directly extracted, this non-boundary information often form many closed curves that are very difficult to be removed by subsequent non-boundary information filters. Thus, to make the non-bubble boundary information as less as possible, for large bubbles, the segmentation algorithm adopts a large sized kernel for the local gray value minima point detection, e.g., 5×5 or 7×7, therefore, the comparison distance is increased. To further remove the speckle noise on a froth surface, the algorithm uses the gradation weighted average gray values in a local area for comparison instead of a single pixel gray value.

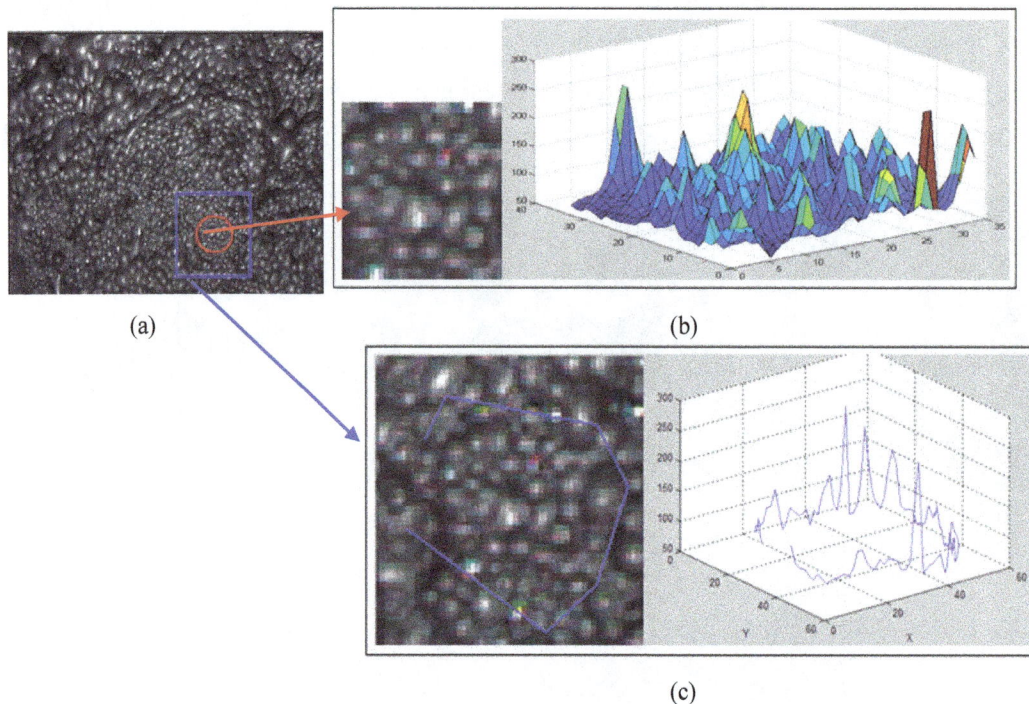

(a)   (b)

(c)

Figure 2. Gray value characteristic of the image of small bubbles. (**a**) A small bubble image; (**b**) **left**: a part of the image (a); **right**: the 3D surface of the left image; (**c**) **left**: small bubbles and the profile line; **right**: the gray level histogram of the profile corresponding to the left image.

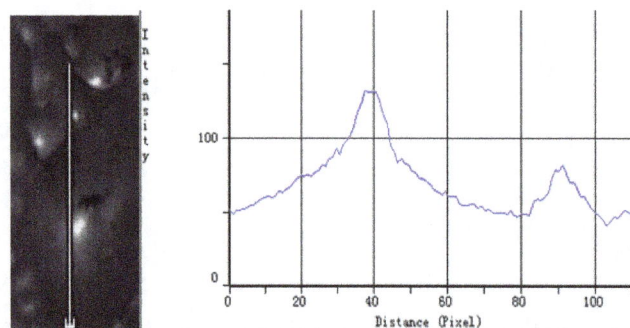

Figure 3. Gray value variation between two bubbles. **Left**: large bubble and the profile line; **Right**: the gray value histogram of the profile corresponding to the left image from bottom to top.

Based on the above idea, an edge detection method based on the improved local gray value minima is designed and developed to extract bubble edges.

We use $f(i, j)$ to denote an original froth image, $g(i, j)$ is for its edge image, and all the values in $g(i, j)$ are set as "0" in initialization.

Step 1, for an image of large bubbles, the algorithm identifies whether the current detecting pixel is a local gray value minima point in the kernel of 5×5. The judgment conditions in four directions are as follows:

$$\begin{cases} f(i, j-1) \geq f(i, j) \leq f(i, j+1) \\ f(i, j+1) \leq w_1 f(i-1, j+2) + w_2 f(i, j+2) + w_3 f(i+1, j+2) \\ f(i, j-1) \leq w_1 f(i-1, j-2) + w_2 f(i, j-2) + w_3 f(i+1, j-2) \end{cases} \tag{1}$$

$$\begin{cases} f(i+1, j-1) \geq f(i, j) \leq f(i-1, j+1) \\ f(i-1, j+1) \leq w_1 f(i-2, j+1) + w_2 f(i-2, j+2) + w_3 f(i-1, j+2) \\ f(i+1, j-1) \leq w_1 f(i+1, j-2) + w_2 f(i+2, j-2) + w_3 f(i+2, j-1) \end{cases} \tag{2}$$

$$\begin{cases} f(i+1, j) \geq f(i, j) \leq f(i-1, j) \\ f(i-1, j) \leq w_1 f(i-2, j-1) + w_2 f(i-2, j) + w_3 f(i-2, j+1) \\ f(i+1, j) \leq w_1 f(i+2, j-1) + w_2 f(i+2, j) + w_3 f(i+2, j+1) \end{cases} \tag{3}$$

$$\begin{cases} f(i-1, j-1) \geq f(i, j) \leq f(i+1, j+1) \\ f(i-1, j-1) \leq w_1 f(i-1, j-2) + w_2 f(i-2, j-2) + w_3 f(i-2, j-1) \\ f(i+1, j+1) \leq w_1 f(i+2, j+1) + w_2 f(i+2, j+2) + w_3 f(i+1, j+2) \end{cases} \tag{4}$$

where $w_i (i = 1, 2, 3)$ are weight coefficients, which are generally inversely proportional to the pixel distance to the detecting pixel, e.g., $w_1 = w_3 = 0.2, w_2 = 0.6$. The four groups of formulae are, respectively, used to detect the local gray values of the minima points in the 0°, 45°, 90°, and 135° directions of the current detecting pixel. In Figure 4, the left diagram shows the four detection directions, where the area closed by the solid curve represents the first comparison region, the area closed by the dotted curve is the second comparison area; the right image shows the detection template in 0° direction, where the shadow parts are the pixels to be used for searching and comparing. The situations in other directions are similar to this.

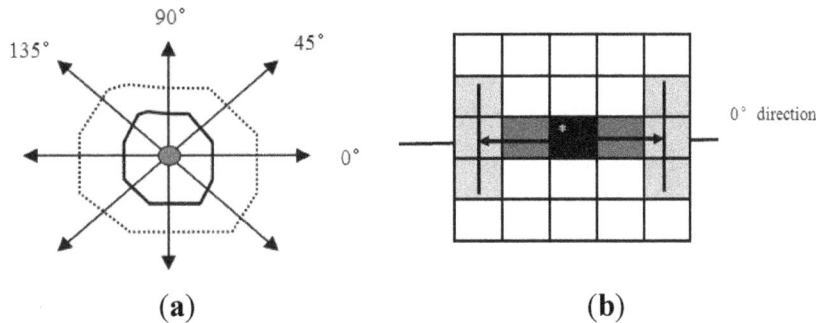

(a)   (b)

Figure 4. Diagram for detection algorithm. (a) The four directions; (b) detection template in 0° direction of a froth image of large bubbles.

For an image of non-large bubbles, we search for the local gray value minima points in a 3×3 kernel, and the above judgment conditions are simplified as:

$$f(i, j-1) \geq f(i, j) \leq f(i, j+1) \tag{5}$$

$$f(i+1, j-1) \geq f(i, j) \leq f(i-1, j+1) \tag{6}$$

$$f(i-1, j) \geq f(i, j) \leq f(i+1, j) \tag{7}$$

$$f(i-1, j-1) \geq f(i, j) \leq f(i+1, j+1) \tag{8}$$

If one of the conditions is satisfactory, the detected point is marked as an edge point, that is to set $g(i, j) = 255$, and its location and direction are also marked; otherwise, go to Step 2.

Step 2, for all images, we identify whether the current detecting pixel is a local gray value minima point in a 5×5 kernel. It is similar to step 1 but with a set of different conditions. This step is applied to extract the edge point in which the gray value is equal to that in another point in its neighborhood region. The judgment conditions are illustrated as:

$$f(i, j-1) > f(i, j) = f(i, j+1) < f(i, j+2) \tag{9}$$

$$f(i+1, j-1) > f(i, j) = f(i-1, j+1) < f(i-2, j+2) \tag{10}$$

$$f(i+1, j) > f(i, j) = f(i-1, j) < f(i-2, j) \tag{11}$$

$$f(i+1, j+1) > f(i, j) = f(i-1, j-1) < f(i-2, j-2) \tag{12}$$

$$f(i, j+1) > f(i, j) = f(i, j-1) < f(i, j-2) \tag{13}$$

$$f(i-1, j+1) > f(i, j) = f(i+1, j-1) < f(i+2, j-2) \tag{14}$$

$$f(i-1, j) > f(i, j) = f(i+1, j) < f(i+2, j) \tag{15}$$

$$f(i-1, j-1) > f(i, j) = f(i+1, j+1) < f(i+2, j+2) \tag{16}$$

where, the first four formulae are utilized to detect the local gray value minima points, respectively, in $0°$, $45°$, $90°$, and $135°$ directions of the current detecting pixel, and the rest four formulae are used in the reverse directions $180°$, $225°$, $270°$ and $315°$, respectively.

If one of the above conditions is met, the detected point is marked as an edge point, *i.e.*, set $g(i, j) = 255$, and its location and direction is also marked.

In accordance with the above-described detection method, each pixel in an image is detected to see if it is a local minima point in a certain direction. Any pixel having the local gray value minima in a certain direction is assigned as an edge point. It is noted that the edge point detection procedure is done after image classification. Figure 5 gives the workflow of the edge detection algorithm. The edge detection results for different classes of bubble images are shown in Figure 6.

As shown in Figure 6, we can see that the significant bubble edges are detected and most of the white spot edges are eliminated for any class of bubble images. Those isolated points, short line segments, and other non-boundary information are removed in the subsequent post-processing procedures.

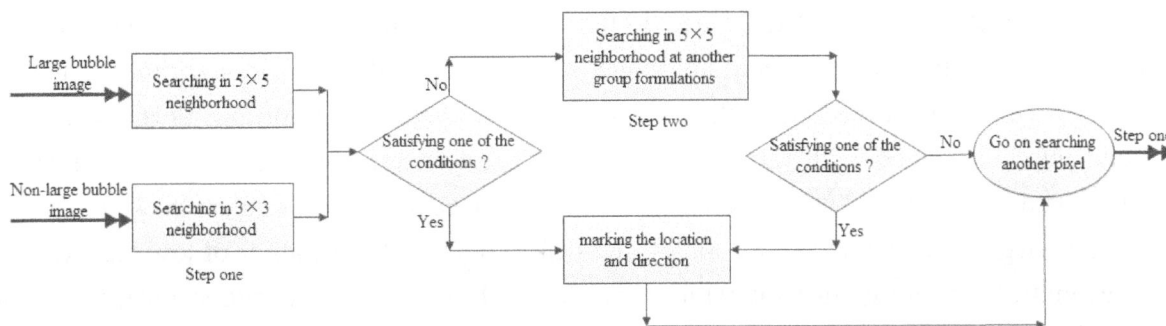

Figure 5. Algorithm workflow of the edge detection based on local gray value minima.

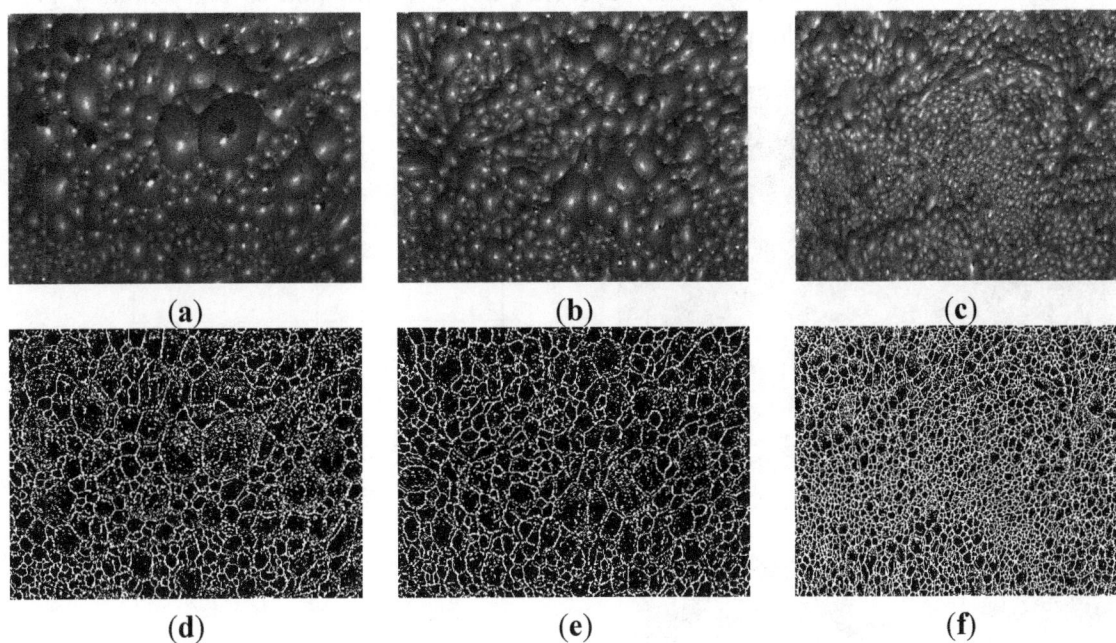

Figure 6. Edge detection results based on the proposed algorithm. (**a–c**) represent the original images of large, middle and small bubbles separately; (**d–f**) are the edge detection results corresponding to the images in (a), (b) and (c), respectively.

2.3. Image Pre-Processing and Classification

In general, there are a lot of noise in an original froth image, which makes the froth surface complicated. The noise might be detected as local gray value minima points to be reserved for they satisfy some conditions. So we need to smooth an original image to alleviate the noise before image segmentation. One simple way to reduce noise is to use a smoothing filter such as a mean filter or a Gaussian filter, and the de-noising procedure can be done before segmenting the classified froth images. Figure 7 shows the images smoothed by a Gaussian filter, from which we can see that the gray value variance in the image is more even after filtering, and it benefits the subsequent edge extraction procedures based on the local gray value minima.

Another preprocessing task is to classify these froth images according to bubble sizes. We have done a lot about the froth image classification with statistical analysis as shown in the reference [15]. Its statistical data have proven the following two facts: One is that the size of the white spot area on a bubble is proportional to the bubble size; the other is that the number of white spots in an image is inversely proportional to the average bubble size. In this study, we do image classification based on the latter. In order to extract the white spots in an image, many traditional threshold algorithms, such as Otsu threshold [24], can be used. As well known, the Otsu threshold algorithm is an automatic threshold segmentation method with the good performance for images having a obvious threshold between objects and background, but it is ineffective for the images having the unclear background, for example, for the images of small and tiny bubbles, the white spots are too small or of low gray values, therefore, it cannot effectively extract the white spots well. Moreover, after image segmentation, the region labeling and area size calculation must be carried out to calculate the number and the size distribution of white spots. To alleviate the problems of the Otsu threshold algorithm and to further simplify the image classification, this paper uses a corner detection method to conveniently get the number of white spots to do the bubble image classification.

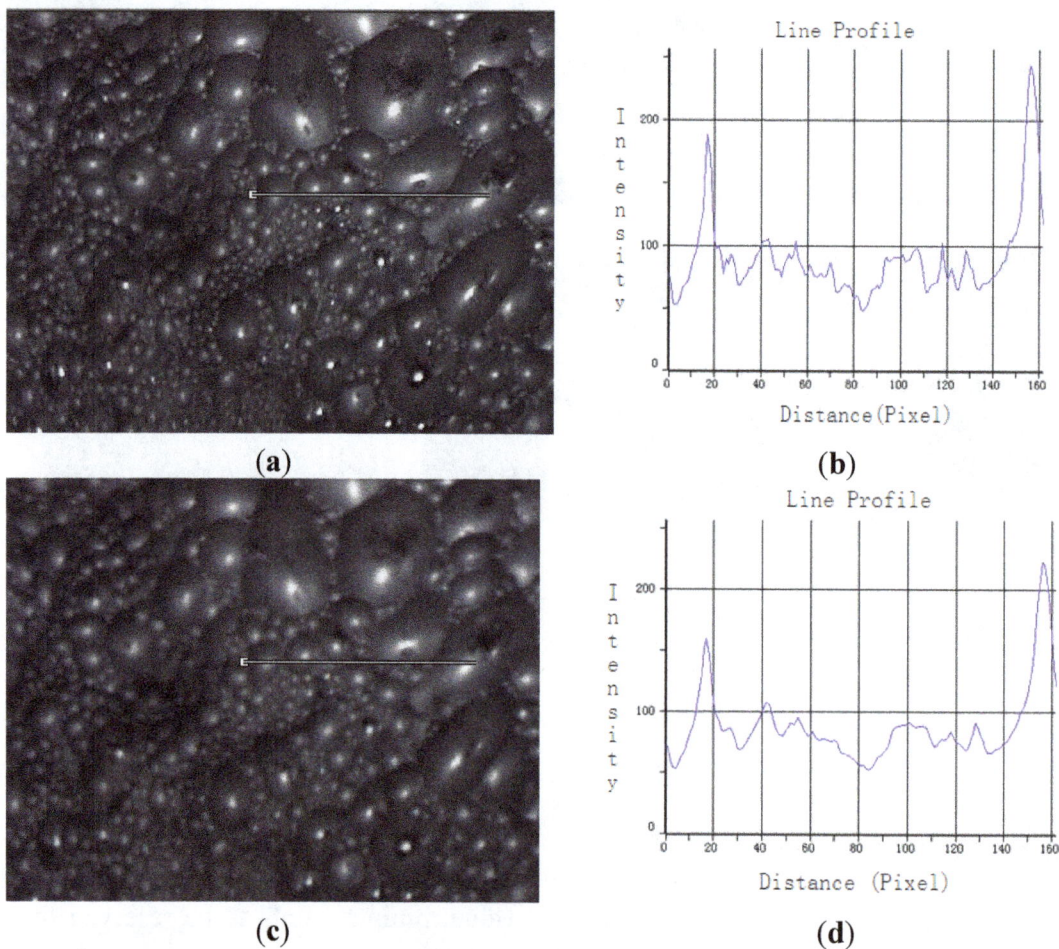

(a)

(b)

(c)

(d)

Figure 7. Gray value variance comparison before and after Gaussian filtering. (a) Original froth image and a profile line; (b) Gray value histogram of the profile corresponding to the image (a) from left to right; (c) Result smoothed by the Gaussian filter and profile line at the same position; (d) Gray value histogram of the profile corresponding to the image (c) from left to right.

A corner point contains the information of the image gray changes. Generally, it is believed that the corner point is the point of a maximum curvature value or a drastic change of the brightness in an image. These points not only retain the important image features but also effectively reduce the amount of data, which can improve the computation speed. A corner point plays a very important role in motion estimation, target tracking, image matching, and others in computer vision. In a froth image, the white spots of bubbles are the areas with a drastic intensity change, which is consistent with the definition of the corner point. Thus, the white spots can be obtained through a corner point detection method, and the number of corner points represents the number of white spots.

At present, the corner detection algorithms can be divided into three types: they are based on gray-scale images, the binary image and the contour curve, respectively. Harris corner detection algorithm is a gray level image corner detection method based on a template [25]. The algorithm detects the gray value change in the neighborhood of the detected pixel, and defines the pixel as a corner point when the change is large enough.

We now give a brief mathematical expression of the Harris corner detection algorithm. Denoting the image intensities by I, and the window function by $w(x,y)$, the window function is operated on the image pixel by pixel to detect the gray level change. The change E produced by a shift (u,v) is given by:

$$E(u,v) = \sum_{x,y} w(x,y)[I(x+u,y+v)-I(x,y)]^2 = \sum_{x,y} w(x,y)[I_x u + I_y v + O(u^2,v^2)]^2 \qquad (17)$$

where, I_x, I_y specify the first gradient results in x, y directions, respectively. It uses the Gaussian equation as the window function:

$$w(x,y) = \exp[-\frac{(x^2+y^2)}{2\sigma^2}] \qquad (18)$$

The change E, for the small shift (u,v), can be concisely written as:

$$E(u,v) \cong \sum_{x,y} w(x,y)[I_x u + I_y v]^2 = \sum_{x,y} w(x,y)[u,v]\begin{bmatrix} I_x^2 & I_x I_y \\ I_x I_y & I_y^2 \end{bmatrix}\begin{bmatrix} u \\ v \end{bmatrix} = [u,v]M\begin{bmatrix} u \\ v \end{bmatrix} \qquad (19)$$

where, M is a 2×2 symmetric matrix, and it can be written as:

$$M = \sum_{x,y} w(x,y)\begin{bmatrix} I_x^2 & I_x I_y \\ I_x I_y & I_y^2 \end{bmatrix} \qquad (20)$$

where, M can be made as a diagonalization matrix:

$$M \to R^{-1}\begin{bmatrix} \lambda_1 & 0 \\ 0 & \lambda_2 \end{bmatrix} R$$

where λ_1 and λ_2 are the eigenvalues of M, they are proportional to the principal curvatures of the local auto-correlation function, and form a rotationally invariant description of M.

Define the following formulation for the corner response function R:

$$R = \lambda_1 \lambda_2 - k(\lambda_1 + \lambda_2)^2 \qquad (21)$$

where, k is an empirical constant, usually k is in the range of 0.04–0.06. R is positive in the corner region, negative in the edge region, and small in the flat region.

The Harris corner point detection algorithm is to compare the corner response function R with a threshold value given in advance. If R is greater than the threshold, then the pixel is marked as a corner point. In the improved Harris corner detection algorithm, the corner point determination condition is changed as: in the two eigenvalues, if the smaller one is greater than a given threshold, the pixel is determined as a corner point. In the mean time, the improved algorithm sets another tolerance distance parameter; in the given tolerance distance it only retains a strong corner point. By setting the reasonable tolerance distance parameter, the improved Harris corner detection algorithm is more suitable for being used to extract the white spots for image classification, and the calculation is simple.

The extraction results of the corner points both in an image of large bubbles and in an image of small bubbles are shown in Figure 8. As we can see, most of the white spots are successfully marked as the corner points. Although there exist some false detection points and some white spots are missed, which only account for a small proportion, it does not affect obtaining the accurate image classification result.

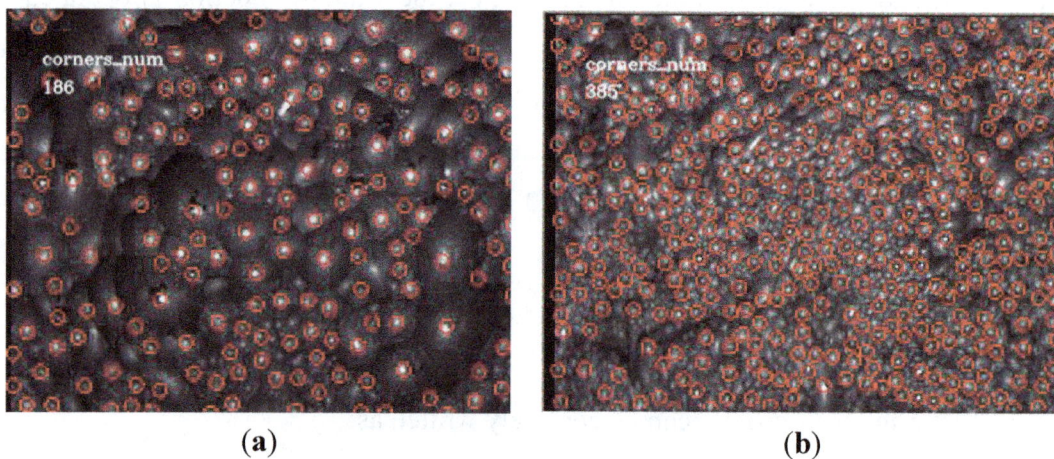

(a)   (b)

Figure 8. Improved Harris corner point detection. (**a**) Result of the large bubbles with tolerance distance 15; (**b**) Result of the small bubbles with tolerance distance 11.

As an illustrative example, we choose six bubble images randomly to do the corner point detection. The six images, of size 384×288, are easily distinguished as the images of large, middle and small bubbles, and two of them are of large bubbles, two are of middle and the rest are of small bubbles. Table 1 shows the basic data and classification information, and the corresponding corner point number distribution is shown in the left of Figure 9.

From Table 1, we see that, with the tolerance distance decreasing, the changes of the numbers of the corner points in the different bubble images are growing more and more. With tolerance distance 40, the difference of the numbers of the three bubble images is little; with tolerance distance 8, the difference is great. With the distance 11, the corner point numbers in the images of small and middle bubbles are both above 200 and the numbers in the images of large bubbles are still below 200.

To verify the above general discipline, with the threshold 5 and the tolerance distance 11 pixels, we do the further test on 10 images of large bubbles and 20 images of middle, small and tiny bubbles, the specific data are shown on the right of Figure 9, where, the red dots represent the number of the corner points in the images of large bubbles, the green dots represent the images of middle and small bubbles, and the blue dots are for the images of tiny bubbles. The data distribution illustrates that the mentioned classification rule is feasible. Based on the above data analyses, we obtain the classification rules as follows: with the threshold 5 and the tolerance distance 11 pixels, if the number of corner points is above 200, the image belongs to that of non-large bubbles (the images of small, tiny and middle bubbles); otherwise, it belongs to that of large bubbles.

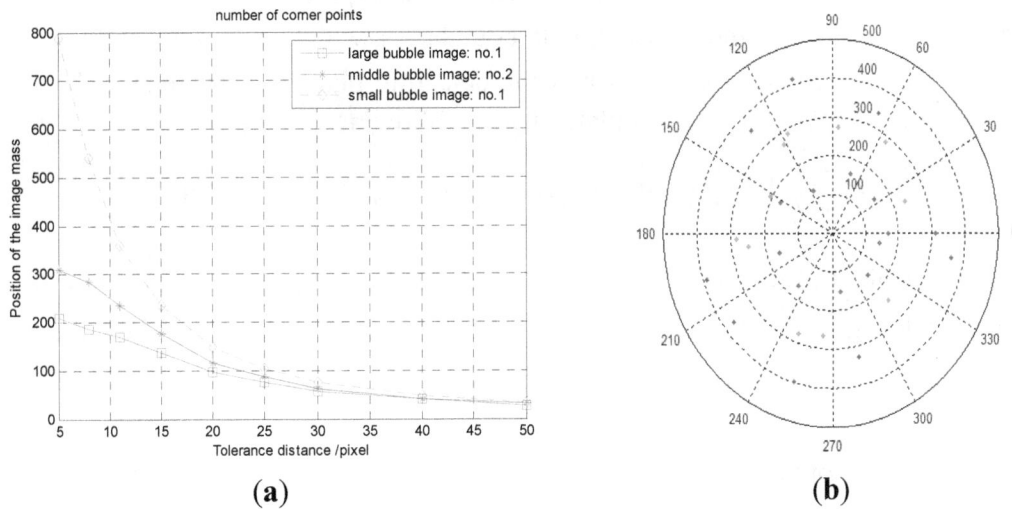

Figure 9. Corner point numbers (white spot numbers) distribution. (a) Corner point number distribution from Table 1; (b) Corner point number distribution of 30 different types of bubble images.

Table 1. Classification analysis based on the number of corner points (threshold of corner point detection algorithm is 5).

Tolerance Distance (Pixel)	Large Image No. 1	Large Image No. 2	Middle Image No. 1	Middle Image No. 2	Small Image No. 1	Small Image No. 2
5	210	200	269	310	787	677
8	187	176	249	284	539	470
11	169	164	213	236	359	372
15	138	136	162	176	234	213
20	97	106	113	115	148	134
25	75	80	80	88	104	103
30	58	56	63	63	77	74
40	41	38	42	41	49	47
50	27	30	28	31	35	32

2.4. Post-Processing after Image Segmentation

After the image classification and the local gray value minima detection, most significant bubble edges exist in the magnitude gradient image. But the result image may not be satisfactory, because there are some isolated points, short line segments and small closed curves caused by noise, and some bubble edges are discontinuous. In order to overcome the problems, we have to do a number of post operations. Our post-processing subroutines include noise filtering, thinning, endpoint detection and gap linking, and region merging.

The isolated points and some short line segments with 2–3 pixels can easily be removed by some common filtering operators. For the images of small and middle bubbles, the non-boundary information is little, and after filtering the edge image for 3–4 times, the isolated points, the short line segments and other non-boundary points can almost be removed. For the images of large bubbles, the filtering steps should be done more times to remove the noise effectively.

Before the edges are further thinned into of a width of one pixel, the simple noise filtering is performed. The last task is to make the closed bubble boundaries, which is done by endpoint detection and gap linking. We use the connection number of the detecting pixel to determine if the pixel is an endpoint. The connection number, N_c, is calculated by the following formula:

$$N_c = \sum_{k=0,2,4,6} [(1-f(x_k)) - (1-f(x_k))(1-f(x_{k+1}))(1-f(x_{k+2}))] \tag{22}$$

where x_k denotes No. k pixel of the eight-neighborhood of the detecting pixel, as shown in Figure 10; and $f(x_k)$ represents the gray value of pixel x_k, where we define: $f(x_k) = 1$ when the pixel is a white point (edge point), and $f(x_k) = 0$ when the pixel is an non-edge point; and when $x_k = x_8$, we define $x_8 = x_0$.

For each edge point, the connection number N_c is calculated. When $N_c = 1$, the detected edge point is marked as an endpoint, and both its direction and location are marked.

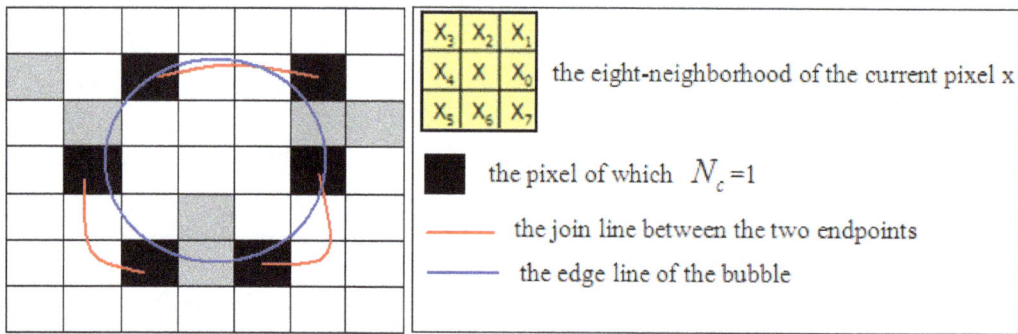

Figure 10. Gap linking diagram between the two endpoints and the representation of the pixels in the eight-neighborhood of the detecting pixel x.

When all the endpoints are marked, the next step is to link the endpoints based on the principle of the similar direction and the nearest distance. We search for another endpoint in its 5×5 kernel as the detected endpoint as center. When a new endpoint is found, and its direction is the same to that of the detected endpoint, the new endpoint is linked along the direction to the current endpoint. When the direction is different, we choose the endpoint with the shortest distance to link along the direction of the current endpoint. When there is no endpoint to be found, then it skips the current endpoint and processes the next endpoint.

Through this procedure, there are some short line segments left. We use a line threshold to remove these false edges. The algorithm of Region merging is also used to remove some complex closed curves.

3. Experimental Results and Discussions

According to the above analyses, the workflow of the proposed method is given in Figure 11. It includes the following key steps:

(1) Applying the improved Harris corner detection method to extract the white spots and calculate the total number of white spots. When the number is less than 200, the image is that of large bubbles; otherwise, it belongs to that of non-large bubbles.

(2) Using the Gaussian filter to smooth the classified bubble image.

(3) Taking advantage of the improved local gray value minima detection method to identify the minima point for each pixel to generate the binary edge image, where a 5×5 template is used for the images of large bubbles, and a 3×3 template for the images of non-large bubbles.

(4) Carrying out the non-boundary information filter processing for the binary edge image to remove the isolated points and short line segments.

(5) Adopting endpoint detection and gap linking, region merging and other operations to obtain the last edge image.

(6) Overlapping the last edge image on the original bubble image.

Based on the above procedure, it can automatically complete the classification, edge detection, segmentation and bubble delineation in a froth image.

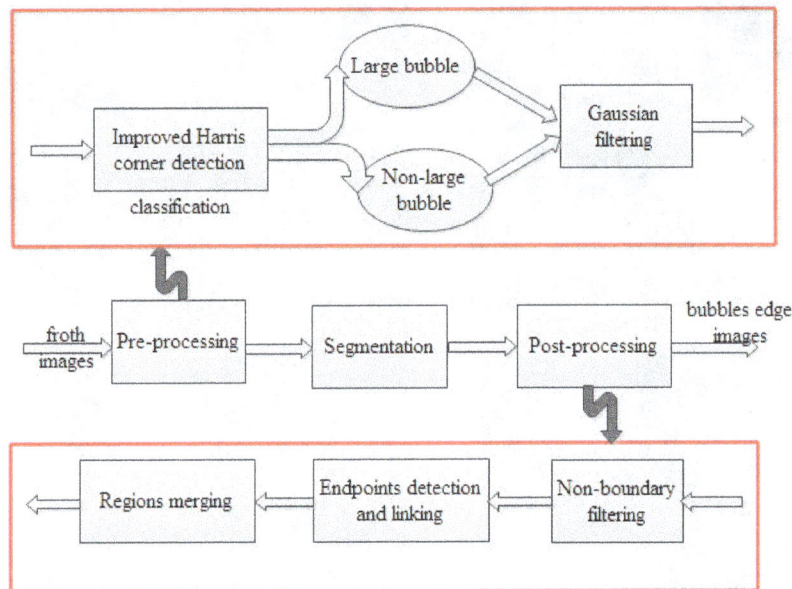

Figure 11. Working procedure of the proposed method.

Figures 12e and 13e give the result images only after the edge detection processing, respectively for the image of large bubbles and the image of non-large bubbles. For the comparison, we also test other edge detection algorithms including Valley-edge [15], Sobel and Canny [26,27]. In Figure 12, the images in (b–d) are the detection results of an image of large bubbles by using Sobel, Canny and Valley-edge respectively, and in Figure 13, the images in (b–d) are the corresponding detection results of an image of non-large bubbles.

As we can see in Figures 12 and 13, the Sobel edge detection operator can only obtain some parts of the edges of bubbles regardless of which class of images, and what is more, the bubble edges are much weaker than the white spot edges. The same result is also obtained by the Canny operator, which lost most of the bubble edges. The above results illustrate that the classic edge detection algorithms fail to segment the bubble images, therefore the result edge (or gradient) images are difficult to be used for the further bubble delineation. Comparing to the results of Sobel and Canny operators, the Valley-edge detection algorithm and the proposed algorithm both produce the better results, where most significant bubble edges are detected and most white spot edges are disregarded. On the contrast, the proposed algorithm can obtain more bubble edge information than the Valley-edge detection algorithm.

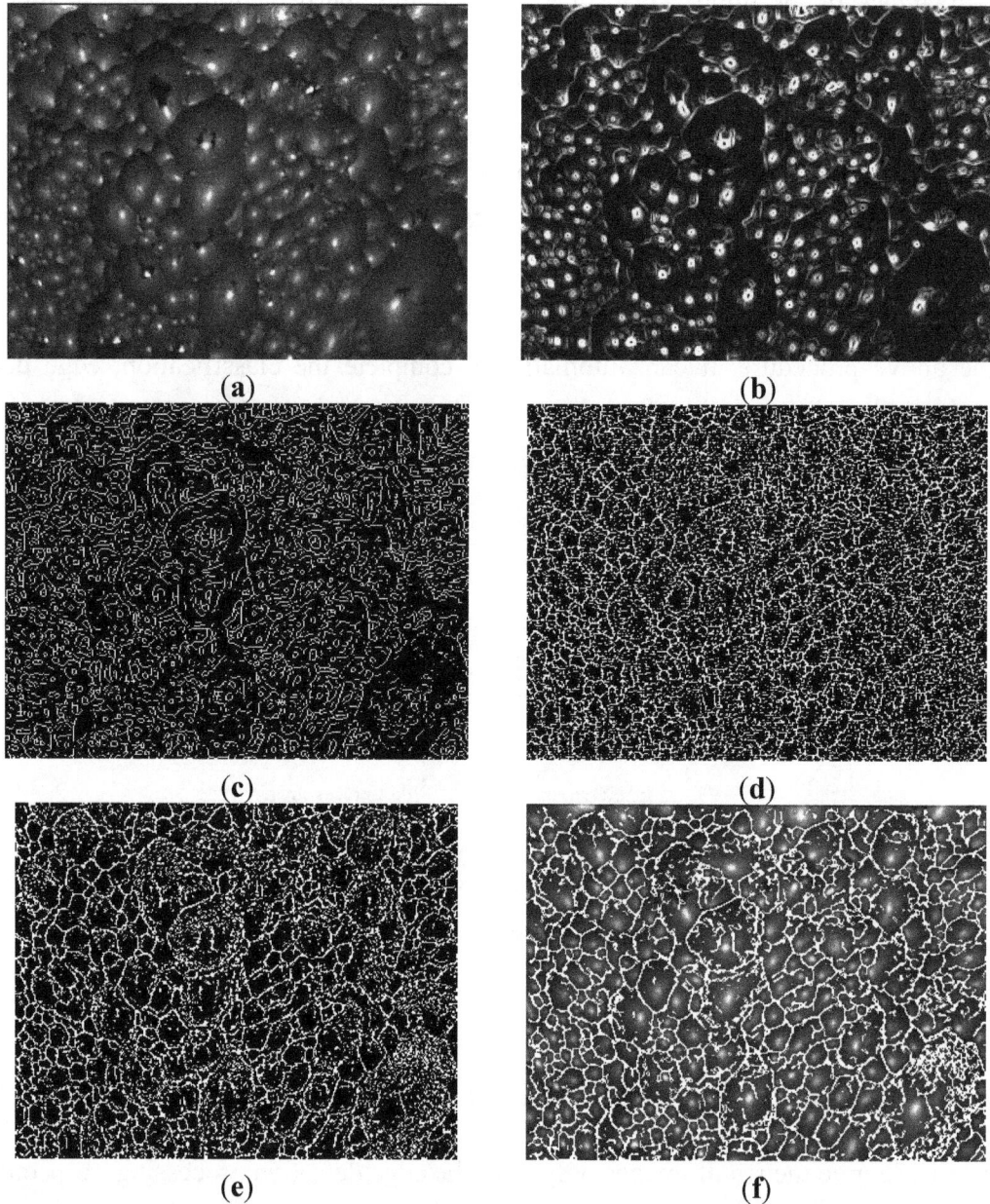

Figure 12. Comparison between the proposed edge detection algorithm and others (for the image of large bubbles). (**a**) Original image; (**b**) Sobel result; (**c**) Canny result; (**d**) Valley-edge result; (**e**) New algorithm result; (**f**) Result of overlap (e) and (a).

After edge detection by the improved local gray value minima algorithm, the post-processing procedures can be done on the froth edge images to obtain the final good image segmentation results. The final segmentation results by the proposed algorithm are shown in the sixth row in Figure 14. We can see that the segmentation results are promising for all classes of froth images, compared with the other image segmentation algorithms based on the gray value similarity. In Figure 14, the four images in the first row are the original images, respectively, representing the images of large, middle, small and mixed-size bubbles; the four images in the second and the third rows show the optimal and Otsu global thresholding results, respectively, where most white spots are delineated, then most bubble edges are lost; the four images in the fourth and fifth rows are the segmentation results of the region growing algorithm with different thresholds, and we can see that the results are also not good. These algorithms based on local or global gray value similarity fail to produce satisfactory segmentation results because the gray values are different both in the interior of a bubble and in different areas in an image.

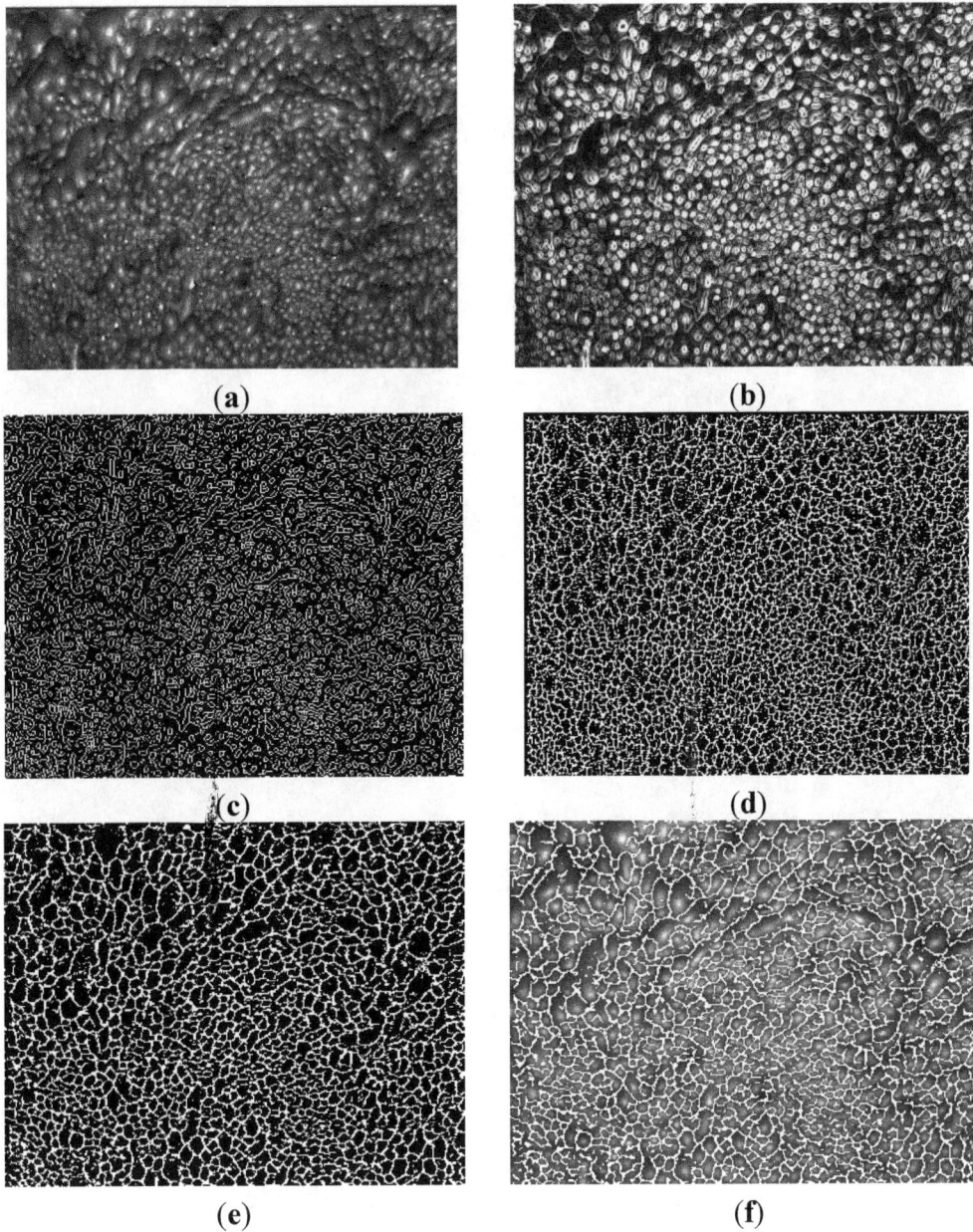

Figure 13. Comparison between the proposed edge detection algorithm and others (for the image of non-large bubbles). (**a**) Original image; (**b**) Sobel result; (**c**) Canny result; (**d**) Valley-edge result; (**e**) New algorithm result; (**f**) Result of overlap (e) and (a).

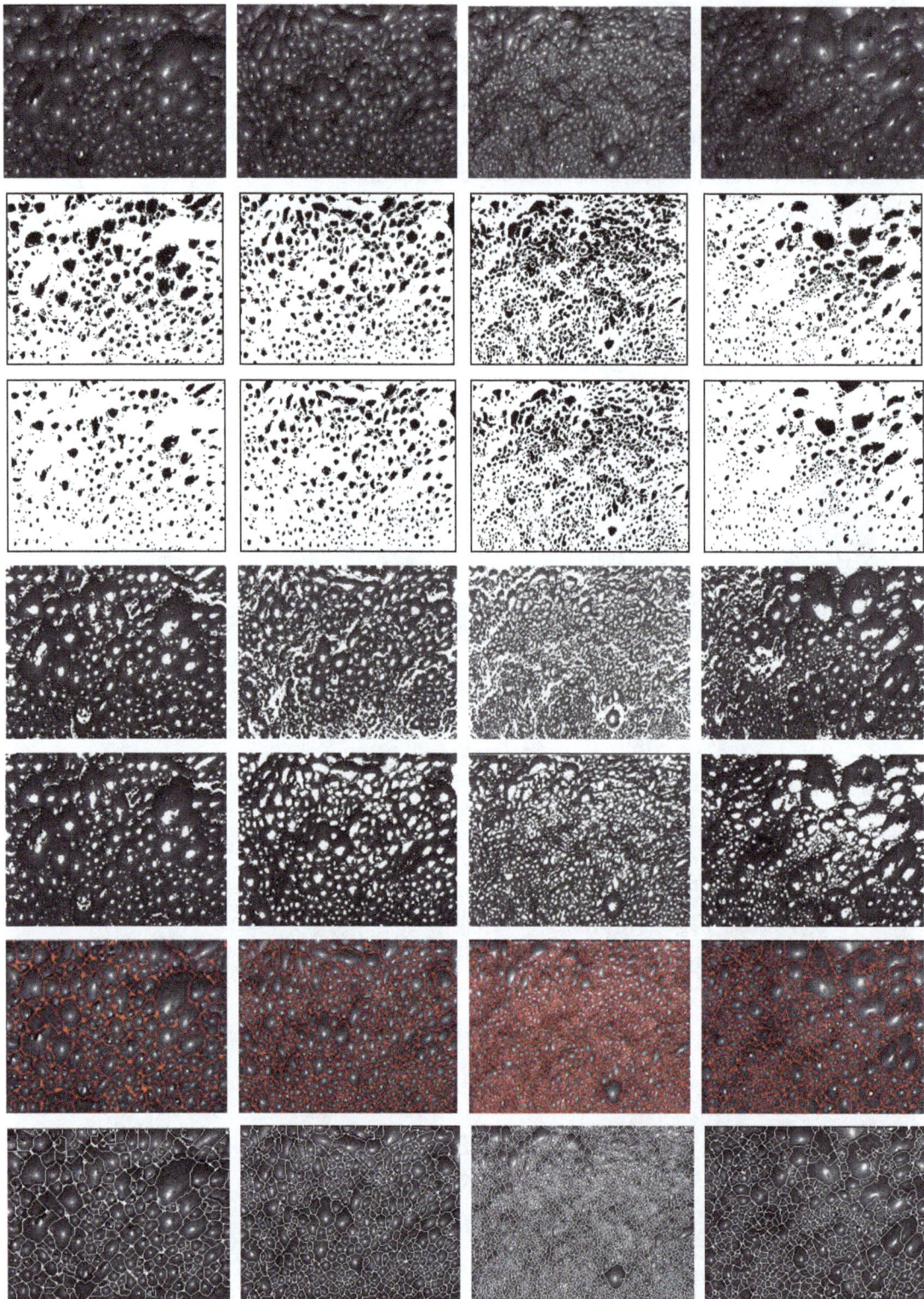

Figure 14. Comparison between the proposed algorithm and the algorithms based on gray value similarity: The first row is for the original images; the second row is for the segmentation results of the optimal threshold algorithm; the third row is for the results of Otsu threhsolding; the fourth and the fifth rows are for the results of the region growing algorithm with the threshold 30 and 35 separately; the sixth row is for the results of the proposed algorithm; the last row is for the results of manual delineation.

Our image segmentation algorithm produces the better results as shown in the four images of the sixth row in Figure 14, where, most bubbles are delineated correctly in each class of images. Even for the small bubbles between the large bubbles and the single large bubble in the small bubbles, the segmentation results are satisfactory although the under-segmentation problem still exists. We also note that it encounters the over-segmentation problem for the images of large bubbles in the mixed-size bubbles, and the black holes in the interior of a large bubble lead to the same problem.

In order to provide a quantitative comparison, we delineate the rest of bubbles manually based on the rough segmentation result of the Watershed algorithm and count the number of bubbles. The manual results are shown in the last row in Figure 14. Table 2 gives the data about the number of bubbles. The numbers are counted based on the results of our algorithm and manual separately.

Table 2. Bubbles numbers comparison between our algorithm and manual segmentation.

Froth Image Type	Large	Middle	Small	Mixed-Size
Manual segmentation	231	451	1537	500
Our algorithm	201	450	1799	472

We also make the comparison between the new algorithm and the Watershed algorithm; the results are shown in Figure 15, where the original images are the same as the image (a) in Figures 12 and 13, respectively, representing the images of large bubbles or non-large bubbles. The commercial software, Image-pro Plus (version 6.0) (Media Cybernetics, Bethesda, MD, USA), is used to achieve the froth image segmentation result based on the Watershed algorithm. We set the gray value threshold in the Watershed algorithm, *i.e.*, we separate objects using the gray levels (the preflooding level parameter) of the objects on the original image. The preflooding level parameter can be used to separate only slightly touching objects, avoiding separation of large objects.

Generally, the preflooding level can be set in a small value for small sized objects, and in a large value for the large sized objects. In Figure 15, images (a) and (c) are the Watershed segmentation results for the image of large bubbles, where the value of preflooding level is set as 20 and 40 separately; and the images in (b) and (d) are the segmentation results for the images of non-large bubbles, where the value of preflooding level is set as 1 and 5, respectively. When using the Watershed algorithm to segment a froth image, it is difficult to set a right adaptable parameter (preflooding level) to obtain a good result. For the image of large bubbles, a too small parameter value will cause the over-segmentation problem, as it can be seen in images (a) and (c); for the image of non-large bubbles, a too big value will lead to under-segmentation problem and also over-segmentation problem in some large bubbles, as shown in images (b) and (d). These results show that the segmentation results of the Watershed algorithm mainly depend on the appropriate parameter setting by manual. However, the parameter setting is hard and time-consuming for the reason that the parameter values differ greatly on different froth images. Obviously, the Watershed algorithm is difficult to be directly applied in an online froth analysis and monitoring system with a right parameter. On the contrary, the proposed method is a fully automatic processing procedure including image classification and image segmentation, in which it needs no manual operation. For the same images, the processing speed is tens to hundred times faster than the Watershed algorithm. Normally, for those froth images of size

384×288 pixels, it needs 0.05–0.1 s for processing one image on a PC (Inter Core 2.9 GHz), which makes the new method applicable in an online system.

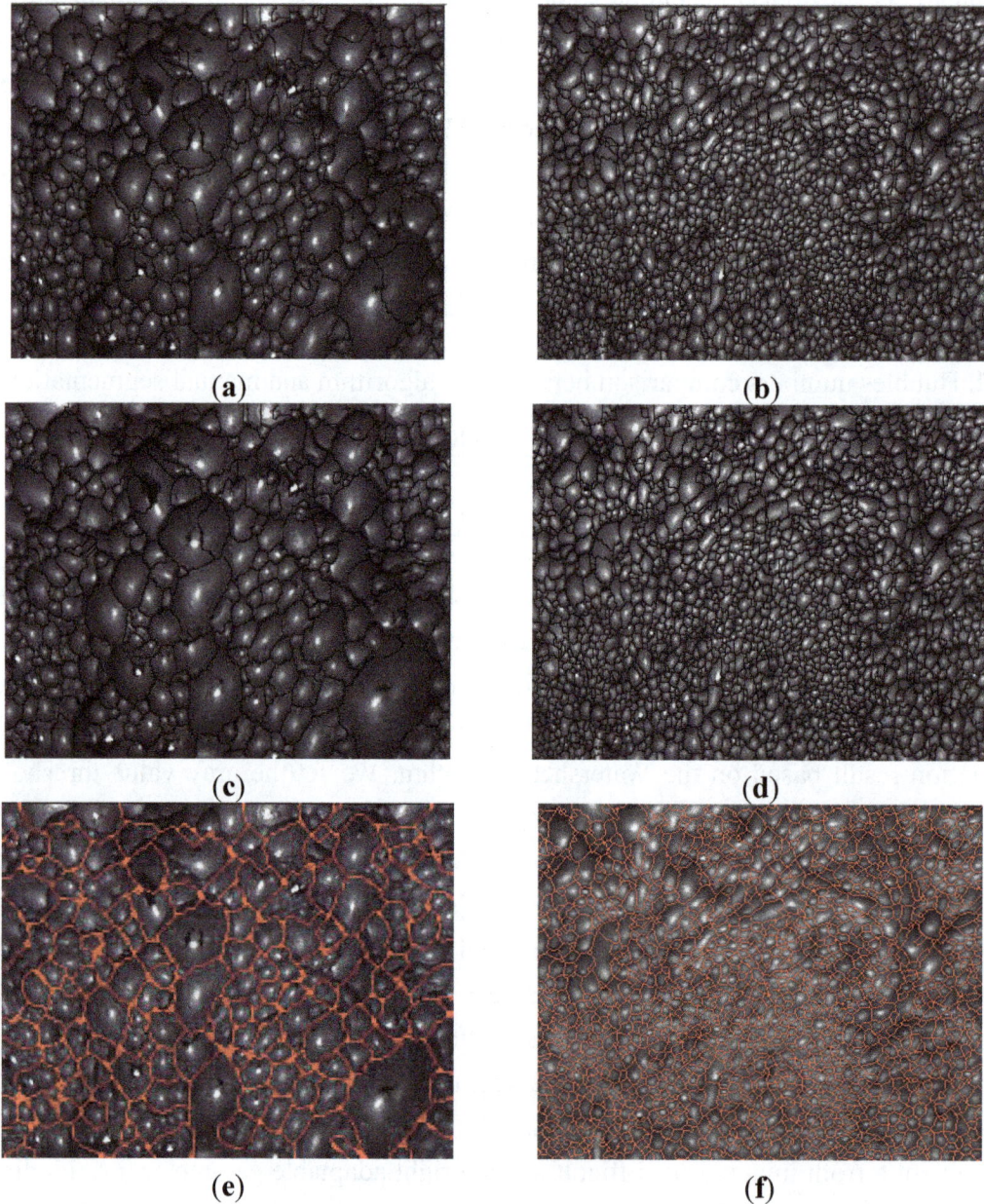

Figure 15. Comparison between the proposed algorithm and the Watershed algorithm. (**a**) and (**c**) are for the Watershed segmentation results with preflooding level = 20, 40 separately for the image of large bubbles; (**b**) and (**d**) are for the Watershed segmentation results with preflooding level = 1, 5 separately for the image of non-large bubbles; (**e**) is for the image segmentation result by the proposed algorithm for the image of large bubbles; (**f**) is for the image segmentation result by the proposed algorithm for the image of non-large bubbles.

4. Conclusions

This paper proposes a froth delineation method that is the combination of image classification and image segmentation. The corner detection algorithm is used to complete the image classification, and the improved local gray value minima algorithm is applied to achieve the image segmentation. The experimental results show that our method can produce good image segmentation results automatically and the processing speed is fast. The segmentation algorithm is to search the edge point located in a certain direction with gray value minima through the inequality comparison; therefore it is not affected by uneven illumination, bubble 3D geometry and other negative factors. In addition, the new image segmentation algorithm only do the logic comparative judgment without the complex numerical calculation, so the computing speed is fast and the method can meet the real-time requirements of the flotation production control in Jin Dong lead zinc flotation plant. The further improvement of the image segmentation algorithm is for the images of the mixed-size bubbles with many black holes. We will compare the proposed image segmentation algorithm at the industrial scale with the existing algorithms including the Watershed algorithm (used in SmartFroth) and others.

Acknowledgments

This research is financially supported by the National Natural Science Fund in China (grant No. 61170147) and the Science and Technology Development Fund in Fuzhou University (grant No. 2014-XY-31).

Author Contributions

Liangqin Chen was in charge of design, implementation and scientific writing of this research under the supervision of Weixing Wang. Final manuscript handling during the publication process was done by Weixing Wang.

Conflicts of Interest

The authors declare no conflict of interest.

References

1. Farrokhpay, S. The significance of froth stability in mineral flotation—A review. *Adv. Colloid Interface Sci.* **2011**, *166*, 1–7.
2. Moolman, D.W.; Eksteen, J.J.; Aldrich, C.; van Deventer, J.S.J. The significance of flotation froth appearance for machine vision control. *Int. J. Miner. Process.* **1996**, *48*, 135–158.
3. Lin, B.; Recke, B.; Knudsen, J.K.H.; Jorgensen, S.B. Bubble size estimation for flotation processes. *Miner. Eng.* **2008**, *21*, 539–548.
4. Yang, C.; Xu, C.; Mu, X.; Zhou, K. Bubble size estimation using interfacial morphological information for mineral flotation process monitoring. *Trans. Nonferr. Met. Soc. China* **2009**, *19*, 694–699.
5. Liu, J.; Gui, W.; Tang, Z.; Yang, C.; Zhu, J.; Li, J. Recognition of the operational statuses of reagent addition using dynamic bubble size distribution in copper flotation process. *Miner. Eng.* **2013**, *45*, 128–141.
6. Xu, C.; Gui, W.; Yang, C.; Zhu, H.; Lin, Y.; Shi, C. Flotation process fault detection using output PDF of bubble size distribution. *Miner. Eng.* **2012**, *26*, 5–12.

7. Runge, K.; McMaster, J.; Wortley, M.; La Rosa, D.; Guyot, O. A correlation between Visiofroth™ measurements and the performance of a flotation cell. In *Ninth Mill Operators' Conference Proceedings 2007*; Nelson, M., Smith, R., Eds.; The Australasian Institute of Mining and Metallurgy: Melbourne, Australia, 2007; pp. 79–86.

8. Jahedsaravani, A.; Marhaban, M.H.; Massinaei, M. Prediction of the metallurgical performances of a batch flotation system by image analysis and neural networks. *Miner. Eng.* **2014**, *69*, 137–145.

9. Wang, Y.; Neethling, S.J. Simulating realistic froth surfaces. *Miner. Eng.* **2006**, *19*, 1069–1076.

10. Wang, Y.; Neethling, S.J. The relationship between the surface and internal structure of dry foam. *Colloids Surf. A* **2009**, *339*, 73–81.

11. Bennet, A.J.R.; Chapman, W.R.; Dell, C.C. Studies in Froth Flotation of Coal. In *Third International Coal Preparation Congress*; Annales des Mines de Belgique: Brussels, Belgium, 1958.

12. Morar, S.H.; Bradshaw, D.J.; Harris, M.C. The use of the froth surface lamellae burst rate as a flotation froth stability measurement. *Miner. Eng.* **2012**, *36–38*, 152–159.

13. Zanin, M.; Wightman, E.; Grano, S.R.; Franzidis, J.-P. Quantifying contributions to froth stability in porphyry copper plants. *Int. J. Miner. Process.* **2009**, *91*, 19–27.

14. Neethling, S.J.; Lee, H.T.; Cilliers, J.J. Simple relationships for predicting the recovery of liquid from flowing foams and froths. *Miner. Eng.* **2003**, *16*, 1123–1130.

15. Wang, W.; Bergholm, F.; Yang, B. Froth delineation based on image classification. *Miner. Eng.* **2003**, *16*, 1183–1192.

16. Citir, C.; Aktas, Z.; Berber, R. Off-line image analysis for froth flotation of coal. *Comput. Chem. Eng.* **2004**, *28*, 625–632.

17. Shen, X.; Tang, Z.; Xu, J.; Gui, W. Based on DCIWPSO in application of valley-edge detection froth image segmentation. *Appl. Res. Comput.* **2010**, *27*, 3564–3566.

18. Yang, C.; Yang, J.; Mu, X.; Zhou, K.; Gui, W. A segmentation method based on clustering pre-segmentation and high-low scale distance reconstruction for colour froth image. *J. Electron. Inf. Technol.* **2008**, *30*, 1286–1290.

19. Aldrich, C.; Marais, C.; Shean, B.J.; Cilliers, J.J. Online monitoring and control of froth flotation systems with machine vision: A review. *Int. J. Miner. Process.* **2010**, *96*, 1–13.

20. Shean, B.J.; Cilliers, J.J. A review of froth flotation control. *Int. J. Miner. Process.* **2011**, *100*, 57–71.

21. Vincent, L.; Soille, P. Watersheds in digital spaces: An efficient algorithm based on immersion simulations. *IEEE Trans. Pattern Anal. Mach. Intell.* **1991**, *13*, 583–598.

22. Tarabalka, Y.; Chanussot, J.; Benediktsson, J.A. Segmentation and classification of hyperspectral images using watershed transformation. *Pattern Recognit.* **2010**, *43*, 2367–2379.

23. Wang, W.; Li, Y.; Chen, L. Bubble delineation on valley edge detection and region merge. *J. China Univ. Min. Technol.* **2013**, *42*, 1060–1065.

24. Otsu, N. A threshold selection method from gray-level histogram. *Automatica* **1975**, *11*, 23–27.

25. Harris, C.; Stephens, M. A combined corner and edge detector. In Proceedings of the 4th Alvey Vision Conference, Manchester, UK, 31 August–2 September 1988; pp.147–151.

26. Gonzalez, R.C.; Woods, R.E. *Digital Image Processing*, 2nd ed.; Prentice Hall: Upper Saddle River, NJ, USA, 2002.

27. Canny, J. A computational approach to edge detection. *IEEE Trans. Pattern Anal. Machine Intell.* **1986**, *6*, 679–698.

10

Submarine Tailings Disposal (STD)—A Review

Bernhard Dold

SUMIRCO (Sustainable Mining Research & Consult EIRL), Casilla 28, San Pedro de la Paz 4130000, Chile; E-Mail: bernhard.dold@gmail.com

Abstract: The mining industry is a fundamental industry involved in the development of modern society, but is also the world's largest waste producer. This role will be enhanced in the future, because ore grades are generally decreasing, thus leading to increases in the waste/metal production ratio. Mine wastes deposited on-land in so-called tailings dams, impoundments or waste-dumps have several associated environmental issues that need to be addressed (e.g., acid mine drainage formation due to sulphide oxidation, geotechnical stability, among others), and social concerns due to land use during mining. The mining industry recognizes these concerns and is searching for waste management alternatives for the future. One option used in the past was the marine shore or shallow submarine deposition of this waste material in some parts of the world. After the occurrence of some severe environmental pollution, today the deposition in the deep sea (under constant reducing conditions) is seen as a new, more secure option, due to the general thought that sulphide minerals are geochemically stable under the reduced conditions prevailing in the deep marine environment. This review highlights the mineralogical and geochemical issues (e.g., solubility of sulphides in seawater; reductive dissolution of oxide minerals under reducing conditions), which have to be considered when evaluating whether submarine tailings disposal is a suitable alternative for mine waste.

Keywords: tailings; acid mine drainage; waste management; marine pollution; solubility; reductive dissolution; sulphide; iron oxide; oceanography; ore deposit

1. Introduction

Mining was, is, and will also in the future be a fundamental industry involved in the development of human society. There is no doubt that mining had and will always have a negative environmental impact, as it is a destructive activity. There is also no doubt that humanity cannot progress without metals. Environmental impacts will continue to increase because the ratio of waste/element produced by mining operations is, and will increase in the future, as high-grade ores become rarer and as low-grade ores are exploited. There is little to question about the value of mining to society, but questions remain about how to best proceed to exploit our resources. Due to the increasing volume of waste associated with low-grade ores and the associated environmental impacts of this waste, mining has increasingly competed with other land uses, making environmental and social issues more and more prominent. Nowadays, base metal mining mainly exploits sulphide mineral ores (*i.e.*, minerals which form in the earth's crust under reducing conditions, for example porphyry deposits; Cu, Mo, Au). This is mainly true for the exploitation of metals like Cu, Zn, Pb, Ni, Mo, Au, Ag, while Al, Fe, and Rare Earth Elements (REE) are extracted mainly from oxide ores (Table 1). Coal mining also exploits a resource that was formed under reducing environment, and therefore which can contain sulphide minerals like pyrite. Sulphide minerals are the primary component responsible for the principal environmental impact of mining, the formation of acid mine drainage (AMD). AMD occurs, when sulphidic mine waste is exposed to atmospheric oxygen and water, as occurs in waste rock dumps and mine tailings storage areas and in underground workings or on the walls of an open pit mine. Due to the problems of AMD formation and the geotechnical instability of mine tailings (fine milled material like sludge) impoundments (reviewed elsewhere in this Special Issue [1]), the mining industry is searching for alternative means of waste management, as this is one of the main issues confronting mining operations. Due to the aforementioned problems of on-land deposition of tailings, one option, which has recently regained atention is submarine tailings disposal (STD). In order to evaluate which waste management option is most suitable for a particular site, many parameters have to be considered. Submarine tailings disposal seems to be an attractive option for preventing AMD because sulphide minerals in the mine waste should be geochemically stable in the long-term due to reduced redox conditions in this environment. Although from a geochemical standpoint this idea is sound, the reality (*i.e.*, nature) is more complex. This review draws attention to several parameters that are often mentioned but poorly documented in the literature [2,3], but are important to consider. The review is not comprehensive and the examples selected mainly highlight important processes occurring in these systems. For this purpose, we discuss the history and the lessons learned from mine waste deposition and the formation of our potential resource (the ore deposit) to understand the biogeochemical interactions occurring between the seawater and the deposited mine waste.

Table 1. Simplified overview of the principal elements extracted by mining and their associated ore deposit type and with the associated environmentally problematic elements and their mineral assemblage. Only the principal minerals and elements are shown; this list can be much more extensive depending on the ore deposit. Hypogene: hydrothermal formation; Supergene: liberation and enrichment processes due to meteorization; AMD: acid mine drainage; REE: Rare Earth Elements.

Target Elements	Principal Ore Deposit Type	Principal Economic Mineral Assemblage	Associated Non-Economic Minerals and Polluting Elements	Environmental Problems Associated to Mineralogy	Environmental Problems Associated to the Process
Cu, Zn, Pb, Ni, Au, Ag, Sb	Hydrothermal (hypogene), e.g., porphyry types, epithermal veins, high-sulfidation	Chalcopyrite, bornite, chalcocite-digenite, covellite sphalerite, galena	Pyrite, Enargite, tetrahedrite-tennantite, e.g., As, Se, Th, Cr, Cu, Zn, Ni, Cd, Pb	AMD, geochemical and geotechnical stability of tailings impoundments	Flotation reagents, cyanide, geochemical and geotechnical stability of tailings impoundments
Fe	Hydrothermal and Laterites (Supergene)	Hematite—magnetite (goethite-ferrihydrite)	e.g., As, Se, Th, Cr, Cu, Zn, Ni, Cd, Pb	Release of associated elements under reducing conditions (reductive dissolution)	Flotation reagents, geochemical and geotechnical stability of tailings impoundments
Al	Bauxites (Supergene)	Gibbsite, boehmite, diaspore	Goethite, hematite, kaolinite, e.g., As, Se, Th, Cr, Cu, Zn, Ni, Cd, Pb		geochemical and geotechnical stability of tailings impoundments. Alkaline tailings sludge (NaOH digestion)
REE	Clay deposits (Supergene) and Pegmatites	e.g., bastnäsite, monazite, allanite, loparite	Clay minerals and Fe and Al hydroxides. Radioactive elements associated	Release of radioactivity	Desorption reagents like $(NH_4)_2SO_4$
Coal	Coal and Lignite	Coal, Lignite, anthrazite	Pyrite, sulfur, volatiles	AMD	Extensive pit lakes

2. A Little Historical Background

When the flotation process was developed at the beginning of the 20th century, the mining process was revolutionized due to the ability to exploit low-grade ores. This also increased considerably the volume of tailings, the waste material remaining after the ore is passed through crushing, milling, and flotation stages in order to extract the target minerals. Other waste materials associated with a mine include overburden (the soils and rocks overlying the ore without economic mineralization), and the country waste rock extracted in the mining process, which are without economic value at the time of mining (but can contain considerable amounts of sulphide minerals). Both types of materials are deposited in so-called waste-dumps, or in stock-piles if there is any remaining economic value, mostly as a coarse grained material called run-of-mine (ROM) [4]. Nowadays, the mining industry is one of the largest waste producers in the world. Consequently, sustainable waste management is in general the most relevant task for humanity to solve in the future. For the mining industry, it represents the most important environmental problem, due to the enormous volume and space it occupies [4–6].

Historically, tailings were deposited close to the mining operation in natural depressions, lakes, or were sent via gravity into rivers, where they sometimes end up at the shoreline of the sea. Some historic operations include Chañaral, Chile [7–9], Ite Bay, Perú [10], Cerro de Pasco, Perú [11], and Bahía Portman, Spain [12–14], while Freeport-Grasberg, Indonesia [2,3,15] still operates this way (Figure 1). Sulphide oxidation and AMD formation in riverine, lake, or marine shore tailings deposits have damaged water resources (superficial and groundwater, and the sea) and the image of mining in the society, and the deposition of these materials in waterbodies or the sea is forbidden in most countries. (e.g., the lawsuit of the village of Chañaral against the Chilean National Copper Corporation (CODELCO) [7].

Nowadays, most exploited metalliferous ore deposits are sulphide deposits, whereas shallow-depth and surface-exposed oxide deposits have been exploited since roman times in Europe, or since Pre-Colombian times in South and Central America. The newer ore deposits are usually located deeper below the Earth's surface and usually are discovered with geophysical methods, unlike the more easily located surface oxide deposits of the past (Figure 2). Therefore, nowadays, as well in the future, the main sources of metals are and will be sulphide ore deposits.

Sulfide minerals form and are stable under reducing conditions generally deep in the Earth's crust or below water, without dissolved oxygen, and in some cases occurring in the presence of reducing agents like organic matter such as peat, or in the deep-sea. Recognizing the stability of sulfide minerals under these conditions, there was increased interest in the 1970s to deposit sulfide mine tailings at depth in the sea, to lessen the potential for sulfide oxidation. With this idea in mind, several operations around the world realized marine tailings disposal.

Riverine, lake, and marine disposal of tailings has a long history, like for example the cases of Chañaral, Chile (Potrerillo-El Salvador, 1938–1975) and Bahía de Ite, Peru (Cuajone y Toquepala, 1960–1997), which are mainly riverine deposits with subsequent formation of marine shore tailings deposits [7,10]. About 40 years ago, the so-called "Deep Sea Tailings Disposal" started with the first operation in 1971 (Atlas mine, Filipinas; Island Copper Mine, Canada) and 1972 (Jordan River Mine, Canada; Black Angel Mine; Greenland). We will see in the following that the expression "Deep Sea" is relative (it can range between a few tens to several thousands of meters of depth). Most scientific studies are published using the term Submarine Tailings Disposal (STD) 41, Submarine Tailings Deposition (STD) 10, Deep Sea Tailings Disposal (DSTD) 18, Deep Sea Tailings Placement (DSTP) 5, and Sub-Sea Tailings Deposition (SSTD). In this review, the most generic term Submarine Tailings Disposal (STD) is used. There are relatively few peer-reviewed papers published (the numbers behind the terms represent the hits in Scopus [16], a scientific publication database) on this topic. This is possibly due to the high costs for oceanographic investigations. Most of the information is produced in the frame of Environmental Impact Studies (EIS) for the permitting process and often there is no public access to the data.

Figure 1. Locations of coastal areas impacted by tailings deposition (including shore deposition, shallow, and deep-sea disposal). In brackets are the economically exploited metals. Adapted from Koski [3].

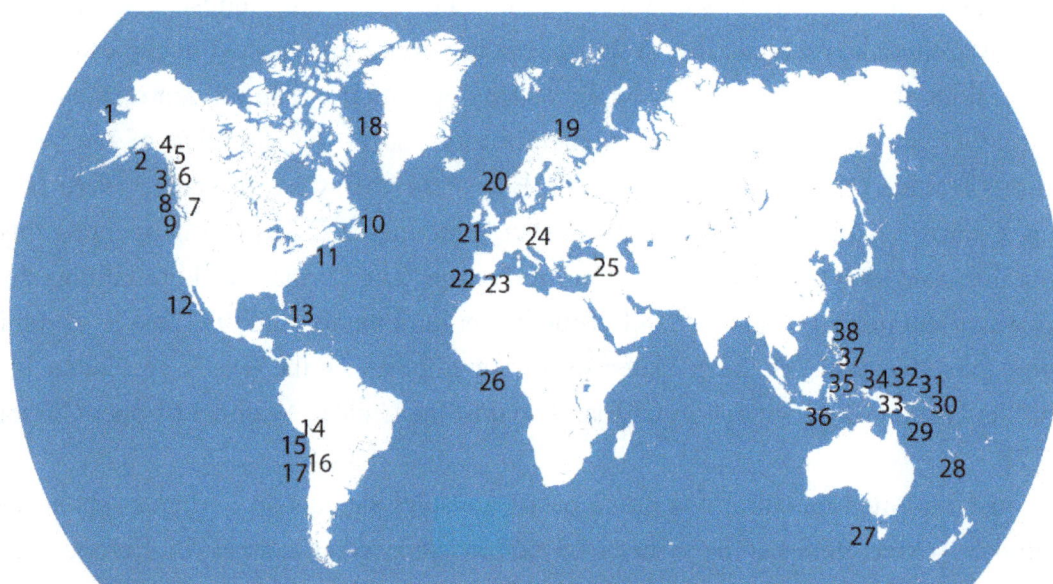

1. Nome Pacer, USA (Au)	20. SØrfjord, Norway (Cu, Pb, Zn)
2. Prince William Sound, USA (Cu)	21. St. Ives bay, Hayle estuary, England (Sn)
3. Bokan Mountain, USA (U)	22. Rio Tinto and Odiel estuaries, Spain (Cu)
4. Klag Bay, USA (Au, Ag)	23. Portman Bay, Spain (Pb, Zn)
5. Salt Chuck mine, USA (Cu, Ag, Au)	24. Grado and Marano lagoons, Italy (Hg)
6. Kisault mine, Canada (Mo)	25. Cayeli Bakir mine, Turkey (Cu, Zn)
7. Britannia mine, Canada (Cu, Zn)	26. Gulf of Benin, Togo (P)
8. Island Copper mine, Canada (Cu, Zn)	27. Macquarie Harbor (estuary), Australia (Cu)
9. Jordan River mine, Canada (Cu)	28. Southwest Lagoon, New Caledonia (Ni)
10. Little Bay, Tilt Cove mines, Canada (Cu)	29. Misima Island, Papua New Guinea (Au)
11. Callahan mine, USA (Zn, Cu)	30. Bougainville Island, Papua New Guinea (Au)
12. Boleo, Lucifer mines, Mexico (Cu, Mn)	31. Lihir Island, Papua New Guinea (Au)
13. Levisa Bay, Cuba (Ni)	32. Simberi Island, Papua New Guinea (Au)
14. Ite Bay, Peru (Cu, Mo)	33. Ok Tedi mine, Papua New Guinea (Cu, Au)
15. Michilla mine, Chile (Cu)	34. Grasberg mine, Indonesia (Au, Cu)
16. Ensenada Chapaco, Chile (Fe-oxides)	35. Buyat Bay, Indonesia (Cu, Au)
17. Chañaral Bay, Chile (Cu)	36. Benete Bay, Indonesia (Cu, Au)
18. Black Angel mine, Greenland (Zn, Pb)	37. Atlas mine, Philippines (Cu)
19. Synvaranger mine, Norway (Fe-oxides)	38. Marinduque Island, Philippines (Cu)

Figure 2. Schematic model of a sulphide ore deposit (porphyry), with the zones, where sulphides, oxides, and sulphates/chlorides may dominate the mineral assemblage. This zoning is a strong simplification and the different mineral groups may coexist. On the left, the different tailings disposal options in water bodies and their dominant geochemical regimes are highlighted.

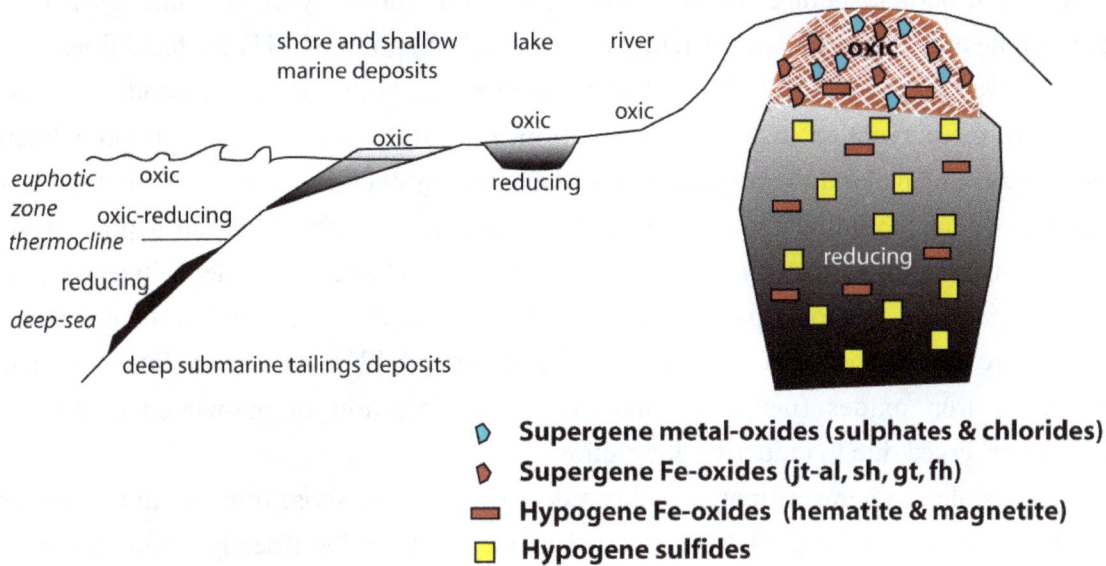

- ◗ **Supergene metal-oxides (sulphates & chlorides)**
- ◗ **Supergene Fe-oxides (jt-al, sh, gt, fh)**
- ▬ **Hypogene Fe-oxides (hematite & magnetite)**
- ☐ **Hypogene sulfides**

Recently, there has been a movement developing against STD around the world, most actively by Earthwork and MiningWatch Canada, who published two reviews about the locations and the mining companies using STD [17–19]. Government and non-government organizations have also been re-evaluating this issue recently [2].

The following is a summary of the evolution of the tailings disposal techniques and their associated environmental problems:

(1) At the beginning of the implementation of flotation, the tailings were deposited by gravity in the closest location available, like rivers, lakes, and in some cases, they reached the sea, depositing the tailings as marine shore tailings deposits. Examples in South America are the valley of El Salado and the Bay of Chañaral in Chile or the river Locumba and the tailings deposited in the Bay of Ite, Perú [7,10]. If the treatment plant was located at the coast, the tailings were directly deposited into the sea; for example, tailings from the mine Michilla and the deposition of tailings from an iron oxide pellet plant at Ensenada Chapaco [20,21]. Sulphide minerals in these types of tailings deposits are exposed to an oxidizing environment during transport and in final deposition, allowing them to oxidize and produce mine drainage and the associated pollution of water resources by the released elements. In the case of the direct deposition of the tailings into the sea, there have been visible effects of increase turbidity and high sedimentation rates in the deposition area, directly impacting the marine fauna [20]. In both cases, the effects were easily visible and the increased pressure of public opinion resulted in a change of the tailings management practice;

(2) In the case of Chañaral, a lawsuit against CODELCO (1989) forced the company to stop tailings deposition in the bay of Chañaral and to build an alternative tailings impoundment called Pampa Austral [7] on-land, which is currently the final disposal facility for the tailings of

the El Salvador mine. The tailings in the valley El Salado and in the Bay of Chañaral, as well as the tailings of the beaches in the north coast of Punta Palitos are in the same condition as directly after deposition and remain as a source for metals pollution to the environment due to sulphide oxidation. In the case of the bay of Ite, Southern Peru Copper Corporation completed a clean-up of the Locumba valley and remediation of tailings deposited at the Bay of Ite, and constructed the on-land tailings impoundment Quebrada Honda, where the tailings from the mines Cuajone and Toquepala are currently deposited. The option of STD for the tailings in the Bay of Ite was evaluated [22], but not implemented. In both cases, the pending tasks to complete are the prevention of AMD formation in the active on-land tailings impoundments, and in the case of Chañaral the remediation of the whole system. In the case of Michilla mine, the sea disposal of the tailings was halted and the tailings were removed and deposited in an on-land tailings impoundment. In the case of the Ensenada Chapaco, the deposition point was changed in 1994 due to problems with turbidity and sedimentation in the intertidal zone, and the tailings are now deposited at between 35 and 130 m depth [21]. As these tailings are mainly composed of iron oxides (magnetite and hematite), liberation of associated contaminant elements is observed due to reductive dissolution;

(3) Due to the visible problems of marine tailings deposition at the shore line and in the euphotic zone, deep submarine tailings disposal started in the early 1970s (ideally <150 m depth to ensure being below the euphotic zone) in Atlas Mine, Filipinas [23], Island Copper Mine (50 m), Canada, Jordan River Mine, Canada and Black Angel Mine, Greenland. The cases of Island Copper Mine and Black Angel Mine have shown the difficulties of predicting the behavior of the oceanographic system and the importance of mineralogical and geochemical characterization of the tailings before marine deposition. In the latter case, soluble and oxide minerals liberated significant amounts of contaminants into the sea [24], similar to the situation of Ensenada Chapaco [21], and clearly showed that oxide minerals should not be deposited in a reducing environment.

As this historic overview shows, inappropriate tailings management and a lack of mineralogical, geochemical, and oceanographic characterization of the systems can lead to environmental damage, and has resulted in a shift by the mining industry back to on-land deposition in tailings impoundments. With the increasing pressures of society on the on-land deposition, the submarine tailings disposal option is nowadays again being considered.

Commonly Mentioned Advantages of STD Are:

- Prevention of acid mine drainage: Reducing environment and lower concentrations of dissolved oxygen limit or prevent sulphide oxidation and any acidity produced through sulphide oxidation will be neutralized by the buffer capacity of marine water;
- Tailings are more geotechnically stable and the possibility of catastrophic failure of tailings dams (on land tailings dam heights may reach several hundred meters), especially in areas with high seismic activity and high rainfall is eliminated [25];
- Minimal land surface is used. This is a strong argument in Norway where, due to the Fjord topography, on-land space for the tailings deposition is very limited;
- Less long-term maintenance required after deposition compared to on-land disposals.

Commonly Mentioned Disadvantages and Risks of STD Are:

- Smothering benthic organisms and physical and geochemical alteration of bottom habitat;
- Reduced number of species and biodiversity of marine communities;
- Risk of liberation of toxic elements from the tailings to the seawater;
- Bioaccumulation of metals through the food chains and ultimately into fish consumed by humans, with associated human health risks;
- The water content of the tailings cannot be recovered; this is especially critical in dry climates;
- The deposited tailings cannot be recovered (possible loss of valuable resources);
- Larger footprint on the seabed than on land;
- Potential toxicity of the flotation reagents used on the marine ecosystem;
- Plume sharing and dispersal of the fine particles throughout the sea;
- Relocation of the tailings in different compartments of the marine ecosystem due to upwelling and currents.

3. Revision of the Legislation Framework and Practice of STD around the World

The London Convention (1972) and its updated version from 1996, the London Protocol (in place since 2006) on the prevention of marine pollution by waste disposal or other materials, are the principal international instruments for the protection of the oceans from anthropogenic pollution. Currently, 42 countries ratified the London Protocol including the significant mining countries such as Australia, Canada, China, and Sweden. This list also includes the main countries which practice STD, like Philippines (9 May 2012), Chile (26 September 2011) and Norway (16 December 1999), but Indonesia and Papua New Guinea (PNG) did not sign. PNG did sign the London Convention (signed by 87 countries) [26].

The United States also did not sign the London Protocol, instead regulating their water quality with the "Clean Water Act", which has been in place since 1991. This being said, most of the developed mining countries do not permit disposal of mine waste into the sea or allow it only in very exceptional cases. Additionally, the UN and World Bank have shown to be very critical of STD and have recommended not using this option.

The London Convention lists materials that are not allowed to be deposited into the sea. In contrast, in the London Protocol, all deposition is forbidden, except for a "reverse list" of some waste types. For example, it allows deposition of "inert inorganic geological material" into the sea. A requirement for following the London Protocol is therefore to show that a material is geological, inorganic, and inert. By definition, nearly all minerals and composites of minerals like rocks form part of the geological–geochemical element cycles. There are only some minerals which could be classified in a geological sense as nearly inert (e.g., quartz, rutile, zircon), but there exist great differences in the solubility and kinetics of dissolution of the different mineral groups in certain geochemical conditions, as explained above (see Table 2).

4. Mineralogical and Geochemical Processes Associated with Ore Deposits and Mine Waste Deposition

The following subsection analyzes the geochemical processes that occur in the different tailings systems to assist in the prediction and prevention of environmental contamination associated with each of the tailings management options.

4.1. Distribution of Minerals and Associated Elements in an Ore Deposit

Understanding the genesis of ore deposits is vital for the exploration of new exploitable mineral deposits. Although the general formation process of some ore deposit types, like porphyry copper deposits, is quiet well understood [27], there remains significant contention in explaining the co-existence of sulphide and iron oxide minerals in the same deposit type as they represent two different geochemical systems (reducing and oxidizing) [28–32]. Although the genesis of this phenomenon is not crucial for final waste management, the fact that the ore body contains both mineral groups (*i.e.*, hypogene sulphides and oxides groups) is a critical factor in determining appropriate waste management. In an ore forming environment, high concentrations of all elements may be involved in the mineralizing process, and there are many trace elements incorporated in the different minerals. Consequently, the final stability of these host minerals is key for the stability of the environmental hazardous elements associated with the final waste material of the mining process. For example, sulphides like pyrite (FeS_2), the most abundant iron sulphide mineral and main mineral responsible for AMD formation, can contain significant amounts of arsenic and other toxic elements. Oxide minerals like magnetite can also contain high concentrations of toxic elements [33].

Coming back to the original hypothesis that sulphide minerals are stable in a reducing environment, but if in a sulphide mineral assemblage there are also associated oxides like for example magnetite, hematite, or goethite, these oxide minerals will not be stable under a reducing environments and will suffer reductive dissolution, resulting in a liberation of the associated elements into the environment.

4.2. Sulphide Oxidation and the Formation of Acid Mine Drainage (AMD)

To better understand why the STD option is again being considered as an option for tailings disposal by the mining industry, we review some key aspects about the formation of acid mine drainage. This chapter is taken from Dold [5] for the convenience of the reader and more details on this issue can be found in this reference and in another review in this Special Issue [1]. The problem of sulphide oxidation and the associated acid mine drainage (AMD), or more generally acid rock drainage (ARD), as well as the dissolution and precipitation processes of metals and minerals, has been a major focus of investigation over the last 50 years [34–38]. There has been less interest in the mineralogical and geochemical interactions taking place within the tailings and waste dumps itself [38,39], despite these reactions being essential for understanding the parameters controlling acid mine drainage formation and for developing effective prevention methods. The primary mineralogical composition of a waste material has a strong influence on the oxidation processes. This has been well illustrated [40–42] and has shown that reaction rates are dependent on the sulphides being oxidized, by the type of oxidant (e.g., Fe(III) or O_2) and the presence of Fe(III) hydroxide coatings. Kinetic-type weathering experiments indicate the importance of trace elements in the stability of individual sulphides. Where different sulphides are in contact with each other, electrochemical processes are likely to occur and influence the reactivity of the sulphides [43].

Most mining operations are surrounded by piles, dumps, or impoundments containing pulverized material or waste from the benefaction process, which are known as tailings, waste rock dumps, stockpiles, or leach dumps or pads. Waste rock dumps contain generally material with low ore grade, which is mined but not milled (run of mine, ROM). These materials typically contain large concentrations of sulphide minerals, which may undergo oxidation, producing a major source of metal and acid contamination [11].

The most common sulphide mineral is pyrite (FeS_2) which oxidizes in several steps including the formation of the meta-stable secondary products ferrihydrite ($5Fe_2O_3 \cdot 9H_2O$), schwertmannite (between $Fe_8O_8(OH)_6SO_4$ and $Fe_{16}O_{16}(OH)_{10}(SO_4)_3$), and goethite ($FeO(OH)$), and the more stable secondary minerals jarosite ($KFe_3(SO_4)_2(OH)_6$), and hematite (Fe_2O_3) depending on the geochemical conditions [35,38,41,44–47]. Pyrite oxidation may be considered to take place in three major steps: (1) oxidation of sulphur (Equation (1)); (2) oxidation of ferrous iron (Equation (2)); and (3) hydrolysis and precipitation of ferric complexes and minerals (Equation (4)). The kinetics of each reaction are different and depend on the conditions prevailing in the tailings.

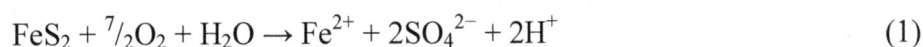

$$FeS_2 + \tfrac{7}{2}O_2 + H_2O \rightarrow Fe^{2+} + 2SO_4^{2-} + 2H^+ \tag{1}$$

$$Fe^{2+} + \tfrac{1}{4}O_2 + H^+ \leftrightarrow Fe^{3+} + \tfrac{1}{2}H_2O \tag{2}$$

reaction rates strongly increased by microbial activity (e.g., *Acidithiobacillus* spp. or *Leptospirillum* spp.)

$$FeS_2 + 14Fe^{3+} + 8H_2O \rightarrow 15Fe^{2+} + 2SO_4^{2-} + 16H^+ \tag{3}$$

Equation (1) describes the initial step of pyrite oxidation in the presence of atmospheric oxygen. Once ferric iron is produced by oxidation of ferrous iron (Equation (2)), oxidation by ferric iron will be the primary oxidant (Equation (3)) of pyrite. Pyrite oxidation by ferric iron is strongly accelerated by microbial activity particularly at low pH [36,48,49]. Under abiotic conditions, the rate of oxidation of pyrite by ferric iron is controlled by the rate of oxidation of ferrous iron, which decreases rapidly with decreasing pH. Below about pH 3, the oxidation of pyrite by ferric iron is about 10–100 times faster than by oxygen [50].

It has been known since the 1960s that microorganisms like *Acidithiobacillus ferrooxidans* or *Leptospirillum ferrooxidans* obtain energy by catalyzing the oxidation of Fe^{2+} to Fe^{3+} [51] and in doing so may increase the rate of reaction (2) up to a factor of about 100 over abiotic oxidation [52]. More recent results show that a complex community of microorganism is responsible for sulphide oxidation [48,53–56]. Nordstrom and Southam [57] state that the initiating step of pyrite oxidation does not require an elaborated sequence of different geochemical reactions that dominate at different pH ranges. *Acidithiobacillus* spp. form nano-environments to grow on sulphide mineral surfaces. These nano-environments can develop thin layers of acidic water that do not affect the bulk pH of the surrounding water. With progressive oxidation, the nano-environments may change to microenvironments. Evidence of acidic micro-environments in the presence of near-neutral pH for the bulk water can be inferred from the presence of jarosite (this mineral forms at pH around 2) in certain soil horizons where the current water pH is neutral [58]. Barker *et al.* [59] observed microbial colonization of biotite and measured pH in microenvironments of surrounding living microcolonies. The solution pH decreased from near-neutral at the mineral surface to 3–4 around microcolonies living within confined spaces at interior colonized cleavage planes.

When mine water, rich in ferrous and ferric iron, reaches the surface it will fully oxidize, hydrolyze and may precipitate to ferrihydrite (fh), schwertmannite (sh), goethite (gt), or jarosite (jt) depending on pH-Eh conditions, and the availability of key elements such as potassium and sulphur (Figure 2). Jarosite, schwertmannite and ferrihydrite are meta-stable with respect to goethite [46].

The hydrolysis and precipitation of iron hydroxides (and to a lesser degree, jarosite) will produce most of the acid in this process. If pH is less than about 2, ferric hydrolysis products like $Fe(OH)_3$ are not stable and Fe^{3+} remains in solution:

$$Fe^{3+} + 3H_2O \leftrightarrow Fe(OH)_3(s) + 3H^+ \tag{4}$$

Note that the net reaction of complete oxidation of pyrite, hydrolysis of Fe^{3+} and precipitation of iron hydroxide (sum of reactions 1, 2 and 4) produces four moles of H^+ per mole of pyrite (in case of $Fe(OH)_3$ formation (Equation (5)), $i.e.$, pyrite oxidation is the most efficient producer of acid among the common sulphide minerals (net reaction 5; Table 2). Nevertheless, it is important to be aware that the hydrolysis of $Fe(OH)_3$ is the main acid producer ($^3/_4$ of moles of H^+ per mol pyrite):

$$FeS_2 + {}^{15}/_4O_2 + {}^7/_2H_2O \rightarrow Fe(OH)_3 + 2SO_4^{2-} + 4H^+ \tag{5}$$

The process of sulphide oxidation concerns all sulphide minerals once exposed to oxidizing conditions (e.g., chalcopyrite, bornite, molybdenite, arsenopyrite, enargite, among others). In this process, different amounts of protons are liberated by the oxidation of the different sulphide minerals [5] and the metals or other harmful elements contained by these sulphides are liberated into the environment.

Table 2. Summary of the most reactive mineral groups and their stability in the environment, associated minerals, and trace elements.

	Sulphides	Oxides & Hydroxides	Carbonates	Sulphates & Chlorides
Stable conditions	reducing (below water cover or in the earth crust with presence of reducing agents like organic matter, or in the Earth crust)	oxidizing (under atmospheric conditions or in the presence of oxidizing agents like dissolved oxygen)	neutral to alkaline	Oxidizing and dry conditions
Unstable conditions	oxidizing (exposure to the atmosphere) or in presence of oxidizing agents like Fe^{3+}, => acid rock drainage (ARD, AMD). There are some indications that high concentrations of Cl could also enhance their dissolution	reducing (below water cover and in the presence of reducing agents like organic matter, sulphate or Cl, for example in lakes with high productivity, or in the sea)	Acid, CO_2	Contact with water (humidity, fog, rain)
Minerals	Pyrite (FeS_2), Chalcopyrite ($CuFeS_2$), Arsenopyrite (FeAsS), Bornite (Cu_5FeS_4), Chalcocite-digenite (Cu_2S), Covelite (CuS), Molibdenite (MoS_2)	Hematite, Magnetite, Goethite, Jarosite-Alunite, Schwertmannite, Ferrihydrite, MnO_2 Birnesite	Calcite, Dolomite, Siderite, Ankerite	Halite, Gypsum, Chalcantite, eriochalcite, $etc.$
Associated trace elements	As, Cr, Se, V, Cu, Zn, Ni, Pb, Cd, REE's	As, Cr, Se, V, Cu, Zn, Ni, Cd, Pb	Fe, Sr, Co, REE's	Metals cations like Cu, Zn, Ni, $etc.$, and anions like SO_4 and Cl

4.3. Stability of Sulphides in Reducing Environment

There is evidence that sulphide minerals are not as stable in marine reducing environments as originally thought. Both in nature [60–62] and in the laboratory, the liberation of heavy metals from the sulphides under reducing conditions [63,64] has been observed. It is not clear which are the associated processes and the controlling parameters, or if it is a problem of some pre-oxidation of the used samples suffering reductive dissolution of trace amounts of oxide minerals. However, as there is contradictory information of the influence of marine water in these processes, more systematic research is needed to understand the long-term stability of sulphide minerals in marine reducing conditions. There are also indications that high Cl concentration might increase the corrosion of sulphides.

4.4. Reductive Dissolution of Oxide Minerals

A metal oxide or hydroxide forms and is stable in oxidizing conditions; therefore, these minerals are unstable under reducing conditions and/or in sediments with high concentrations of silica and/or organic matter [65]. Under these conditions, oxides will suffer reductive dissolution [10,66–69], which will also liberate associated trace elements, which might be toxic to the environment [70–73]. Reductive dissolution, which is catalyzed by microorganisms that use the organic matter as electron source and Fe(III) as an electron sink, is recognized as one of the most archaic forms of bacterial metabolism [66], and is a key process in the biogeochemical cycle of iron in the oceans [74,75]. Reductive dissolution of oxides has only recently been used in a new biohydrometallurgical process for the exploitation of Ni-laterites [73,76].

Two different types of iron oxides in a wider sense can be distinguished: (1) those from hypogene origin, *i.e.*, hydrothermal formation; and (2) those which form due to supergene processes, resulting from meteorization processes of the primary mineralogy, for example during sulphide oxidation (Figure 2 and Table 1). The most relevant hypogene oxides are principally iron oxides like magnetite and hematite, while the supergene Fe(III) hydroxides or oxyhydroxide sulphates are goethite, ferrihydrite, jarosite-alunite, schwertmannite, and in hot climates also hematite can result as a dehydration product of the former minerals and form a gossan [5]. The reductive dissolution of these minerals is a key source for the limiting element iron for the phytoplankton production in the southern oceans [75]. The instability of magnetite in reducing marine sediments is well known due to the problem of paleomagnetic record in marine sediments [77,78]. Therefore, the liberation of iron as a limiting nutrient is not a problem for the ocean, however these iron oxides can contain other associated trace elements, which can be toxic to the environment, like for example Cu, Ni, Zn, Cd, Pb, V, As, Cr, Se, among others [33].

4.5. Dissolution of Sulphates and Chlorides

Sulphate and chloride minerals are very soluble (meaning they dissolve easily in contact with water) and they form from highly concentrated solutions like brines, most often due to evaporation, like in salt lakes in desert climates or in hydrothermal solutions. Exposed to water they dissolve rapidly into their respective ions and liberate the element into the aqueous phase (Table 1). They generally form in oxidizing environments, and in climates with high evaporation. They are also called efflorescent salts [79], due to their tendency to effloresce above the surface due to capillary transport of the solution towards the surface were the water evaporates and the ions subsequent precipitate when the mineral reached supersaturation. With the next rain or fog, these minerals re-dissolve and the elements again enter the hydrogeological cycle [7,8,39].

Therefore, to stabilize the groups of reactive minerals, which can have environmental hazardous elements associated (sulphides, oxides, carbonates, sulphates-chlorides), it can be said as a simplification, that the sulphides have to be maintained under reducing conditions, the oxides under oxidizing conditions, the carbonates neutral, and the sulphates and chlorides dry. The problem for the waste management starts when these mineral groups are found together in one ore deposit and have environmental hazardous elements associated (see Table 2); unfortunately, this is normally the case in an ore deposit [39].

This overview on the mineral groups, which might liberate elements into the environment, makes clear that in many parts of an ore deposit, several of the mineral groups can co-exist in one rock type and therefore makes the waste management options very limited. However, in order to be able to predict the behavior of the material in different geochemical conditions, only a detailed mineralogical (quantitative and qualitative) and geochemical analysis gives the necessary information to prevent so called "un-predictable" environmental hazards, like that which has often occurred in the past—as for example at Black Angel mine [24].

5. The Receiving Environment

In most of the feasibilities studies of STD, the main part of the budget is used by the oceanographic study and the mineralogical study is often very weak or inexistent. As was hopefully clearly expressed above, the first step must be the detailed mineralogical and geochemical characterization of the future tailings to be deposited. This study will show whether the tailings are suitable to deposit into a marine environment, which is argued to be a reducing environment. Then, this receiving environment has to be studied in detail, in order to ensure the stability of the geochemical environment needed to ensure the deposited minerals are non-reactive and stable in place. As mentioned above, mine tailings can be a complex mixture of different mineral groups, which are stable in different geochemical conditions. Thus, these mineral assemblages can produce problems not only in the final waste deposit, but also in the mining process itself, as for example by way of solution contamination or decreasing the recovery amount [6]. Therefore, in addition to the mineralogical and geochemical characterization of the tailings material to be deposited in a certain mine waste deposit type or site, the geochemical characterization of the deposition site and long-term stability also has to be evaluated. In STD, this implies that detailed oceanographic studies, including baseline studies of the mineralogy and geochemistry of the sediments, their pore water geochemistry, microbiology and macro fauna, organic matter composition, studies of the complete water column, and long-term studies to predict El Niño-Southern Oscillation (ENSO) influences and to prevent upwelling, amongst other possible site-specific requests, have to be completed.

6. The Use of STD around the World

There are principally four countries, which use STD nowadays: Indonesia [15], Papua New Guinea, Philippines and Norway [2,3] (of which Philippines and Norway signed the London Protocol). There is one active STD operation in Turkey [80] and one in Chile (Ensenada Chapaco) [20]. In most cases, the deposition is from Cu-Au sulphide mineralization (see Table 3). There exist only two operations, which deposit iron oxide ores (Sydvarangar, Norway and Ensenada Chapaco, Chile). Several ceased operations exist in Canada, United States, and Greenland [3,24,81].

Table 3. Summary of the project and sites where tailings disposal into water bodies occurs (lakes, rivers and the sea (STD)); Adapted from [18].

Body of Water	Mines and Location	Type of Ore	Type of Dumping
Basamuk (Astrolabe) Bay, Bismarck Sea	Ramu Nickel and Yandera mines, Papua New Guinea	nickel-cobalt; copper-gold	Marine (proposed)
Norwegian Fjords	Kirkenes, Kvannevann, Stjernøya, Hustadmarmor, Skaland, Engebøfjellet, & Repparfjorden	iron oxides, industrial minerals, titanium, copper	Marine (proposed & actual)
Canadian lakes	Across Canada	gold, nickel, copper, copper-gold, copper-zinc, iron, diamonds	Lakes (proposed & actual)
South American Pacific Coast	Ensenada Chapaco, Chile	iron oxides	Marine
Senunu Bay	Batu Hijau mine, Indonesia	copper-gold	Marine
Luise Harbor	Lihir mine, Papua New Guinea	gold	Marine
Pigiput Bay	Simberi mine, Papua New Guinea	gold	Marine
Black Sea	Cayeli Bakir, Turkey	copper-zinc	Marine
Otomina and Ajkwa Rivers, Arafura Sea	Grasberg mine, West Papua	copper-gold	River & Marine shore
Porgera River, Fly River system	Porgera mine, Papua New Guinea	gold	River
Ok Tedi River, Fly River system	Ok Tedi mine, Papua New Guinea	copper-gold	River
Auga River	Tolukuma mine, Papua New Guinea	copper-gold	River
Lower Slate Lake	Kensington mine, USA	gold	Lake

7. The Experience with STD around the World

7.1. The Chañaral Case, Chile (Ceased)

The Chañaral case represents one of the first riverine and subsequent marine shore tailings depositions (1938–1975) and was extensively studied to understand the effects on the associated ecological compartments (mainly the sea and the population of Chañaral). There are scientific studies available on the mineralogical-geochemical, hydrogeological processes occurring in the tailings deposit [7,8], microbiological [9], public health [82], and the impact on the marine environment [83–86].

It can be summarized that the deposition of tailings at the shoreline of the Bay of Chañaral and the beaches to the north towards the National Park Pan de Azúcar resulted in the oxidation of the sulphides and the formation of an acid oxidation zone on the surface of the tailings. This together with the climatic conditions in the Atacama desert resulted in Aeolian transport of mainly Cu, Zn, and Ni in water-soluble mineral form (eriochalcite; $CuCl_2 \cdot 2H_2O$) towards the village of Chañaral [7,8].

The levels of copper (20.2 ± 11.5 µg/L), mercury (2.2 ± 2.3 µg/L) and lead (2.1 ± 7 µg/L) in urine of the population of Chañaral were higher than the reported concentrations of the non-exposed population. Additionally, the concentrations were higher than in international studies of exposed populations. The levels of the people from Chañaral, which had higher levels than normal, were: 44.8% for copper, 29.4% for total arsenic, 21.1% for nickel, 16.9% for inorganic arsenic, 9.3% for mercury, and 8.3% for lead [82].

Towards the sea, a flux of dissolved elements present as oxyanions like As and Mo and heavy metals like Cu, Ni, Zn by colloidal transport was detected [7]. The liberation of arsenic from typical Fe(III) hydroxides from the oxidation zone (jarosite, schwertmannite, goethite, ferrihydrite) was confirmed by laboratory studies of the liberated arsenic that is adsorbed and co-precipitated in these minerals in seawater [87]. As a result, the incorporation of the heavy metals and metalloids such as arsenic in the flora and fauna in the associated marine environment was established [84–86].

7.2. The Ite Bay, Peru (Ceased Since 1997)

A similar case is the Ite Bay, Peru marine shore tailings deposit, where tailings from two porphyry copper deposits (Cuajone and Toquepala) were sent from 1960 to 1997 via gravity down the Locumba River towards the Ite Bay, where they finally formed the shore tailings deposit. The difference from Chañaral is that here enough water was available to start a successful remediation [10,88]. Also, the tailings in the Locumba River valley were removed by a clean-up initiated by Southern Peru Copper Corporation.

7.3. Island Copper Mine, Canada (Ceased Since 1994)

Island Copper, Canada and Atlas Mine, Philippines were the first operations in the world to implement STD in 1971. Island Copper was in operation until 1994. All studies in this case were carried out by Poling and Ellis [81,89,90]. Although, their book title "Underwater Tailing Placement at Island Copper Mine: A Success Story" [90] gives the impression that this is a good example of how STD should be done. However, the data on this case of the same authors presented in earlier papers indicate that there was displacement of the tailings at lower depths due to upwelling and unpredicted currents. Tailings with thickness >20 cm have shown minor biodiversity in benthic organisms and higher copper concentrations were registered in mollusks [81]. Also, the Canadian Department of the Environment released in 1996 a report that examined decades of environmental monitoring data at the Island Copper mine site and concluded that the sea floor showed widespread and permanent alteration by tailings. As a result of this, the Canadian site-specific regulations were repealed in 2002 and the marine disposal of tailings was banned [2].

7.4. Black Angel Mine, Greenland (Ceased since 1990)

The Black Angel Mine (Pb-Zn) in Greenland operated between 1973 and 1990 using seawater in the flotation process and STD into the fjord Affarlikassa "A". After one year of operation, high Pb and Zn concentrations were detected in the Fjord Qaumarujuk, outside of fjord "A" and later Cd contamination was also detected. Geochemical, mineralogical and oceanographic studies have shown that the tailings contained significant amounts of oxides and soluble Pb-Zn minerals, which were not stable at the deposition site. There was also a temporal stratification detected in the fjords, which allowed dispersion of the fine particles in the water column between the fjords [24,91,92]. This case shows the problem of soluble oxide minerals in deep-sea deposition. Similar conditions occurred in the Chilean case Ensenada Chapaco (dissolution and reductive dissolution). One, two, and three decades since operation at Black Angel ceased, independent scientific studies showed that the pollution persists [93–98] and is having significant long-term effects on the benthic macro fauna through the presence of heavy metals [99,100]. The study of Josefson *et al.* [100] showed that 33 years after deposition ceased at the Black Angel Mine, degradation of the integrity of the benthic community was directly correlated with persisting, high Pb concentrations in the sediment (>200 mg/kg).

7.5. Kitsault Molybdenum Mine, Canada (Ceased since 1982)

The Kitsault Molybdenum Mine reopened temporarily in 1981–1982 after operation between 1967 and 1972. Consultants originally proposed STD similar to that at Island Copper Mine, but shortly after, new oceanographic instrumentations revealed that water currents, which were originally not predicted, were present. Between 1983 and 1989, the after-closure monitoring program detected that the tailings pore water was enriched in Mo, possibly liberating this element into the marine environment. Additionally, a biological impact on the benthic community of the fjord was observed in the deposition zone [101–103].

7.6. Marcopper Mine, Marinduque, Philippines (Ceased)

The Marcopper Mine in the Philippines experienced some accidents, in which tailings were dispersed into the sea. As a result, studies have shown increased concentrations of different metals in the sediments as well as in the pore water, suggesting toxicity in the marine environment [61,62].

7.7. Ensenada Chapaco, Huasco, Chile (Iron Oxide Ore, Active)

In the Bay of Enenada Chapaco, Huasco, Chile, an iron pellet plant had deposited since 1978 the tailings of several iron oxide deposits (mainly from Algarrobo and Los Colorados) from the Chilean Iron Belt directly into the sea, first directly at the shore and since 1994 at 35 m depth. There is only one scientific publication about the effect of the Ensenada Chapaco tailings deposition on the benthic community [20]. This study investigated the impact of 16 years of tailings disposal in the intertidal zone (1978–1994) on the macro benthic community. Data on the macro fauna between 20 and 50 m depth obtained four months before the intertidal deposition was ceased show the community suffered significantly in terms of abundance, species richness, diversity and high dominance, and caused deep changes in community structure due to the tailings deposition. The complete disappearance of opportunistic species like poliquetos from the families Capetellidae, Spionidae and/or Cirratulidae are attributed to the turbidity and the high sedimentation levels in the bay.

In 1994, the deposition point was changed to 35 m depth, resulting in deposition of the tailings down to a depth of 130 m. There is some information on the impact of the deposited tailings in the recently published Environmental Impact Study (EIS) of an expansion project to deep-sea tailings deposition between 200 and 800 m depth ("Actualización del Sistema de Depositación de Relaves de

Planta de Pellets"), information, which is public and available online [21]. The baseline study for the EIS suggests that hematite-magnetite in the submarine deposited tailings (35–130 m depth) undergo reductive dissolution and release associated trace elements to the seawater [21].

7.8. Sydvarangar, Norway (Iron Oxide Ore, Active)

Sydvarangar is one of two operations that dispose tailings from iron oxide ores into a marine environment (deposition point at 28 m depth). No scientific study on the environmental impact is available and no information is given if there is reductive dissolution occurring. On the Internet, there are some presentations with limited information on this case. The main concern is the potential contamination by the flotation reagents used (Magnafloc 1707).

8. Conclusions

There is a wide consensus among industry and stakeholders that riverine and shore marine disposal should be generally banned due to the high risk of environmental contamination, but there are still some mines operating using these methods, for example Grasberg, Indonesia. At the moment, there are several mines using STD as their tailings management option, but most are in Indonesia, Philippines, Papua New Guinea and Norway. Disposal depths vary from 20 to several hundred meters and the final deposition depth of the tailings in some cases reaches 3000–4000 m. For new projects, deep-sea submarine tailings disposal is typically considered. Forty-two countries have signed the London Protocol, among which nearly all are important mining countries, and therefore the marine deposition of waste is effectively forbidden, with the exception of inorganic, inert geological material. Most of the STD operations are sulphide mines exploiting Cu, Zn, Pb, Mo, Au, and Ag. Only two operations deposit iron oxide tailings into the sea (Ensenada Chapaco, Chile and Sydvarangar, Norway), with some indication of reductive dissolution and subsequent liberation of trace metals into the marine environment in the case of Ensenada Chapaco. There are some cases where in three decades of post-closure the marine ecosystem has not yet recovered. Therefore, it can be concluded that tailings from metal mines are usually not composed of inert geological minerals, and only detailed mineralogical and geochemical studies can give light to the question of whether this material will remain stable in submarine disposal.

In all operations, the following risks of STD have to be evaluated before a decision can be taken:

(1) Liberation of elements associated with oxide minerals due to reductive dissolution under reducing conditions and/or dissolution of soluble sulphate or chloride minerals (e.g., Ensenada Chapaco, Chile and Black Angel Mine, Greenland). Therefore, this type of mineral is not suitable for marine disposal;

(2) Liberation of metals from sulphides. In some studies, the liberation of heavy metals from sulphides in reducing marine conditions is reported. However, due to the lack of detailed mineralogy and geochemistry, it is not clear if this is due to liberation from sulphide minerals under reducing marine conditions, or if these samples contained some oxide minerals. There is a need for more detailed research to address these questions;

(3) Smothering of the benthic fauna due to the high sedimentation rates at the point of tailings deposition. Some studies report a fast repopulation of the tailings by the benthic community after deposition has ceased [104], while other studies present data which indicate that benthic community structures suffered strongly from contamination and open the way to opportunistic species, with a reduction in biodiversity [100,105]. Toxicity effects were detected in some species [62,94,106];

(4) Failure of the tailings tubing (technical problems, tides, earth quakes, tsunamis or storm events) and subsequent dispersion of the tailings in the sea without any control and associated pollutions [61,62];

(5) Dispersion of the tailings to greater distances than predicted by models and increase of turbidity and plume shearing (e.g., Misima Mine, Papua New Guinea);

(6) Re-deposition of the tailings in more shallow levels, due to upwelling and unforeseen currents (e.g., Island Copper Mine, Canada; Bahía Portman, Spain). This will increase the risk of sulphide oxidation in the euphotic zone;

(7) Toxicity of the flotation reagents. Little is known about the toxicity of the flotation reagents in the marine environment;

(8) A resource will be lost. Considering that the efficiency of the exploitation of an ore deposit ranges today between 80% and 90%, 10%–20% of the resource will end up as waste. With new metal recovery technology, old tailing impoundments can in some cases be exploited as mines. This has to be considered in the context that once the tailings are deposited into the deep-sea, this potential resource will be lost forever and future generations will not have the option to re-exploit the resource with improved techniques.

Summarizing these points, the conditions which enable a secure submarine disposal of sulphide tailings are: stable reducing condition; no upwelling or other currents which could remobilize the tailings; and the tailings must contain only sulphide minerals and insoluble silicate minerals, without associated contaminant trace elements.

In order to be able to predict the behavior of the tailings in the marine setting, the first step is to execute a thorough mineralogical and geochemical study of the future tailings that will be deposited in the marine environment. This must include qualitative characterization and quantification of all minerals present in the tailings (e.g., silicates, sulphide, oxides, sulphates-chlorides), and a study of the trace element concentrations associated with the different mineral groups, *i.e.*, sulphides, oxides, soluble minerals with the methodology described by Dold and Weibel [6]. Once this analysis provides the conclusion that the mineral assemblage is suitable for deposition in a reducing environment, then, but only then, is it worth starting an extensive oceanographic study to search for a suitable site in the marine environment. The reason to follow this order is that a mineralogical and geochemical study is more cost-effective and faster than an oceanographic study. Most of the environmental hazards experienced by STD would have been possible to predict and to prevent if good mineralogical and geochemical data were available at the time of decision-making.

Acknowledgments

I would like to thank the reviewers and Jeff Bain for thorough corrections and helpful comments.

Conflicts of Interest

The author declares no conflict of interest.

References

1.   Dold, B. Evolution of Acid Mine Drainage formation in sulphidic mine tailings. *Minerals* **2014**, *4*, 621–641.

2.   Vogt, C. International Assessment of Marine and Riverine Disposal of Mine Tailings. In Proceedings of the 34th Meeting of the London Convention and the 7th Meeting of the London Protocol, London, UK, 1 November 2012.

3.   Koski, R.A. Metal dispersion resulting from mining activities in coastal environments: A pathways approach. *Oceanography* **2012**, *25*, 170–183.

4.   Dold, B. Sustainability in metal mining: From exploration, over processing to mine waste management. *Rev. Environ. Sci. Biotechnol.* **2008**, *7*, 275–285.

5.   Dold, B. Basic Concepts in Environmental Geochemistry of Sulphide Mine-Waste Management. In *Waste Management*; Kumar, S., Ed.; InTech: Rijeka, Croatia, 2010; pp. 173–198.

6.   Dold, B.; Weibel, L. Biogeometallurgical pre-mining characterization of ore deposits: An approach to increase sustainability in the mining process. *Environ. Sci. Pollut. Res.* **2013**, *20*, 7777–7786.

7.   Dold, B. Element flows associated with marine shore mine tailings deposits. *Environ. Sci. Technol.* **2006**, *40*, 752–758.

8.   Bea, S.A.; Ayora, C.; Carrera, J.; Saaltink, M.W.; Dold, B. Geochemical and environmental controls on the genesis of efflorescent salts on coastal mine tailings deposits: A discussion based on reactive transport modeling. *J. Contam. Hydrol.* **2010**, *111*, 65–82.

9.   Korehi, H.; Blöthe, M.; Sitnikova, M.A.; Dold, B.; Schippers, A. Metal mobilization by iron- and sulfur-oxidizing bacteria in a multiple extreme mine tailings in the Atacama Desert, Chile. *Environ. Sci. Technol.* **2013**, *47*, 2189–2196.

10. Dold, B.; Diaby, N.; Spangenberg, J.E. Remediation of a marine shore tailings deposit and the importance of water-rock interaction on element cycling in the coastal aquifer. *Environ. Sci. Technol.* **2011**, *45*, 4876–4883.

11. Dold, B.; Wade, C.; Fontbote, L. Water management for acid mine drainage control at the polymetallic Zn-Pb-(Ag-Bi-Cu) deposit of Cerro de Pasco, Peru. *J. Geochem. Explor.* **2009**, *100*, 133–141.

12. Benedicto, J.; Martinez-Gomez, C.; Guerrero, J.; Jornet, A.; Rodriguez, C. Metal contamination in Portman Bay (Murcia, SE Spain) 15 years after the cessation of mining activities. *Cienc. Mar.* **2008**, *34*, 389–398.

13. Martinez-Sanchez, M.J.; Navarro, M.C.; Perez-Sirvent, C.; Marimon, J.; Vidal, J.; Garcia-Lorenzo, M.L.; Bech, J. Assessment of the mobility of metals in a mining-impacted coastal area (Spain, Western Mediterranean). *J. Geochem. Explor.* **2008**, *96*, 171–182.

14. Cesar, A.; Marín, A.; Marin-Guirao, L.; Vita, R.; Lloret, J.; Del Valls, T.A. Integrative ecotoxicological assessment of sediment in Portman Bay (southeast Spain). *Ecotoxicol. Environ. Saf.* **2009**, *72*, 1832–1841.

15. Rusdinar, Y.; Edraki, M.; Baumgartl, T.; Mulligan, D.; Miller, S. Long term performance of hydrogeochemical riverine mine tailings deposition at Freeport Indonesia. *Mine Water Environ.* **2013**, *32*, 56–70.

16. Scopus. Available online: http://www.scopus.com/ (accessed on 12 June 2014).

17. Submarine Tailings Disposal Toolkit. Available online: http://www.miningwatch.ca/files/ 01.STDtoolkit.intr_.pdf (accessed on 12 June 2014).

18. MiningWatch Canada Webpage. Troubled Waters: How Mine Waste Dumping is Poisoning Our Oceans, Rivers, and Lakes. Available online: http://www.miningwatch.ca/news/troubled-waters-how-mine-waste-dumping-poisoning-our-oceans-rivers-and-lakes (accessed on 12 June 2014).

19. Moran, R.; Reichelt-Brushett, A.; Young, R. Out of sight, out of mine: Ocean dumping of mine wastes. *World Watch* **2009**, *22*, 30–34.

20. Lancellotti, D.A.; Stotz, W.B. Effects of shoreline discharge of iron mine tailings on a marine soft-bottom community in northern Chile. *Mar. Pollut. Bull.* **2004**, *48*, 303–312.

21. RESCAN EIA "Actualización del Sistema de Depositación de Relaves de Planta de Pellets". Available online: http://www.sea.gob.cl (accessed on 23 June 2014).

22. *RESCAN Offshore Tailings Disposal. Environmental Impcat Asessment*; Report to Southern Peru Copper Corporation: Phoenix, AZ, USA, 1992.

23. Diegor, W.G.; Momongan, P.C.; Diegor, E.J.M. Notes on the geochemical dispersion of elements in the beach sediments along the coast west of the Atlas Mines, Visayas, Philippines. *J. Asian Earth Sci.* **1997**, *15*, 285–294.

24. Poling, G.W.; Ellis, D.V. Importance of geochemistry: The Black Angel lead-zinc mine, Greenland. *Mar. Georesour. Geotechnol.* **1995**, *13*, 101–118.

25. Azam, S.; Li, Q. Tailings dam failures: A review of the last one hundred years. *Geotech. News* **2010**, *28*, 50–53.

26. London Convention and Protocol. Available online: http://www.imo.org/OurWork/Environment/ SpecialProgrammesAndInitiatives/Pages/London-Convention-and-Protocol.aspx (accessed on 12 June 2014).

27. Sillitoe, R.H. Porphyry Copper Systems. *Econ. Geol.* **2010**, *105*, 3–41.

28. Sun, W.D.; Liang, H.Y.; Ling, M.X.; Zhan, M.Z.; Ding, X.; Zhang, H.; Yang, X.Y.; Li, Y.L.; Ireland, T.R.; Wei, Q.R.; Fan, W.M. The link between reduced porphyry copper deposits and oxidized magmas. *Geochim. Cosmochim. Acta* **2013**, *103*, 263–275.

29. Pokrovski, G.S. Use and misuse of chemical reactions and aqueous species distribution diagrams for interpreting metal transport and deposition in porphyry copper systems: Comment on Sun *et al.* (2013) "The link between reduced porphyry copper deposits and oxidized magmas", *Geochim. Cosmochim. Acta 103*, 263–275. *Geochim. Cosmochim. Acta* **2014**, *126*, 635–638.

30. Richards, J.P. Discussion of Sun *et al.* (2013): The link between reduced porphyry copper deposits and oxidized magmas. *Geochim. Cosmochim. Acta* **2014**, *126*, 643–645.

31. Sun, W.D.; Huang, R.F.; Liang, H.Y.; Ling, M.X.; Li, C.Y.; Ding, X.; Zhang, H.; Yang, X.Y.; Ireland, T.; Fan, W.M. Magnetite-hematite, oxygen fugacity, adakite and porphyry copper deposits: Reply to Richards. *Geochim. Cosmochim. Acta* **2014**, *126*, 646–649.

32. Sun, W.D.; Zhang, C.C.; Liang, H.Y.; Ling, M.X.; Li, C.Y.; Ding, X.; Zhang, H.; Yang, X.Y.; Ireland, T.; Fan, W.M. The genetic association between magnetite-hematite and porphyry copper deposits: Reply to Pokrovski. *Geochim. Cosmochim. Acta* **2014**, *126*, 639–642.

33. Nystroem, J.O.; Henriquez, F. Magmatic features of iron ores of the Kiruna type in Chile and Sweden; ore textures and magnetite geochemistry. *Econ. Geol.* **1994**, *89*, 820–839.

34. Sato, M. Oxidation of sulphide ore bodies; II. Oxidation mechanisms of sulphide minerals at 25 °C. *Econ. Geol.* **1960**, *55*, 1202–1231.

35. Nordstrom, D.K. Aqueous Pyrite Oxidation and the Consequent Formation of Secondary Iron Minerals. In *Acid Sulphate Weathering*; Kittrick, J.A., Fanning, D.S., Eds.; Soil Science Society of America: Madison, WI, USA, 1982; pp. 37–56.

36. Moses, C.O.; Kirk Nordstrom, D.; Herman, J.S.; Mills, A.L. Aqueous pyrite oxidation by dissolved oxygen and by ferric iron. *Geochim. Cosmochim. Acta* **1987**, *51*, 1561–1571.

37. Blowes, D.W.; Reardon, E.J.; Jambor, J.L.; Cherry, J.A. The formation and potential importance of cemented layers in inactive sulphide mine tailings. *Geochim. Cosmochim. Acta* **1991**, *55*, 965–978.

38. Jambor, J.L. Mineralogy of Sulphide-Rich Tailings and Their Oxidation Products. In *Short Course Handbook on Environmental Geochemistry of Sulphide Mine-Waste*; Jambor, J.L., Blowes, D.W., Eds.; Mineralogical Association of Canada: Quebec, QC, Canada, 1994; Volume 22, pp. 59–102.

39. Dold, B.; Fontboté, L. Element cycling and secondary mineralogy in porphyry copper tailings as a function of climate, primary mineralogy, and mineral processing. *J. Geochem. Explor.* **2001**, *74*, 3–55.

40. Rimstidt, J.D.; Chermak, J.A.; Gagen, P.M. Rates of Reaction of Galena, Spalerite, Chalcopyrite, and Asenopyrite with Fe(III) in Acidic Solutions. In *Environmental Geochimistry of Sulphide Oxidation*; Alpers, C.N., Blowes, D.W., Eds.; American Chemical Society: Washington, DC, USA, 1994; Volume 550, pp. 2–13.

41. Rimstidt, J.D.; Vaughan, D.J. Pyrite oxidation: A state-of-the-art assessment of the reaction mechanism. *Geochim. Cosmochim. Acta* **2003**, *67*, 873–880.

42. Huminicki, D.M.C.; Rimstidt, J.D. Iron oxyhydroxide coating of pyrite for acid mine drainage control. *Appl. Geochem.* **2009**, *24*, 1626–1634.

43. Kwong, Y.T.J. *Prediction and Prevention of Acid Rock Drainage from a Geological and Mineralogical Perspective*; Canada Centre for Mineral and Energy Technology: Ottawa, ON, Canada, 1993; p. 47.

44. Dutrizac, J.E.; Jambor, J.L. Jarosites and their application in hydrometallurgy. *Rev. Mineral. Geochem.* **2000**, *40*, 405–452.

45. Cornell, R.M.; Schwertmann, U. *The Iron Oxides*; 2nd ed.; Wiley-VCH: Weinheim, Germany, 2003.

46. Bigham, J.M.; Schwertmann, U.; Traina, S.J.; Winland, R.L.; Wolf, M. Schwertmannite and the chemical modeling of iron in acid sulphate waters. *Geochim. Cosmochim. Acta* **1996**, *60*, 2111–2121.

47. Jerz, J.K.; Rimstidt, J.D. Pyrite oxidation in moist air. *Geochim. Cosmochim. Acta* **2004**, *68*, 701–714.

48. Ehrlich, H.L. *Geomicrobiology*; Marcel Dekker: New York, NY, USA, 1996.

49. Nordstrom, D.K.; Jenne, E.A.; Ball, J.W. Redox Equilibria of Iron in Acid Mine Waters. In *Chemical Modeling in Aqueous Systems*; American Chemical Society: Washington, DC, USA, 1979; Volume 93, pp. 51–79.

50. Ritchie, A.I.M. Sulphide Oxidation Mechanisms: Controls and Rates of Oxygen Transport. In *Short Course Handbook on Environmental Geochemistry of Sulphide Mine-Waste*; Jambor, J.L., Blowes, D.W., Eds.; Mineralogical Association of Canada: Quebec, QC, Canada, 1994; Volume 22, pp. 201–244.

51. Bryner, L.C.; Walker, R.B.; Palmer, R. Some factors influencing the biological and non-biological oxidation of sulphide minerals. *Trans. Soc. Min. Eng. AIME* **1967**, *238*, 56–65.

52. Singer, P.C.; Stumm, W. Acid mine drainage: The rate-determining step. *Science* **1970**, *167*, 1121–1123.

53. Johnson, D.B. Biodiversity and ecology of acidophilic microorganisms. *FEMS Microbiol. Ecol.* **1998**, *27*, 307–317.

54. Johnson, D.B. Importance of Microbial Ecology in the Development of New Mineral Technologies. In *Biohydrometallurgy and the Environment toward the Mining of the 21st Century*; Amils, R., Ballester, A., Eds.; Elsevier: Amsterdam, The Netherlands, 1999; Volume 9, pp. 645–656.

55. Norris, P.R.; Johnson, D.B. Acidophilic Microorganisms. In *Extremophiles: Microbial Life in Extreme Environments*; Horikoshi, K., Grant, W.D., Eds.; John Wiley and Sons: New York, NY, USA, 1998; pp. 133–154.

56. Johnson, D.B.; Hallberg, K.B. The microbiology of acidic mine waters. *Res. Microbiol.* **2003**, *154*, 466–473.

57. Nordstrom, D.K.; Southam, G. Geomicrobiology of Sulphide Mineral Oxidation. In *Geomicrobiology*; Banfield, J.F., Nealson, K.H., Eds.; Mineralogical Society of America: Chantilly, VA, USA, 1997; Volume 35, pp. 361–390.

58. Carson, C.D.; Fanning, D.S.; Dixon, J.B. Alfisols and Ultisols with Acid Sulphate Weathering Features in Texas. In *Acid Sulphide Weathering*; Kittrick, J.A., Fanning, D.S., Hossner, L.R., Eds.; Soil Science Society of America: Madison, WI, USA, 1982; Volume 10, pp. 127–146.

59. Barker, W.W.; Welch, S.A.; Chu, S.; Banfield, J.F. Experimental observations of the effects of bacteria on aluminosilicate weathering. *Am. Mineral.* **1998**, *83*, 1551–1563.

60. Sherwood, J.E.; Barnett, D.; Barnett, N.W.; Dover, K.; Howitt, J.; Ii, H.; Kew, P.; Mondon, J. Deployment of DGT units in marine waters to assess the environmental risk from a deep sea tailings outfall. *Anal. Chim. Acta* **2009**, *652*, 215–223.

61. David, C.P. Heavy metal concentrations in marine sediments impacted by a mine-tailings spill, Marinduque Island, Philippines. *Environ. Geol.* **2002**, *42*, 955–965.

62. Carr, R.S.; Nipper, M.; Plumlee, G.S. Survey of marine contamination from mining-related activities on Marinduque Island, Philippines: Porewater toxicity and chemistry. *Aquat. Ecosyst. Health Manag.* **2003**, *6*, 369–379.

63. Lin, H.K.; Walsh, D.E.; Chen, X.; Oleson, J.L. Release of heavy metals from sulphide flotation tailings under deepwater discharge environments. *Miner. Metall. Process.* **2009**, *26*, 174–178.

64. Rzepka, P.; Walder, I.F.; Bozecki, P.; Rzepa, G. Sub-sea tailings deposition leach modeling. *Mineral. Mag.* **2013**, *77*, 2107, doi:10.1180/minmag.2013.077.5.18.

65. Florindo, F.; Roberts, A.P.; Palmer, M.R. Magnetite dissolution in siliceous sediments. *Geochem. Geophys. Geosyst.* **2003**, *4*, 1053, doi:10.1029/2003GC000516.

66. Weber, K.A.; Achenbach, L.A.; Coates, J.D. Microorganisms pumping iron: Anaerobic microbial iron oxidation and reduction. *Nat. Rev. Microbiol.* **2006**, *4*, 752–764.

67. Lovley, D.R. Dissimilatory Fe(III) and Mn(IV) reduction. *Microbiol. Rev.* **1991**, *55*, 259–287.

68. Lovley, D.R. Dissimilatory metal reduction. *Annu. Rev. Microbiol.* **1993**, *47*, 263–290.

69. Lovley, D.R. Microbial Fe(III) reduction in subsurface environments. *FEMS Microbiol. Rev.* **1997**, *20*, 305–313.

70. Stone, A.T.; Morgan, J.J. Reductive Dissolution of Metal Oxides. In *Aquatic Surface Chemistry*; Stumm, W., Ed.; John Wiley and Sons: Hoboken, NJ, USA, 1987; pp. 221–254.

71. Ribeta, I.; Ptacek, C.J.; Blowes, D.W.; Jambor, J.L. The potential for metal release by reductive dissolution of weathered mine tailings. *J. Contam. Hydrol.* **1995**, *17*, 239–273.

72. Larsen, O.; Postma, D. Kinetics of reductive bulk dissolution of lepidocrocite, ferrihydrite, and goethite. *Geochim. Cosmochim. Acta* **2001**, *65*, 1367–1379.

73. Hallberg, K.B.; Grail, B.M.; Plessis, C.A.D.; Johnson, D.B. Reductive dissolution of ferric iron minerals: A new approach for bio-processing nickel laterites. *Miner. Eng.* **2011**, *24*, 620–624.

74. Raiswell, R.; Canfield, D.E. The Iron Biogeochemical Cycle Past and Present. *Geochem. Perspect.* **2012**, *1*, 1–2.

75. Dold, B.; Gonzalez-Toril, E.; Aguilera, A.; Lopez-Pamo, E.; Bucchi, F.; Cisternas, M.-E.; Amils, R. Acid rock drainage and rock weathering in Antarctica—Important sources for iron cycling in the Southern Ocean. *Environ. Sci. Technol.* **2013**, *47*, 6129–6136.

76. Du Plessis, C.A.; Slabbert, W.; Hallberg, K.B.; Johnson, D.B. Ferredox: A biohydrometallurgical processing concept for limonitic nickel laterites. *Hydrometallurgy* **2011**, *109*, 221–229.

77. Hepp, D.A.; Mörz, T.; Hensen, C.; Frederichs, T.; Kasten, S.; Riedinger, N.; Hay, W.W. A late Miocene-early Pliocene Antarctic deepwater record of repeated iron reduction events. *Mar. Geol.* **2009**, *266*, 198–211.

78. Rowan, C.J.; Roberts, A.P.; Broadbent, T. Reductive diagenesis, magnetite dissolution, greigite growth and paleomagnetic smoothing in marine sediments: A new view. *Earth Planet. Sci. Lett.* **2009**, *277*, 223–235.

79. Nordstrom, D.K. Advances in the hydrogeochemistry and microbiology of acid mine waters. *Int. Geol. Rev.* **2000**, *42*, 499–515.

80. Berkun, M. Submarine tailings placement by a copper mine in the deep anoxic zone of the Black Sea. *Water Res.* **2005**, *39*, 5005–5016.

81. Ellis, D.V.; Pedersen, T.F.; Poling, G.W.; Pelletier, C.; Horne, I. Review of 23 years of STD: Island Copper Mine, Canada. *Mar. Georesour. Geotechnol.* **1995**, *13*, 59–99.

82. Cortes, S. Percepción y Medición del Riesgo a Metales en una Población Expuesta a Residuos Mineros. Ph.D. Thesis, Universidad de Chile, Santiago de Chile, Chile, 2009. (In Spanish)

83. Castilla, J.C.; Nealler, E. Marine environmental impact due to mining activities of El Salvador copper mine, Chile. *Mar. Pollut. Bull.* **1978**, *9*, 67–70.

84. Castilla, J.C. Environmental impact in sandy beaches of copper mine tailings at Chanaral, Chile. *Mar. Pollut. Bull.* **1983**, *14*, 459–464.

85. Lee, M.R.; Correa, J.A.; Castilla, J.C. An assessment of the potential use of the nematode to copepod ratio in the monitoring of metals pollution. The Chañaral case. *Mar. Pollut. Bull.* **2001**, *42*, 696–701.

86. Farina, J.M.; Castilla, J.C. Temporal variation in the diversity and cover of sessile species in rocky intertidal communities affected by copper mine tailings in Northern Chile. *Mar. Pollut. Bull.* **2001**, *42*, 554–568.

87. Alarcon, R.; Gaviria, J.; Dold, B. Liberation of adsorbed and co-precipitated arsenic from jarosite, schwertmannite, ferrihydrite, and goethite in seawater. *Minerals* **2014**, *4*, 603–620.

88. Diaby, N.; Dold, B. Evolution of geochemical and mineralogical parameters during *in situ* remediation of a marine shore tailings deposit by the implementation of a wetland cover. *Minerals* **2014**, *4*, 578–602.

89. Ellis, D.V. Effect of mine tailings on the biodiversity of the seabed: Example of the Island Copper Mine, Canada. *Seas Millenn. Environ. Eval.* **2000**, *3*, 235–246.

90. Poling, G.; Ellis, D.J.; Murray, J.W.; Parsons, T.R.; Pelletier, C. *Underwater Tailing Placement at Island Copper Mine: A Success Story*; Society for Mining, Metallurgy, and Exploration: Englewood, CO, USA, 2002.

91. Asmund, G. Environmental studies in connection with mining activity in Greenland. *Gronland Geol. Unders. Rapp.* **1986**, *128*, 13–22.

92. Asmund, G.; Broman, P.G.; Lindgren, G. Managing the environment at the Black Angel Mine, Greenland. *Int. J. Surf. Min. Reclam. Environ.* **1994**, *8*, 37–40.

93. Elberling, B.; Asmund, G.; Kunzendorf, H.; Krogstad, E.J. Geochemical trends in metal-contaminated fiord sediments near a former lead-zinc mine in West Greenland. *Appl. Geochem.* **2002**, *17*, 493–502.

94. Elberling, B.; Knudsen, K.L.; Kristensen, P.H.; Asmund, G. Applying foraminiferal stratigraphy as a biomarker for heavy metal contamination and mining impact in a fiord in West Greenland. *Mar. Environ. Res.* **2003**, *55*, 235–256.

95. Perner, K.; Leipe, T.; Dellwig, O.; Kuijpers, A.; Mikkelsen, N.; Andersen, T.J.; Harff, J. Contamination of arctic Fjord sediments by Pb-Zn mining at Maarmorilik in central West Greenland. *Mar. Pollut. Bull.* **2010**, *60*, 1065–1073.

96. Søndergaard, J.; Asmund, G.; Johansen, P.; Elberling, B. Pb isotopes as tracers of mining-related Pb in lichens, seaweed and mussels near a former Pb-Zn mine in West Greenland. *Environ. Pollut.* **2010**, *158*, 1319–1326.

97. Søndergaard, J.; Asmund, G.; Johansen, P.; Rigit, F. Long-term response of an arctic fiord system to lead-zinc mining and submarine disposal of mine waste (Maarmorilik, West Greenland). *Mar. Environ. Res.* **2011**, *71*, 331–341.

98. Søndergaard, J.; Bach, L.; Asmund, G. Modelling atmospheric bulk deposition of Pb, Zn and Cd near a former Pb-Zn mine in West Greenland using transplanted Flavocetraria nivalis lichens. *Chemosphere* **2013**, *90*, 2549–2556.

99. Loring, D.H.; Asmund, G. Heavy metal contamination of a Greenland fjord system by mine wastes. *Environ. Geol. Water Sci.* **1989**, *14*, 61–71.

100. Josefson, A.B.; Hansen, J.L.S.; Asmund, G.; Johansen, P. Threshold response of benthic macrofauna integrity to metal contamination in West Greenland. *Mar. Pollut. Bull.* **2008**, *56*, 1265–1274.

101. Pedersen, T.F.; Ellis, D.V.; Poling, G.W.; Pelletier, C. Effects of changing environmental rules: Kitsault molybdenum mine, Canada. *Mar. Georesour. Geotechnol.* **1995**, *13*, 119–133.

102. Odhiambo, B.K.; Macdonald, R.W.; O'Brien, M.C.; Harper, J.R.; Yunker, M.B. Transport and fate of mine tailings in a coastal fjord of British Columbia as inferred from the sediment record. *Sci. Total Environ.* **1996**, *191*, 77–94.

103. Pedersen, T.F. Early diagenesis of copper and molybdenum in mine tailings and natural sediments in Rupert and Holberg inlets, British Columbia. *Can. J. Earth Sci.* **1985**, *22*, 1474–1484.

104. Kline, E.R.; Stekoll, M.S. Colonization of mine tailings by marine invertebrates. *Mar. Environ. Res.* **2001**, *51*, 301–325.

105. Burd, B.J. Evaluation of mine tailings effects on a benthic marine infaunal community over 29 years. *Mar. Environ. Res.* **2002**, *53*, 481–519.

106. Reichelt-Brushett, A. Risk assessment and ecotoxicology limitations and recommendations for ocean disposal of mine waste in the Coral Triangle. *Oceanography* **2012**, *25*, 40–51.

Permissions

The contributors of this book come from diverse backgrounds, making this book a truly international effort. This book will bring forth new frontiers with its revolutionizing research information and detailed analysis of the nascent developments around the world.

We would like to thank all the contributing authors for lending their expertise to make the book truly unique. They have played a crucial role in the development of this book. Without their invaluable contributions this book wouldn't have been possible. They have made vital efforts to compile up to date information on the varied aspects of this subject to make this book a valuable addition to the collection of many professionals and students.

This book was conceptualized with the vision of imparting up-to-date information and advanced data in this field. To ensure the same, a matchless editorial board was set up. Every individual on the board went through rigorous rounds of assessment to prove their worth. After which they invested a large part of their time researching and compiling the most relevant data for our readers.

The editorial board has been involved in producing this book since its inception. They have spent rigorous hours researching and exploring the diverse topics which have resulted in the successful publishing of this book. They have passed on their knowledge of decades through this book. To expedite this challenging task, the publisher supported the team at every step. A small team of assistant editors was also appointed to further simplify the editing procedure and attain best results for the readers.

Apart from the editorial board, the designing team has also invested a significant amount of their time in understanding the subject and creating the most relevant covers. They scrutinized every image to scout for the most suitable representation of the subject and create an appropriate cover for the book.

The publishing team has been an ardent support to the editorial, designing and production team. Their endless efforts to recruit the best for this project, has resulted in the accomplishment of this book. They are a veteran in the field of academics and their pool of knowledge is as vast as their experience in printing. Their expertise and guidance has proved useful at every step. Their uncompromising quality standards have made this book an exceptional effort. Their encouragement from time to time has been an inspiration for everyone.

The publisher and the editorial board hope that this book will prove to be a valuable piece of knowledge for researchers, students, practitioners and scholars across the globe.

List of Contributors

Brandon Reynolds
Ecosystem Science and Management, University of Wyoming, 1000 E. University Ave, Laramie, Y 82071, USA

K. J. Reddy
Ecosystem Science and Management, University of Wyoming, 1000 E. University Ave, Laramie, WY 82071, USA

Morris D. Argyle
Chemical Engineering, Brigham Young University, Provo, UT 84602, USA

Krishnamoorthy Arumugam
Department of Earth and Environmental Sciences, University of Michigan, 1100 North University Avenue, 2534 C.C. Little, Ann Arbor, MI 48109-1005, USA

Udo Becker
Department of Earth and Environmental Sciences, University of Michigan, 1100 North University Avenue, 2534 C.C. Little, Ann Arbor, MI 48109-1005, USA

Ugo Bardi
Dipartimento di Scienze della Terra, Università di Firenze, Via G. La Pira 4, 50121 Firenze, Italy

Stefano Caporali
Dipartimento di Chimica, Università di Firenze, Via della Lastruccia 3, 50019 Sesto Fiorentino, Consorzio Interuniversitario Nazionale per la Scienza e Tecnologia dei Materiali, Via Giusti 9, 50123 Firenze, Italy

Maria Mäkitalo
Department of Civil, Environmental and Natural Resources Engineering, Luleå University of Technology, SE-97187 Luleå, Sweden

Christian Maurice
Department of Civil, Environmental and Natural Resources Engineering, Luleå University of Technology, SE-97187 Luleå, Sweden
Ramböll Sverige AB, Kyrkogatan 2, Box 850, SE-97126 Luleå, Sweden

Yu Jia
Department of Civil, Environmental and Natural Resources Engineering, Luleå University of Technology, SE-97187 Luleå, Sweden

Björn Öhlander
Department of Civil, Environmental and Natural Resources Engineering, Luleå University of Technology, SE-97187 Luleå, Sweden

Nigel J. Cook
School of Chemical Engineering, University of Adelaide, Adelaide, SA 5005, Australia

Barbara Etschmann
School of Chemical Engineering, University of Adelaide, Adelaide, SA 5005, Australia
School of Geosciences, Monash University, Clayton, VIC 3800, Australia
South Australian Museum, North Terrace, Adelaide, SA 5000, Australia

Cristiana L. Ciobanu
School of Chemical Engineering, University of Adelaide, Adelaide, SA 5005, Australia

Kalotina Geraki
Diamond Light Source, Harwell Science and Innovation Campus, Didcot, Oxon OX11 0QX, UK

Daryl L. Howard
Australian Synchrotron, 800 Blackburn Rd., Clayton, VIC 3168, Australia

Timothy Williams
The Monash Centre for Electron Microscopy, Monash University, Clayton, VIC 3800, Australia

Nick Rae
School of Geosciences, Monash University, Clayton, VIC 3800, Australia
Australian Synchrotron, 800 Blackburn Rd., Clayton, VIC 3168, Australia

Allan Pring
South Australian Museum, North Terrace, Adelaide, SA 5000, Australia
School of Chemical and Physical Sciences, Flinders University, GPO Box 2100, Adelaide, SA 5000, Australia

Guorong Chen
Key Laboratory for Ultrafine Materials of Ministry of Education, School of Materials Science and Engineering, East China University of Science and Technology, Shanghai 200237, China

Bernt Johannessen
Australian Synchrotron, 800 Blackburn Rd., Clayton, VIC 3168, Australia

Joël Brugger
School of Geosciences, Monash University, Clayton, VIC 3800, Australia
South Australian Museum, North Terrace, Adelaide, SA 5000, Australia

Bernhard Dold
SUMIRCO (Sustainable Mining Research & Consult EIRL), Casilla 28, San Pedro de la Paz 4130000, Chile

Yi Wai Chiang
School of Engineering, University of Guelph, Guelph, ON N1G 2W1, Canada
Department of Microbial and Molecular Systems, KU Leuven, Leuven 3001, Belgium

Rafael M. Santos
Department of Chemical Engineering, KU Leuven, Leuven 3001, Belgium

Aldo Van Audenaerde
Department of Chemical Engineering, KU Leuven, Leuven 3001, Belgium

Annick Monballiu
Laboratory for Microbial and Biochemical Technology (Lab µBCT), KU Leuven @ Brugge-Oostende (Kulab), Oostende 8400, Belgium

Tom Van Gerven
Department of Chemical Engineering, KU Leuven, Leuven 3001, Belgium

Boudewijn Meesschaert
Department of Microbial and Molecular Systems, KU Leuven, Leuven 3001, Belgium

Jeffery A. Greathouse
Geochemistry Department, Sandia National Laboratories, P.O. Box 0754, Albuquerque, NM 87185-0754, USA

Karen L. Johnson
School of Engineering and Computing Sciences, Durham University, South Road, Durham DH1 3LE, UK

H. Christopher Greenwell
Department of Earth Sciences, Durham University, South Road, Durham DH1 3LE, UK

Weixing Wang
School of Physics & Information Engineering, Fuzhou University, 350108 Fuzhou, China
Royal Institute of Technology, 100 44 Stockholm, Sweden

Liangqin Chen
School of Physics & Information Engineering, Fuzhou University, 350108 Fuzhou, China

Bernhard Dold
SUMIRCO (Sustainable Mining Research & Consult EIRL), Casilla 28, San Pedro de la Paz 4130000, Chile